Lecture Notes in Mathematics

Edited by A. Dold and B. Eckmann

859

Logic Year 1979–80

The University of Connecticut, USA

Edited by M. Lerman, J. H. Schmerl, and R. I. Soare

Springer-Verlag
Berlin Heidelberg New York 1981

Editors

Manuel Lerman
James H. Schmerl
Department of Mathematics, The University of Connecticut
Storrs, CT 06268, USA

Robert I. Soare
Department of Mathematics, The University of Chicago
Chicago, IL 60637, USA

AMS Subject Classifications (1980): 03-06, 03C30, 03C45, 03C60, 03C65, 03C75, 03D25, 03D30, 03D55, 03D60, 03D65, 03D80, 03F30, 03G30

ISBN 3-540-10708-8 Springer-Verlag Berlin Heidelberg New York
ISBN 0-387-10708-8 Springer-Verlag New York Heidelberg Berlin

Library of Congress Cataloging in Publication Data Main entry under title: Logic year 1979–80, the University of Connecticut. (Lecture notes in mathematics; 859) Bibliography: p. Includes index. 1. Logic, Symbolic and mathematical--Congresses. I. Lerman, M. (Manuel), 1943-. II. Schmerl, J. H. (James Henry), 1940-. III. Soare, R. I. (Robert Irving), 1940-. IV. Series: Lecture notes in mathematics (Springer-Verlag); 859. QA3.L28 vol. 859 [QA9.AI] 510s [511.3] 81-5628 AACR2
ISBN 0-387-10708-8 (U.S.)

Printing and binding: Beltz Offsetdruck, Hemsbach/Bergstr.
2141/3140-543210

PREFACE

Each year the Mathematics Department of the University of Connecticut sponsors a special year which is an intense concentration in a specific area of Mathematics. The year 1979-80 was devoted to Mathematical Logic, with special emphasis on recursion theory and model theory. Visiting scholars from other institutions, either for the whole year or for one of the two semesters, formed the core of this successful year. Stephen Simpson (Pennsylvania State University) and David Kueker (University of Maryland) were visitors for the entire year; Richard Shore (Cornell University) and Robert Soare (University of Chicago) visited just for the fall semester; and Michael Morley (Cornell University) and Joram Hirschfeld (Tel-Aviv University) visited just for the spring semester. Visiting graduate students included: Klaus Ambos, Stephen Brackin, Steven Buechler, David Cholst, Peter Fejer, David Miller, Charles Steinhorn, and Galen Weitkamp.

The highlight of the year was the Conference on Mathematical Logic, which took place November 11-13, 1979, at Storrs. There were 80 logicians in attendance. Included on the program were ten invited hour addresses, twenty contributed fifteen minute talks, and two papers presented by title.

This volume represents the proceedings both of the Logic Year and also of the Conference. Almost all of the papers included herein have been based either on talks presented at the Conference or on presentations made to one of the various seminars, including the joint University of Connecticut - Yale - Wesleyan logic seminars, that were regularly held during the course of the year.

The Logic Year and the Conference could not have been so successful without the greatly appreciated assistance and cooperation of many organizations and individuals. We thank the National Science Foundation for financial support under grant MCS 79-03308; we thank the Research Foundation for additional financial assistance; we thank the University of Connecticut Office of Conferences, Institutes and Administrative Services for their able handling of the organization of the Conference; we thank our consulting editors Steve Simpson, Richard Shore and David Kueker for their expertise; and finally we thank all of those individuals who by attending the Conference contributed to making it an outstanding event.

M. Lerman

J. Schmerl

R. Soare

CONFERENCE PROGRAM

I. Invited Addresses

1. Harrington, Leo: Building Arithmetical Models of Peano Arithmetic.
2. Jockusch, Carl: Some Easy Constructions of r. e. Sets.
3. Macintyre, Angus: Primes in Non-standard Models of Arithmetic.
4. Makkai, Michael: The Category of Models of a Theory.
5. Millar, Terrence: Topics in Recursive Model Theory.
6. Morley Michael: Theories with few Models.
7. Moschovakis, Yiannis: Ordinal Games and Recursion Theory.
8. Nerode, Anil: Recursive Model Theory and Constructive Algebra.
9. Sacks, Gerald: The Limits of Recursive Enumerability.
10. Vaught, Robert: Infinitary Languages and Topology.

II. Contributed Papers

1. Baldwin, John: Why Superstable Theories are Super.
2. Byerly, Robert E.: An Invariance Notion in Recursion Theory.
3. Cherlin, Gregory: Real Closed Rings.
4. DiPaola, Robert: The Theory of Partial α-Recursive Operators, Effective Operations, and Completely Recursively Enumerable Classes.
5. Epstein, Richard, Hass, Richard, and Kramer, Richard: A Hierarchy of Sets and Degrees Below 0'.
6. Fejer, Peter A.: Structure of r.e. Degrees.
7. Glass, Andrew: On Elementary Types of Automorphism Groups of Linearly Ordered Sets.
8. Kaufmann, Matt: On Existence of Σ_n End Extensions.
9. Kierstead, Henry and Remmel, Jeffrey B.: On the Degrees of Indiscernibles in Decidable Models.
10. Kolaitis, Phokion G.: Spector-Gandy Theorem for Recursion in E and Normal Functionals.
11. Kranakis, Evangelos: On Σ_n Partition Relations.
12. Maass, Wolfgang: R.E. Generic Sets.
13. Manaster, Alfred: Recursively Categorical, Topologically Dense, Decidable Two Dimensional Partial Orderings.
14. Odifreddi, Piergiorgio: Strong Recudibilities.
15. Posner, David: The Upper Semilattice of Degrees $\leq 0'$ is Complemented.
16. Slaman, Theodore A. and Sacks, Gerald E.: Inadmissible Forcing.
17. Smith, Kay Ellen: Boolean-Valued Models and Galois Theory for Commutative Regular Rings.
18. Smith, Rick: A Survey of Effectiveness in Field Theory.

CONFERENCE PROGRAM (CONT.)

II. Contributed Papers (Cont.)

19. Srebrny, Marian: Measurable Uniformization.

20. Watnick, Richard: Recursive and Constructive Linear Orders.

III. Papers Presented by Title

1. Calude, Christian: Category Methods in Computational Complexity.

2. Miller, David: The Degree of Semirecursive η-hyperhypersimple sets.

TABLE OF CONTENTS

CONFERENCE PARTICIPANTS

Ambos, Klaus
Baldwin, John T.
Barnes, Robert
Bohorquez, Jaime
Brackin, Stephen H.
Brady, Stephen
Bruce, Kim
Buechler, Steven A.
Byerly, Robert
Cherlin, Greg
DiPaola, Robert
Cholst, David
Dorer, David
Dougherty, Dan
Epstein, Richard L.
Fejer, Peter
Fisher, Edward R.
Friedman, Sy D.
Glass, A. M. W.
Gold, Bonnie
Griffor, Edward
Harrington, Leo
Hay, Louise
Hodes, Harold
Homer, Steven
Hoover, D. N.
Hrbacek, Karel
Jockusch, Carl, Jr.
Joseph, Debra
Kanamori, A.
Kaufmann, Matt
Keisler, H. Jerome
Kierstead, Henry A.
Kolaitis, Phokion G.
Kramer, Richard L.
Kranakis, Evangelos
Krause, Ralph M.
Kueker, David
Landraitis, Charles
Lerman, Manuel
Lin, Charlotte
Maass, Wolfgang
Macintyre, Angus
Makkai, Mihaly
Manaster, Alfred
Mansfield, R. B.
Marker, Dave
Mate, Attila
McKenna, Kenneth
Millar, Terrence
Miller, David P.
Morley, Michael
Moschovakis, Y. N.
Nerode, Anil
Odell, David A.
Odifreddi, George
Posner, David B.
Powell, William C.
Sacks, Gerald
Schmerl, James
Scowcroft, Philip
Shamash, Josephine
Shoenfield, Joseph
Shore, Richard A.
Simpson, Stephen G.
Slaman, Theodore
Smith, Carl
Smith, Kay
Smith, Rick
Smith, Stuart T.
Soare, Robert
Srebrny, Marian
Stob, Michael
Van den Dries, Lou
Vaught, Robert
Weiss, Michael
Weitkamp, Galen
Watnick, Richard
Weaver, George
Welaish, Jeffrey
Wood, Carol

DEFINABILITY AND THE HIERARCHY OF STABLE THEORIES

John T. Baldwin

It is well known that a theory T is stable if and only if for every A contained in a model of T and every type p in S(A), p is definable over A in the following sense:

> The type p in S(A) is definable over B by the map d if for each formula $\phi(x;y)$ there is a formula $d\phi(y)$ with parameters from B such that for each sequence a in A: $\phi(x;a)$ is in p if and only if $d\phi(a)$ holds.

In fact, in [2] we proposed that a slight variant of this property be taken as the definition of a stable theory. There is a natural objection to this proposal; the usual definition of stable, superstable, and totally transcendental theories in terms of the cardinality of the space of types yields immediately the hierarchy: totally transcendental implies superstable implies stable. Is there a similar hierarchy of definability which defines totally transcendental and superstable in terms of "definability of types"? In this paper we provide such a hierarchy. Namely, we will show the following results.

Let $S_n(T)$ denote the collection of n-types over the empty set. We say T is a <u>small</u> theory if for each n, $|S_n(T)| \leq |T|$.

THEOREM 1. The countable small theory T is totally transcendental if and only if for every A contained in a model of T and every p in S(A), there is a finite subset B of A such that p is definable over B.

We will define below the concept "p is definable almost over B".

THEOREM 2. The countable theory T is superstable if and only if for every A contained in a model of T and every p in S(A), there is a finite B contained in A such that p is definable almost over B.

Most of the results in this paper are easy corollaries to theorems in [7]. The main claim to novelty lies in the recognition that a nice hierarchy can be defined in terms of definability. However, our viewpoint is much different from Shelah's. Several notions of rank are central to his development. His results and even some of his definitions depend upon properties of these ranks. In contrast, our development depends only upon the basic properties of forking as developed either along Shelah's line or along that of Lascar-Poizat. With one exception which we will discuss later, the results in this paper hold for uncountable languages with essentially the same proofs. For simplicity of notation, we concentrate on the countable case. The paper is designed to be read by anyone who has read III.1 and III.2 of [7] or [5] or[3].

We follow various notational conventions common in this subject which are explained in these sources. For example, all our constructions take place within a very saturated "monster model". Since it is not usually important to know the length of a finite sequence of variables or elements we write x or a omitting the usual overscore. When the length is important, it is given explicitly.

Section 1. The notion of forking (or more precisely non-forking) provides an explication in a general model theoretic context of the idea of algebraic independence. In particular, if $A \subseteq B$ and $t(c;B)$ does not fork (d.n.f.) over A ($t(c;B)$ denotes the type of c over B) then, intuitively, "c obeys no more relations over B than it does over A". More detailed explanations occur in the three references cited above. More formally, we adopt the following definition.

DEFINITION 1.1. Let $A \subseteq B$ and let c be an arbitrary element. Then, $t(c;B)$ forks over A if there is a formula $\phi(x;y)$ a sequence b from B and sequences b_i for $i < \omega$ such that:

i) $t(b_i;A) = t(b;A)$ for all i.

ii) $\phi(x;b) \in t(c;B)$.

iii) The set $\{\phi(x;b_i): i < \omega\}$ is n-inconsistent for some n. (That is, no more than n of these formulas can be simultaneously satisfied.)

This definition is slightly simpler than the one given in [7] but is equivalent to that definition for stable theories. In fact, the precise definition of forking used is of little importance for this paper. After the next technical lemma where we rely on the definition, we will list the principal properties of forking. In the remainder of the paper (except for 3.4) we will rely not on the definition of forking but only on the properties listed here.

1.2 LEMMA. Let a_i for i in I be a sequence of n element sequences such that $p_i = t(a_i,B)$ d.n.f. over A, where $p = p_i|A$. If D is a ultrafilter on I and a denotes the ultraproduct of the a_i with respect to D, then $q = t(a,B)$ d.n.f. over A.

PROOF. If $\phi(x;b) \in t(a;B)$ then for almost all (with respect to D) i, $\phi(x;b) \in p_i$. But then, since p_i d.n.f. over A, the formula $\phi(x;b)$ does not cause $t(a;B)$ to fork over A. Since this holds for each formula $\phi(x;b)$, $t(a;B)$ d.n.f. over A.

1.3 THEOREM. If T is a stable theory then:

i) If $p \in S(A)$ then p does not fork over A.

ii) If $A \subseteq B \subseteq C$ and $p \in S(C)$ then

a) If p does not fork over A then p d.n.f. over B and $p|B$ d.n.f. over A.

b) If p d.n.f. over B and $p|B$ d.n.f. over A then p d.n.f. over A.

iii) If $A \subseteq B \subseteq C$ and $p \in S(B)$ d.n.f. over A then there exists an extension p' of P in S(C) which d.n.f. over A.

iv) If b is in B and $t(B;C)$ d.n.f. over A then $t(b;C)$ d.n.f. over A.

v) If $A \subseteq B$, $p \in S(A)$ and q is an extension of p in S(B) which does not fork over A and if p is not algebraic over A, then q is not algebraic over B.

1.4 THEOREM (THE SYMMETRY LEMMA). Let T be stable. Then, for any b_0, b_1 and A, $t(b_0, A \cup b_1)$ forks over A iff $t(b_1, A \cup b_0)$ forks over A.

The following result has the same character as those in 1.3 but its proof relies on the symmetry lemma so we list it separately.

1.5 THEOREM. Let A be contained in B. For any c and d

$t(c^\frown d, B)$ d.n.f. over A

if and only if

$t(c, B)$ d.n.f. over A, and

$t(d, B \cup c)$ d.n.f. over $A \cup \{c\}$.

1.6 DEFINITION. Let M be a model of T and $M \subseteq A$, then $p \in S(A)$ is a coheir of $p|M$ if every finite subset of p is satisfiable in M.

This definition is truth functionally equivalent to the definition in [5] and is provably equivalent to the assertion that p does not fork over M.

1.7 THEOREM. Suppose $A \subseteq B$, p is in S(A) and p' is an extension of p to a member of S(B). Then TFAE

i) p' d.n.f. over A

ii) For every pair of models $M \subseteq M'$ with $A \subseteq M$ and $B \subseteq M'$:

(*) there is an extension p_1 of p in S(M) whose coheir on M' extends p'.

iii) There exist a pair of models $M \subseteq M'$ which satisfy (*) and such that $A \subseteq M$, $B \subseteq M'$ and $t(B, M)$ d.n.f. over A.

The following result is immediate in Shelah's development of forking and an early result in the Lascar-Poizat development.

1.8 LEMMA. If p is a type over B and p forks over A there is a finite set B_0 contained in B such that $p|(A \cup B_0)$ forks over A.

1.9 DEFINITION. Let A be a subset of B, then N(B,A) is the subset of S(B) consisting of those members of S(B) which do not fork over A.

Section 2. The notion of forking is designed to provide a canonical extension of a type over a set A to a type over a larger set B. In this section we discuss to what extent the notion "non-forking" can be replaced by the somewhat more intuitive notion "definable". We first review the notion of the definability of a type. If T is stable, every type is definable in the sense mentioned in the introduction.

> The type p in S(A) is definable over B by the map d if for each formula $\phi(x;y)$ there is a formula $d\phi(y)$ with parameters from B such that for each sequence a in A: $\phi(x;a)$ is in p if and only if $d\phi(a)$ holds.

This result is most directly proved by using a rank function to code the length of trees as in [7]. Essentially the same proof, but of a special case and more cumbersome because the rank machinery is not invoked occurs in [1].
Frequently this notion is employed for types over models where it is easy to show [4] that a type can have (up to equivalence) only one definition. Here, however, it is important to consider various definitions of a type since only some of them may have consistent extensions of the following sort.

2.1 DEFINITION. Let p be in S(A) and $A \subseteq B$. If p is defined by d, the d-extension of p on B, denoted d(p,B) is the collection of formulas with parameters from B which satisfy:

$$\phi(x;b) \in d(p,B) \text{ iff } d\phi(b).$$

Note that if d defines $p \in S(A)$ over $C \subseteq A$, there is a d-extension of p to any set B containing C (not just those containing A). In general, the d-extension of p to B may not even be consistent. However, if A is a model M we have the following result.

2.2 LEMMA. Let M be a model of a stable theory and $p \in S(M)$. Suppose that d defines p over M. Then for any A with $M \subseteq A$, $d(p,A)$ is a consistent complete type. In fact, $d(p,A)$ is the unique coheir of p on A.

PROOF. This result is implicit in section 4 of [5] and explicit in [3].

We want to extend this result by requiring not that $M \subseteq A$ but only that the subset of M over which p is defined by d is contained in A.

2.3 LEMMA. If $B \subseteq N$ and $p \in S(N)$ is definable over B by d then p d.n.f. over B.

PROOF. Let $B \subseteq M \subseteq N'$ and suppose $N \subseteq N'$. Let q, q', be the d-extension of p to M, N' respectively. Now q' extends p and by 2.3 is the coheir of q which extends $p|B$ so by lemma 1.7 p d.n.f. over B.

2.4 LEMMA Let $B \subseteq M$ and suppose $p \in S(M)$ is definable over B by d. Then for any A containing B, the d-extension of p on A is a consistent type which does not fork over B.

PROOF. Let M' be a common extension of M and A. By Lemma 2.3 $d(p,M')$ does not fork over B whence by Theorem 1.3 $d(p,A) = d(p,M')|A$ d.n.f. over B.

We have established that, roughly speaking, definable extensions do not fork. The converse is false. For example, if T is the theory of an equivalence relation with exactly two infinite equivalence classes and M is a model of T, the type of a new element in one class does not fork over $\emptyset$ but is not definable over $\emptyset$. In order to obtain a converse, we introduce the following notion.

2.5 DEFINITION. Let A be contained in B and let p be in S(B). Then, p is stationary over A if:

i) p d.n.f. over A and

ii) for every C containing B, p has a unique extension in N(C,B).

We want to show that if $p \in S(A)$ is stationary over A then there is a definition d of p over A such that for any B with $A \subseteq B$, d(p,B) is a consistent complete type which does not fork over A. For this we require some further definitions.

2.6 DEFINITION. The type p splits over A if there exist a,b in dom(p) such that t(a;A)=t(b;A) but for some $\phi(x;y)$, $\phi(x;a)$ is in p while $\phi(x;b)$ is not.

2.7 LEMMA. If $A \subseteq B$, p does not fork over A, and $p|B$ is stationary over A then p does not split over B.

PROOF. Suppose p splits over B, then for some a,b realizing the same type over B, $p|B \cup \{\phi(x;a)\}$ and $p|B \cup \{\sim\phi(x;b)\}$ are both consistent. Let F be an automorphism fixing B and taking a to b. Then $p|B \cup \{\phi(x;b)\}$ is a non-forking extension of $p|B$ (since it is the image under F of $p|B \cup \{\phi(x;a)\}$). But this contradicts the assumption that $p|B$ is stationary.

The proof given here of the following result derives from arguments in [6].

2.8 THEOREM. Suppose $p \in S(A)$ is stationary over A. Then there is a definition, d', of p over A such that for any B containing A and any q in S(B) which extends p and does not fork over A, q is d'(p,B).

Proof. Extend B to a $(|T|+|A|)^+$ saturated model, M. By lemma 2.7 if r denotes the extension of q to M which does not fork over A, r does not split over A. Now r is definable over M, say by d. Let P contain all the parameters which occur in formulas in the range of d. Let X_1={p': p' in S(M) and $d\phi(y)$ in p'} and let X_2={p': p' in S(M) and $\sim d\phi(y)$ in p'}. Now X_1 and X_2 are closed sets which partition S(M). Let X'_1 and X'_2 be the projections of X_1, X_2 on S(A). Then X'_1 and X'_2 form a closed partition of S(A). The only difficulty in this assertion is to show that the two sets are disjoint. So suppose some type p_1 in S(A) is in both X'_1 and X'_2. Then there are extensions p_2, p_3 of p_1 in S(M) containing $d\phi(y)$, $\sim d\phi(y)$ respectively. Moreover, $p_i|P$ is realized in M for i=2,3, say by c and d. But then $\phi(x,c)$ is in r and $\sim\phi(x,d)$ is in r contradicting the fact that p does not split over A. Thus, by compactness, X_1 is definable by some closed formula $\psi(y)$ with parameters from A. Now let $d'\phi$ be ψ. To see d' is the required definition, apply Lemma 2.4 and the definition of stationary.

We have characterized stationary types in terms of definablity. Our next step is to extend this characterization to non-forking types. For this we we require a few more definitions.

2.9 DEFINITION. The formula $\phi(x;b)$ is almost over A if

$$G=\{\phi(x,F(b)):\ F \text{ an automorphism fixing } A\}$$

contains only finitely many inequivalent formulas. The type p is almost over A just if each formula in p is almost over A.

2.10 DEFINITION. The set of finite equivalence relations over A, denoted $FE^m(A)$, is the collection of 2m-ary relations on the monster model which:

i) are definable with parameters from A,

ii) are equivalence relations on the collection of m-tuples from the monster model, and

iii) have only finitely many equivalence classes.

2.11 THEOREM. The formula $\phi(x;b)$, where x is an m-tuple, is almost over A iff there is a finite equivalence relation E in $FE^m(A)$ such that:

$$(x)(y)[E(x;y) \rightarrow [\phi(x;b) \leftrightarrow \phi(y;b)]] .$$

If the conclusion of this theorem holds we say $\phi(x;b)$ depends on the finite equivalence relation E.

Shelah makes the following definition in [7].

2.12 DEFINITION. The type p is definable almost over A if for each formula $\phi(x;y)$ there is a formula $\phi^*(y)$ which is almost over A such that for each sequence b in dom(p):

$$\phi(x;b) \in p \text{ if and only if } \phi^*(b).$$

2.13 THEOREM. Let $A \subseteq B$ and suppose that $p \in S(B)$ d.n.f. over A, then p is definable almost over A.

For this result we must apply an important theorem [5,7].

2.14 THEOREM. (THE FINITE EQUIVALENCE RELATION THEOREM) Let $p \in S(A)$ and $A \subseteq M$. Suppose p_0 and p_1 are distinct extensions of p in S(M) which d.n.f. over A. Then there exists an $R(x;y)$ in $FE^m(A)$ such that:

$$p_0(x) \cup p_1(y) \vdash \neg R(x;y).$$

PROOF OF THEOREM 2.13. It suffices to show that for each formula $\phi(x;y)$ there is a finite equivalence relation over A, $E(u;v)$, and a sequence c such that for all b in B:

$$\phi(x;b) \in p \text{ if and only if } E(b;c).$$

Choose any b in B such that $\phi(x;b)$ is in p and choose c such that $E(c;b)$ for each E in $FE^m(A)$ (where m is the length of b). If the theorem is false, for each E_i in $FE^m(A)$ there is a b_i in B such that $\neg\phi(a,b_i)$ but $E_i(b_i,c)$. Fix a realizing p. Let $p_i = t(b_i, A \cup \{a\})$ and let D be a non-principal ultrafilter on I (an index set for $FE^m(A)$). If b^* denotes the ultraproduct of the b_i mod D, we have by Lemma 1.1 that $t(b^*, A \cup \{a\})$ d.n.f. over A. Since the finite equivalence relations are closed under finite conjunction we also have $E_i(b^*,c)$ for all i. But $t(b;A \cup \{a\})$ and $t(b^*, A \cup \{a\})$ are distinct nonforking extensions of $p|A$. This contradicts 2.14 and proves the theorem.

3. In this section we want to use the technical results connecting forking with definability to show the relation between definability and the spectrum of stability. We begin with Shelah's proof of Lachlan's theorem that an $\aleph_0$ categorical superstable theory is totally transcendental since this suggested our results. The crucial tool here is the finite equivalence relation theorem.

Recall the following definitions.

3.1 DEFINITION. The theory T is stable in λ if for every $A \subseteq M$ a model of T if $|A| \leq |B|$ then $|S(A)| \leq |S(B)|$.

3.2 DEFINITION. The theory T is

i) stable if T is stable in some λ.

ii) superstable if T is stable in λ for all $\lambda \geq \exp(2,|T|)$.

iii) ω-stable (or totally transcendental) if T is countable and stable in $\aleph_0$.

Morley originally defined a notion, totally transcendental, by means of rank which is equivalent for countable theories to ω-stability. We will describe the relation of that notion to this paper below.

3.3 THEOREM. If T is a countable superstable theory and T is ω-categorical then T is ω-stable.

PROOF. If not, there is a model M of T with $|S(M)|>|M|$. Without loss of generality we may fix an integer m such that the number of m-types over M is greater than $|M|$. Since T is superstable, for each p in S(M) there is a finite subset of M over which p does not fork. Since there are only $|M|$ finite subsets of M, the theorem follows if we can show that for any finite A contained in M N(M,A) is also finite. By Ryll-Nardjewski's Theorem, S(A) is finite so it suffices to show that any fixed member r of S(A) has only finitely many extensions in N(M,A). Thus, suppose p(x) and q(x) are distinct members of N(M,A) extending $r \in S(A)$. Then by 2.14, there is an

$E(x;y)$ in $FE^m(A)$ such that $p(x) \cup q(y)$ implies $\neg E(x;y)$. Thus $N(M,A)$ is bounded by the product over the members E of $FE^m(A)$ of $Exp(2,n(E))$ where $n(E)$ denotes the number of equivalence classes of E. But the ω-categoricity of T implies by Ryll-Nardjewski's theorem that the number of formulas with $|A|+2m$ free variables is finite and this number certainly bounds $|FE^m(A)|$. Thus $N(M,A)$ is finite and the theorem follows.

We require three more lemmas before our main results. The key to these results is the observation that in the preceeding proof it would have sufficed to establish that $|N(M,A)|$ was countable. Similarly, to show T is superstable, it suffices to show that for any M and any finite A contained in M, $|N(M,A)| \leq Exp(2,|T|)$.

Matt Kaufmann pointed out that the proof of 3.5 (below) depended on the following lemma which is easy to derive in the Lascar Poizat development of forking but for which I have not found a simple proof in the context set forth here. It is II.3.6 in [3] and follows from section 2 in [5].

3.4 LEMMA. Let A be a subset of the card(A)+ saturated model M and suppose that p_0, p_1 in S(M) extend p in S(A) and neither p_0 nor p_1 fork over A. If $\phi(x;b)$ is in p_0 then for some b' in M with $t(b;A)=t(b',A)$, $\phi(x;b')$ is in p_1.

3.5 LEMMA If T is a countable totally transcendental theory then for any M and any m-type p in S(M), there is a finite subset A_0 of M such that p d.n.f. over A_0 and $p|A_0$ is stationary.

PROOF. Supposing the lemma is false we will construct for each $i \in \omega$, a finite set A_i contained in M such that p d.n.f. over A_i, two sequences a_i, a'_i (with a_i from M) and an E_i in $FE^m(A_i)$ such that $\neg E_i(a_i,a'_i)$ and $t(a_i;A_i)=t(a'_i;A_i)$. Suppose by induction that we have made the first n steps of this construction. Let $A_{n+1}=A_n \cup \{a_n\}$. Then p does not fork over A_{n+1} but $p|A_{n+1}$ is not stationary over A_{n+1}. Let p' extending p and p'' be distinct extensions of $p|A_{n+1}$ which do not fork over A_{n+1}. Then there exists E_{n+1} in $FE^m(A_{n+1})$ such that $p'(x) \cup p''(y) \vdash \neg E_{n+1}(x;y)$. Since M is a model we can choose a_{n+1} in M such that $E_{n+1}(x;a_{n+1})$ is in p'. Applying 3.4 to some $|A|^+$-saturated model, N, containing M (without loss of generality dom(p'') is N) there exists an $a'_{n+1} \in N$ with $t(a'_{n+1};A_{n+1})=t(a_{n+1};A_{n+1})$ and $E(x,a'_{n+1}) \in p''$. But this implies $\neg E_{n+1}(a_{n+1},a'_{n+1})$.

We now show that the A_i, a_i, a'_i constructed above contradict the ω-stability of T. We will define for each $s \in Exp(2,<\omega)$, an automorphism f_s such that there are $Exp(2,\aleph_0)$ types over the countable set $A=\{f_s(a_i)\colon i \in \omega,\ s \in Exp(2,<\omega)\}$. Let $f_{\{0\}}(a_0)=a_0$ and $f_{\{1\}}(a_0)=a'_0$. Suppose that for each $s \in Exp(2,n)$ we have defined f_s. Since $t(a_n;A_n)=t(a'_n;A_n)$, there is an automorphism g_n which fixes A_n and maps a_n to a'_n. Let $f_{s\frown\{0\}}(a_n)=f_s(a_s)$ and let $f_{s\frown\{1\}}(a_n)=f_s(g_n(a_n))$. Now for each $\tau \in Exp(2,\omega)$, let p_τ be the type $\{E_i(x,f_{\tau|i}(a_i))\colon \tau(i)=0\}$. If c realizes $p_{\tau'}$ and d realizes p_τ where $\tau(k) \neq \tau'(k)$ then $\neg E_k(c,d)$, whence the result.

3.6 LEMMA If $p,q \in S(M)$ are definable over $A \subseteq M$ and $p \neq q$, then $p|A \neq q|A$.

PROOF. By 2.4, p and q do not fork over A; whence by 2.11 there is a finite equivalence relation E over A such that: $p(x) \cup q(y) \vdash \neg E(x,y)$. Now suppose that p is defined by d. Write $dE(x,y)(y)$ as $dE(y)$. Then $\neg dE(x)$ is in $q|A$. Moreover, $dE(x)$ is in $p|A$. For, if not, choosing a sequence b_i for $i \in \omega$ such that b_i realizes $d(p;\{A \cup b_j\colon j<i\})$, we have $\neg E(b_i,b_j)$ if $i \neq j$ contradicting the assumption that E has only a finite number of classes.

THEOREM 1. The countable small theory T is totally transcendental if and only if for every A contained in a model of T and every p in S(A), there is a finite subset B of A such that p is definable over B.

PROOF. Suppose first that T is totally transcendental. Let M be a model of T and p be in S(M). Then by Lemma 3.5 there is a finite subset B of M such that p is stationary over A. By Theorem 2.8 p is definable over B. Suppose for any M and any p in S(M), there is a finite subset B of M such that p is definable over B. There are only |M| finite subsets of M and since T is small, there are only $\aleph_0$ types over a finite set. Mapping each element of S(M) to its restriction to a finite set over which it is definable shows (by Lemma 3.6) that there are only |M| types in S(M).

This theorem requires the assumption that T is small. Consider the theory T of infinitely many equivalence relations E_i, each with two classes whose prototypic model has universe $\mathrm{Exp}(2,\aleph_0)$ with $E_i(\sigma, \tau)$ holding iff $\sigma(i)= \tau(i)$. This is the theory of independent or crosscutting equivalence relations. Now T is superstable but not totally transcendental yet every type over the prototypic model is definable over a singleton.

The following lemma is proved by constructing a tree as in the last theorem but the construction is somewhat more complex.

3.7 LEMMA If T is superstable and p ∈ S(A) then there is a finite subset A_0 of A such that p d.n.f. over A_0.

This is theorem III.3.11 of [7]. A somewhat simpler proof will appear in [3] using the simpler definition of forking given here which applies for stable theories.

THEOREM 2. The countable theory T is superstable if and only if for every A contained in a model of T and every p in S(A), there is a finite B contained in A such that p is definable almost over B.

PROOF. Suppose first that T is superstable. Let M be a model of T and p be in S(M). Then, by the definition of superstability there is a finite B contained in M such that p d.n.f. over B. By theorem 2.13 p is definable almost over B. Now let M be a model of T and suppose that for each p there is a finite B such that p is definable almost over B. Since there are only |M| finite subsets of M, we will see that T is stable in every cardinal greater than Exp(2,$\aleph_0$) if we show that there are at most Exp(2,$\aleph_0$) types which are definable almost over a finite set. But, such a type depends on the choice for each formula ϕ of one of a finite number of equivalence classes (of the finite equivalence relation on which ϕ depends) so the theorem follows.

As remarked earlier, the theorems on definability of types hold mutatis mutandis for theories in uncountable languages. The same remark applies to Theorem 2 but not to Theorem 1. That is, if we attempted to define an uncountable totally transcendental theory to be one which is stable in |T|, this could hold for "accidental" reasons. (For example, the theory could be the disjoint union of a theory T with |T| = Exp(2,$\aleph_0$) and categorical in all powers $\geq$ Exp(2,$\aleph_0$) and a countable superstable but not ω-stable theory.) To obtain a result equivalent to the definability condition in Theorem 1, one must define totally transcendental in terms of rank. Thus, Shelah adopted this approach in [7] thereby illustrating Morley's prescience in originally defining totally transcendental in terms of rank rather than the cardinality of Stone spaces.

REFERENCES

[1.] J. T. Baldwin, Countable theories categorical in uncountable power, Ph.D. thesis, Simon Fraser (1971).

[2.] J.T. Baldwin, Conservative extensions and the two cardinal theorem for stable theories, Fund. Math. 88 (1975), pp 7-9.

[3.] J.T. Baldwin, book on stable theories (to appear).

[4.] D. Lascar, Rank and definability in superstable theories, Israel J. of Math. 23 (1976), pp 53-87.

[5.] D. Lascar and B. Poizat, An introduction to forking, J.S.L. 44 (1979), pp 330-351.

[6.] B. Poizat, Théories Instables, (to appear).

[7.] S. Shelah, Classification Theory and the Number of Non-isomorphic Models, North Holland Publishing Co., Amsterdam (1978).

University of Illinois, Chicago Circle

Research for this paper was partially supported by N.S.F. grant MCS 77-01667.

QE RINGS IN CHARACTERISTIC p

C. BERLINE, C.N.R.S., Université Paris VII

G. CHERLIN, Rutgers University(1)

A structure M is said to be QE (for "quantifier eliminable") relative to a language L if the L-theory of M admits elimination of quantifiers. The present paper is a contribution to the classification of QE rings of prime characteristic ; a sequel in preparation will deal with characteristic p^n for $n > 1$, and will contain a fuller account of the classification problem, its history, and the current status of the problem. Suffice it to say here that our understanding of the structure of the Jacobson radical remains unsatisfactory, but the classification is in all other respects complete, due to the combined work of Boffa, Macintyre, Point, Rose, and the present authors.

§1. INTRODUCTION.

R will be a QE ring of prime characteristic p throughout, and J will be its Jacobson radical. When J = (0) it is known that R can be of five possible types [2]:

- A finite field, a product of two isomorphic finite fields, a 2 × 2 matrix ring over a prime field, an algebraically closed field, or an atomless p^n-ring (equivalently, a Boolean power of a finite field over an atomless Boolean algebra).

---

(1) Research supported in part by NSF Grant MCS 76-06484 A01 and by the Alexander - von - Humboldt - Stiftung.

We will show that when $J \neq (0)$ then R is usually just $F_p[J]$, that is the extension of J by an identity element 1 of order p. (There are two exceptional cases, both finite and of characteristic 2.) This result is obtained in §2.

The remaining two sections deal with the structure of J. §3 deals with the case in which J is finite. Here the complete classification is known. In §4 we prove that there are $2^{\aleph_0}$ inequivalent infinite QE nilrings J in each prime characteristic. For reasons of space our further results in the case of infinite J will be published elsewhere. In any case the precise classification eludes us.

§2. $R = F_p[J]$ USUALLY.

We assume from now on that $J \neq (0)$. It is immediate that the ring J admits elimination of quantifiers in the language of rings with 0, but omitting the constant 1. (J is definable in R without parameters, and inherits QE relative to the original language in an obvious sense. The nonzero constants are easily seen to be irrelevant to J in prime characteristic.)

In the next two lemmas we adapt work of Boffa and Point to our present purpose.

LEMMA 1 - For x in J, $x^3 = (0)$.

Proof :

We cite two facts from [2]

1. There is a ring R' elementarily equivalent to R which is algebraic over F_p.
2. Any nilpotent element of R has order at most 3.

Therefore $J' = J(R')$ satisfies the identity : $x^3 = 0$ (cf. [2]), and hence J satisfies the same identity.

BP LEMMA - Let $x_1,\dots,x_k$ and $y_1,\dots,y_k$ be two sets of F_p-linearly independent elements of R such that $x_ix_j = y_iy_j = 0$ for all pairs i,j. Then $(R,\bar{x}) \equiv (R,\bar{y})$.

Proof :

Notice that $1,x_1,\dots,x_k$ and $1,y_1,\dots,y_k$ are linearly independent sets and that there is a ring isomorphism : $\langle 1,\bar{x}\rangle \simeq \langle 1,\bar{y}\rangle$. The claim then follows by QE.

Notation. Set $X = \{x \in R : x^2 = 0, x \neq 0\}$.

LEMMA 2 - If $x^2 = 0$ then $x \in J$.

Proof :

By Lemma 1 $X \cap J \neq \phi$. By the BP Lemma with $k = 1$, $X \subseteq J$, and the claim follows.

LEMMA 3 -

1.1 $XRX = (0)$.

1.2 $XJ = JX = (0)$

2. $J^3 = (0)$.

Proof :

1.1. Notice that for x in X, if $xrx \neq 0$ then x, xrx are linearly independent : if $xrx = nx$ then $(n-xr)x = 0$ and $n-xr$ is invertible (since $xr \in J$ it is nilpotent), and hence $x = 0$, a contradiction. Applying the BP Lemma with $k = 2$ to x, xrx and xrx, x we conclude :

$$x = xrxsxrx$$

for some s. Then $(1-xrxsxr)x = 0$, and we argue as before that $x = 0$, a contradiction.

1.2. If x is in X, a is in J, and $ax \neq 0$ then we argue as above that x and ax are linearly independent. Then by 1.1 the BP Lemma applies with $k = 2$ to x, ax and ax, x, yielding :

$$x = bax$$

for some b in J. Then as above $(1-ba)x = 0$ and we argue that $x = 0$, which is a contradiction. Thus $JX = 0$, and similarly $XJ = 0$.

2. For x in J we have $x^2J = Jx^2 = (0)$ by 1.2 and Lemma 1. Hence for x,y in J we compute :

$$0 = (x+y)^2x = (x^2+xy+yx+y^2)x = xyx.$$

Therefore $(xy)^2 = 0$, so by 1.2 $xyJ = (0)$, as claimed.

LEMMA 4 - Either :

1. For all x in X $xR = F_p.x$; or
2. For all x in X $xR = X \cup \{0\}$.

Proof :

By the BP Lemma for $k = 2$, we find that either :

1. For x,y in X linearly independent, $y \notin xR$; or
2. For x,y in X linearly independent, $y \in xR$.

The claim follows easily.

LEMMA 5 - R/J contains no nontrivial idempotent elements.

Proof :

Notice that for any two nontrivial idempotents e,e' we have :

$$(R,e) \equiv (R,e').$$

Now suppose that $e \neq 0,1$ is an idempotent in R. We consider two cases.

Case 1. $xR = X \cup \{0\}$ for all x in X.

Fix x in X. Replacing e by 1-e if necessary, suppose that $ex \neq 0$. Then replacing x by ex, we suppose that $ex = x$. Then $ey = y$ for all y in $xR = X \cup \{0\}$. Hence by our opening remark, $(1-e)y = y$ for all y in X. But then $x = ex + (1-e)x = 2x$, so $x = 0$, a contradiction.

Case 2. $xR = F_p.x$ for all x in X.

Fix x in X and suppose that $xe \neq 0$, replacing e by (1-e) if necessary. Then $xe = nx$ for some $0 < n < p$. It follows easily that $nx = n^2x$ and hence that $n = 1$, $xe = x$. This holds for all x in X, leading to a contradiction as in the first case.

LEMMA 6 - R/J is a field.

Proof :

Replacing R by an elementarily equivalent ring as in the proof of Lemma 1, we may suppose that R is algebraic over F_p. Then every element of R has a power which is idempotent (for each x of R find integers m,n such that $0 < m < n$ and $x^m = x^n$ and check that $x^{m(n-m)}$ is an idempotent), hence by Lemma 5 all elements of R are either invertible or nilpotent. Thus R is local and R/J is a division ring algebraic over $\mathbb{F}_p$, hence a field.

LEMMA 7 - If $a,b \in R - F_p[J]$ and $\bar{a} = a/J$, $\bar{b} = b/J$ then $(R/J,\bar{a}) \equiv (R/J,\bar{b})$.

Proof :

Suppose $R/J \models \mathcal{Y}(\bar{a})$. Fix x in X and let $y = ax$. Then R satisfies : "$\exists u\ \mathcal{Y}(\bar{u})$ holds in R/J and $u.x = y$". By the BP Lemma for $k = 2$ applied to the pairs x,ax and x,bx, we have also an element v of R such that : $\mathcal{Y}(\bar{v})$ holds in R/J and $v.x = bx$. Since $(v-b)x = 0$ and R/J is a field, therefore $\bar{v} = \bar{b}$, so R/J satisfies $\mathcal{Y}(\bar{b})$, as claimed.

LEMMA 8 - $R/J \simeq F_p$ for $p > 2$. $R/J \simeq F_2$ or F_4 if $p = 2$.

Proof :

As usual we may suppose that R is algebraic over F_p. It follows easily from Lemma 7 that R/J is finite. Let the order of R/J be p^n, and suppose that $n > 1$. By Lemma 7, all elements of $R/J - F_p$ have the same minimal polynomial, which is a separable polynomial of degree n. Hence $p^n - p = n$, and then $p = 2$, $n = 2$.

In the remainder of the present section we will suppose that $p = 2$ and that $R/J \simeq F_4$. We will see that there are only two such exceptional QE rings.

LEMMA 9 - R contains a subfield isomorphic with F_4.

Proof :

By assumption there is an element a in R such that $u = a^2 + a + 1$ is in J. Setting $i = a + u + u^2$ we find that $i^2 + i + 1 = 0$, so $F_2[i] \simeq F_4$.

Notation. We fix i in R so that $i^2 + i + 1 = 0$.

We will prove shortly that $J^2 = 0$.

Notation.

1. $V = X \cup \{0\}$.
2. For x in X, $S(x) \subseteq J/V$ is defined to be the set :

$\{a/V : a^2 = x\}$.

Notice that V and J/V are F_2-vector spaces.

LEMMA 10 - Dim V = 2.

Proof :

For x in X, the map $r \mapsto rx$ induces an F_2-monomorphism from R/J into V. The surjectivity of this maps follows from Lemma 4.

LEMMA 11 - If $r^2 + r + 1 = 0$ then $r = i + x$ or $i + 1 + x$ with $x^2 = 0$.

Proof :

$r = t + a$ with $t = i$ or $i + 1$ and a in J. Since $r^2 + r + 1 = 0$ we find :

(1) $\quad ta + at = a^2 + a.$

Multiplying this on the right by a :

(2) $\quad ta^2 + ata = a^2.$

Multiplying this on the left by t :

(3) $\quad (ta)^2 = t^2a^2 + ta^2 = a^2.$

Expand $(ta+a)^2$ and use (2,3) to get :

(4) $\quad ((t+1)a)^2 = a^2.$

From (3,4) it follows that $(sa)^2 = a^2$ for s in R-J, and by QE that the same applies to any b in J-V, so that if b is in J-V and $y = b^2$, then b, ib, (i+1)b represent distinct elements of S(y). It follows that the cardinality of S(y) is divisible by 3.

Now we need results from §3. Let k = dim J/V. Then we will see (Lemma 7 cf §3) :

1. $k \leq 4$. Also, by QE we deduce easily :

2. For x in X, card(S(x)) = $(2^k-1)/3$.

We may therefore conclude that 2^k-1 is divisible by 9, and hence that k = 0, in which case certainly $a^2 = 0$ after all.

LEMMA 12 - $J^2 = (0)$.

Proof :

Suppose on the contrary that $a \in J$ and $a^2 \neq 0$. It is easy to see that there is then a ring isomorphism :

$$\langle a, ia \rangle \simeq \langle a, ia + a^2 \rangle.$$

It follows then from QE that there is an equation :

$$ra = ia + a^2$$

where $r^2 + r + 1 = 0$. By Lemma 11 we can take $r = i$ or $i + 1$, and thus $a^2 = a$ or 0, a contradiction in either case.

Notation.

1. $A_1 = F_4[x]/(x^2)$.
2. A_2 = algebra freely generated over F_2 by elements i,x satisfying : $i^2 + i + 1 = x^2 = 0$, $xi = (i+1)x$.

LEMMA 13 - R is isomorphic with A_1 or A_2.

Proof :

In view of Lemmas 8-10 and the proof of the latter, R has as an F_2-basis the set : $1,i,x,ix$. To determine the structure of R it remains to determine the value of xi. Clearly $xi \neq x$, so xi is either ix or $(i+1)x$, leading to the two possibilities stated.

LEMMA 14 - A_1 and A_2 are QE rings.

Proof :

It suffices to indicate the treatment of A_2. We use the following criterion for the finite ring A_2 to be QE : every isomorphism between proper subrings A,B of A_2 (containing 1) extends to an automorphism of the ring.

Suppose first that A is generated over F_2 by a single element a, which may be taken to be of form $ci + c'x + c''ix$. If $c = 0$ then $a^2 = 0$ and the image b of a in B has the same form. It is easy to construct an automorphism of A_2 which fixes i and carries a to b, by defining it first as a linear map and verifying that the defining relations of A_2 are preserved.

If $c \neq 0$ we can use an automorphism interchanging i and $i + 1$ and fixing x if necessary to ensure that the image b of a in B is again of the form $di + d'x + d''ix$. Then there is an automorphism of A_2 carrying a to b which fixes J pointwise.

The only other proper subring A of A_2 is the ring $F_2[J]$, as is easily checked. This has automorphisms of order 3, which extend to A_2 by fixing i, as well as automorphisms of order 2, which extend by interchanging i and $i + 1$.

This concludes our discussion of the exceptional cases.

§3. FINITE QE NILRINGS.

The QE rings of prime characteristic which remain to be classified are those of the form $F_p[J]$ where J is a QE ring (without identity) satisfying $J^3 = 0$. In the present section we assume that J is finite. It will turn out that there are only four nontrivial possibilities in this case. The case in which $J^2 = (0)$ is excluded at the outset as trivial.

Example 1. $A_2 = \langle x : 2x = x^3 = 0\rangle$.

Example 2. $B_2 = \langle x,y : 2x = 2y = x^3 = y^3 = 0,\ xy = yx = x^2 + y^2\rangle$.

Example 3. $C_p = \langle x,y : px = py = x^3 = y^3 = 0,\ xy = yx = 0,\ y^2 + tx^2 = 0\rangle$.
(Here t is a nonsquare in F_p, and the isomorphism type of C_p does not depend on the choice of t.)

Example 4. $D_p(t) = \langle x,y : px = py = x^3 = y^3 = 0,\ xy = 0,\ yx = x^2,\ y^2 = tx^2\rangle$.
(Here 1-4t is a non-square).

LEMMA 5 - The rings A_2, B_2, C_p, and $D_p(t)$ all are QE nilrings.

Proof :

We use the criterion given in the proof of Lemma 14 of 2. As this is quite straightforward, we confine ourselves to some remarks used in connection with the rings C_p and $D_p(t)$.

The elements of order two are those of the form ax^2 with a in F_p. It is necessary to check that these elements have square roots. Since $(mx+ny)^2 = Q(m,n).x^2$ with Q a nondegenerate form on F_p, this follows since Q represents all elements of F_p.

The only other point which is not entirely obvious is that the automorphism group acts transitively on elements of the form mx + ny (m or n nonzero). The map $x \mapsto mx + ny$ extends to an automorphism carrying y to :

$$ntx + my \text{ in the case of } C_p \ ;$$

$$ntx + (m+n)y \text{ in the case of } D_p(t).$$

We now consider an arbitrary finite QE nilring J, and we let V be the set of elements of order two in J. Then V is the left and right annihilator of J, and is a vector space over F_p. (We remark that the following lemma does not require the hypothesis that J is finite).

LEMMA 6 - Suppose that for all x,y in J, if $x^2 = y^2$ then $x = \pm y$ modulo V. Then J is isomorphic with A_2 or B_2.

Proof :

Clearly we may suppose that dim V ≥ 2, and fix a,b linearly independent in V. The set : $\{uv : u^2 = a,\ v^2 = b\}$ contains only one or two elements, and it follows easily from QE that this set is contained in $\langle a,b\rangle$. Thus if we fix x,y so that $x^2 = a$ and $y^2 = b$ then we may conclude :

(*) $\qquad xy = ma + nb$

where the pair (m,n) is determined up to sign by a,b. Since $\langle a,b\rangle \simeq \langle b,a\rangle$ we may conclude also that one of the following holds :

(+) $\qquad yx = mb + na$

(-) $\qquad yx = -(mb + na)$.

Now if m = n = 0 then we get a contradiction : xy = yx = 0, so $(x+y)^2 = a+b$, and since $\langle a,b\rangle \simeq \langle a,a+b\rangle$ therefore also $x(x+y) = 0$, yielding $x^2 = 0$.

Furthermore, if $p \neq 2,3$ we get a contradiction as follows. Fix $k \in F_p$, $k^2 \neq 0,1$ and deduce from (*), QE, and $\langle k^2a,b\rangle \simeq \langle a,b\rangle$:

$$kma + knb = kxy = \pm(mk^2a + nb).$$

Then from $km = \pm k^2m$ deduce $m = 0$, and from $kn = \pm n$ deduce $n = 0$.

Next we eliminate the possibility $p = 3$. Here we consider cases, according as (+) or (-) holds.

If (+) holds :

$$(x+y)^2 = (m+n+1)a + (m+n+1)b.$$

Since x,y are linearly independent modulo V, therefore $m+n+1 \neq 0$. Now by (*,+) and QE, we can compute $x(x+y)$ and $(x+y)x$ in two ways each to find :

$$(m+1)a + nb = \pm(m+n(m+n+1))a + n(m+n+1)b$$
$$(n+1)a + mb = \pm(n+m(m+n+1))a + m(m+n+1)b$$

with the same sign in both cases. The solutions are :

if the sign is positive : $m = n = 1$; then $(x+y)^2 = 0$, a contradiction.

if the sign is negative : $m+n = 1$; then $(x-y)^2 = 0$, a contradiction.

If (-) holds :

$$(x+y)^2 = (m-n+1)a + (n-m+1)b \neq 0.$$

We may suppose without loss of generality that $n-m+1 \neq 0$, in which case a calculation like the preceding gives : $m = 1$, $n = -1$. This then yields :

$$(x-y)^2 = -a, \; x(x-y) = b.$$

In other words, $\sqrt{a} \cdot \sqrt{-a} = \pm b$, which contradicts QE since e.g. $\langle a,b\rangle \simeq \langle a,a+b\rangle$.

Now we know that $p = 2$, and hence there is no ambiguity of sign in the choice of m,n. Computing $x.(x+y)$ and comparing coefficients as above we find $m = n = 1$, as in the ring B_2.

If dim $V = 2$ then our analysis is complete. Otherwise select $z^2 = c$ with a,b,c linearly independent and conclude as above : $xz = zx = a+c$, $yz = zy = b+c$. Taking $u = x+y$ we will have $u^2 = a+b$ and we compute easily :

$$a+b+c = u^2 + z^2 = uz = xz + yz = a+b,$$

a contradiction. This completes the analysis.

The essential point in the analysis of the remaining cases is contained in the following result.

Notation. $k = \dim V$.

LEMMA 7 - Dim $J/V \leq 2k$.

Proof :

Let $a_1,\dots,a_k$ be a basis for V and fix $x_1,\dots,x_d$ linearly independent modulo V.

For $c_1,\dots,c_d$ in F_p we have :

$$(\sum c_i x_i)^2 = Q_1(\bar{c})a_1 + \dots + Q_k(\bar{c})a_k$$

where $Q_1,\dots,Q_k$ are quadratic forms. If $d > 2k$ then the Chevalley theorem [4] yields a nontrivial simultaneous zero $\bar{c}$ of the Q_i in F_p, and then $\sum c_i x_i \in V$, a contradiction.

LEMMA 8 - If $k = 1$ then J is isomorphic to A_2, C_p or $D_p(t)$.

Proof :

If $\dim J/V = 1$ it is easy to see that J is isomorphic with A_2 (since every element of V has a square root, other characteristics are ruled out). Thus by Lemma 7 we may now suppose that $\dim J/V = 2$. Fix a in V - (0). By QE, the number of square roots of a in J modulo V is independent of the choice of a, and is therefore p+1. Hence we can choose x,u linearly independent modulo V, such that $x^2 = u^2 = a$. Set xu = ca, ux = c'a.

Notice that $cc' \neq 1$. (Let $Q(m,n) = m^2 + (c+c')mn+n^2$. Then $(mx+nu)^2 = Q(m,n)a \neq 0$ for $(m,n) \neq (0,0)$, hence $(c+c')^2 - 4$ is not a square, and thus $cc' \neq 1$.) Choose $t \neq 0$ so that :

$$n = (t/(1-cc'))^{1/2}$$

exists, set $y = -ncx + nu$, and compute : $y^2 = ta$, $xy = 0$. If $yx = jx^2$ with $j \neq 0$ replace y by $j^{-1}y$, and t by t/j^2. This concludes our analysis.

Finally let us assume : J is not isomorphic with any of the rings A_2, B_2, C_p, or $D_p(t)$. We will arrive below at a contradiction, after a detailed analysis of the situation.

LEMMA 9 - $\dim J/V = 2k$.

Proof :

Let $d = \dim J/V$. Then $d \leq 2k$. Consider the squaring map from nonzero elements of J/V to nonzero elements of V. Then QE implies easily that $p^k - 1$ divides $p^d - 1$. Thus k divides d. If k = d then Lemma 6 applies, contradicting our assumption. Thus d = 2k.

LEMMA 10 - Fix a in V - (0) and let $X = \{x \in J : x^2 = a\}$. Fix r in X, and let $Y = \{x \in X : rx \notin \langle a\rangle\}$. Let $Y/V = \{x/V : x \in Y\}$ and define $h : Y/V \to V - \langle a\rangle$ by $h(y/V) = ry$. Then h is a bijection.

Proof :

h is surjective :

By QE it suffices to show that $rX \not\subseteq \langle a\rangle$. If on the contrary $rX \subseteq \langle a\rangle$, then by QE we have $X^2 \subseteq \langle a\rangle$. Then setting $S = \langle X\rangle$, we have $S^2 \subset \langle a\rangle$, and hence $\dim S/S \cap V \leq 2$

(by the proof of Lemma 7). However the cardinality of $X/V = \{x/V : x \in X\}$ is easily seen (using QE) to be $(p^{2k}-1)/(p^k-1) = p^k + 1 > p^k$, so dim $S/S \cap V > k$. Thus we conclude that $k = 1$, and Lemma 8 applies, contradicting our assumption on J.

h is 1-1 :

By QE the cardinality of $h^{-1}(b)$ is independent of b for b in $V - \langle a \rangle$. Thus if h is not 1-1 then it is at least 2-to-1, and noting that $h(r) = a$ we obtain the inequality :

$$1 + 2(p^k - p) \leq p^k + 1 .$$

This implies that $p^{k-1} \leq 2$, and as the possibility $k = 1$ has been ruled out, we conclude : $p = 2$, $k = 2$, $n = 4$, card$(X/V) = 5$. However this case also leads to a contradiction ; letting $V = 0,a,b,c$ we notice that for x,y in X distinct modulo V, we have :

$$xy = b \text{ or } c,\ yx = b \text{ or } c,\ xy + yx = 0 \text{ or } a,$$

and it follows easily that $X \cup V$ is an additive group, of order 24 and exponent 2, with is impossible.

LEMMA 11 - With the notation of Lemma 10, if $xy \notin \langle a \rangle$ then $yx = xy$.

Proof :

Define $H(r,b,c)$ = "$\exists x \in X \quad rx = b \ \& \ xr = c$". By Lemma 10 if $b \notin \langle a \rangle$ then r,b determine a unique c satisfying H. Hence by QE we have $c \in V \cap \langle r,b \rangle = \langle a,b \rangle$. In other words, for x in Y we have :

$$xr = ma + nrx,$$

with $m,n \in F_p$. Furthermore, applying QE to subrings of the form $\langle r,b \rangle$, we conclude that m,n do not depend on x and r, subject to $rx \notin \langle a \rangle$. Our claim is that $m = 0$ and $n = 1$.

Suppose firstly that $n \neq 0$. Then also $xr \notin \langle a \rangle$, and we can interchange the roles of x and r to conclude :

$$xr = ma + nrx = m(n+1)a + n^2xr,$$

so that $n^2 = 1$, $m(n+1) = 0$. It suffices in this case to prove : $n \neq -1$.

If $n = -1$ then $xy + yx \in \langle a \rangle$ for all $x,y \in X$ and hence $(c_1x+c_2y+c_3z)^2 = Q(c_1,c_2,c_3)a$ with Q a quadratic form, for x,y,z in X. Choosing x,y,z linearly independent we conclude that Q has no nontrivial zero, contradicting [4].

It remains to show that $n \neq 0$. If $n = 0$ then consider the p^k-p elements ry with $y \in Y$. For each such element, $yr \in \langle a \rangle$. Furthermore $r^2 \in \langle a \rangle$. Hence :

$$\text{card } \{x/V \in X/V : xr \notin \langle a \rangle\} \leq p.$$

But Xr contains $V - \langle a \rangle$, so : $p^k-p \leq p$. Since the case $k = 1$ has been excluded, we conclude that $k = 2$, $p = 2$, $d = 4$. By a calculation it can be shown that the multiplication table for X can be cast into the following form (where V = 0,a,b,c ; c = a+b)

$\cdot$	x_1	x_2	x_3	x_4	x_5
x_1	a	b	c	ma	ma
x_2	ma	a	b	ma	c
x_3	ma	ma	a	c	b
x_4	b	c	ma	a	ma
x_5	c	ma	ma	b	a

Another calculation, using this table, shows that x_1,x_2,x_3,x_4 are linearly independent modulo V. Now set $y_1 = x_1 + x_2$, $y_2 = x_2 + x_3$, and compute :

$$y_1^2 = y_2^2 = ma + b, \quad y_1y_2 = b, \quad y_2y_1 = (m+1)a \ .$$

Since n = 0, we should have $y_1y_2 = my_1^2$ or $y_2y_1 = m\,y_1^2$, but by inspection this fails ; this is the desired contradiction.

LEMMA 12 - With the notation of Lemma 10, J = <X>.

Proof :

Set S = <X>. If $x^2 \in \langle a\rangle$ for all $x \in S$ then also $xy + yx \in \langle a\rangle$ for $x,y \in S$ and hence the argument of Lemma 7 applies to prove that $\dim S/S \cap V \leq 2$. But as noted previously, $\dim S/S \cap V > k$, so k = 1, contradicting our assumption on J.

Thus S contains some element y such that $y^2 \notin \langle a\rangle$, and since S is a-definable, it follows from QE that S contains all such elements. A simple computation establishes that $\mathrm{card}(S/S \cap V) > p^{2k-1}$, and hence S cannot be a proper subring of J.

PROPOSITION - The finite QE ring J must be isomorphic to one of the rings A_2,B_2,C_p, or $D_p(t)$.

Proof :

Using the notation of Lemma 10, fix x,y in X linearly independent modulo V so that $xy \in \langle a\rangle$. (If x in X is fixed arbitrarily, it follows easily from Lemma 10 that a suitable element y can be found in X.) By Lemma 11 $yx \in \langle a\rangle$.

Define H(x,y,b,c) = "$\exists r \in X \quad xr = b \ \& \ yr = c$". By Lemma 10 the triple x,y,b determines a unique c satisfying H if $b \in V - \langle a\rangle$, so that QE yields :

$$xr \in V - \langle a\rangle \text{ implies that } yr = ma + nxr$$

for some $m,n \in F_p$ depending only on x,y.

Set $u = y - nx \notin V$. Then $uy = ma$ for the $p^k - p$ elements y/V in X/V such that $xy \notin \langle a\rangle$. Also $ux, uy \in \langle a\rangle$, so :

$$\mathrm{card}(uX - \langle a\rangle) \leq (p+1) - 2 < p^k - p ,$$

so that $uX \subseteq \langle a\rangle$. Thus $uJ = u\langle X\rangle \subseteq \langle a\rangle$. In particular if $u^2 = a'$ and $X' = \{r : r^2 = a'\}$ Then $uX' \subseteq \langle a\rangle$, contradicting $V - \langle a'\rangle \subseteq uX'$. This contradiction completes the proof.

§4. INFINITE QE NILRINGS IN CHARACTERISTIC p.

We will construct $2^{\aleph_0}$ elementarily inequivalent QE nilrings in any prime characteristic p : Other results will appear in [1].

It is helpful to review some general considerations in connection with the construction of QE structures. We make the simplifying assumption that we are dealing with a finite language.

DEFINITION 1 - Let A be a structure.

1. A is uniformly locally finite iff there is a function $f : \mathbb{N} \to \mathbb{N}$ such that for all $a_1, \ldots, a_n$ in A :

$$\mathrm{card}\langle a_1, \ldots, a_n\rangle \leq f(n).$$

2. Sub(A) is the class of finite structures isomorphic with a substructure of A.

Our construction will be based directly on the following general result due to Fraïssé [5] (the usefulness of this result has been emphasized recently by Schmerl, cf.e.g. [6]).

THEOREM 2 - Let K be a collection of finite structures. Then the following are equivalent :

1. K = Sub(A) for some uniformly locally finite QE structure A.
2. K satisfies.
   (i) K is closed with respect to isomorphism and substructure.
   (ii) K is uniformly locally finite in the obvious sense.
   (iii) K has the joint embedding property.
   (iv) K has the amalgamation property.

Remarks on the proof :

I. If K = Sub(A) and A is uniformly locally finite then knowledge of K is equivalent to the universal part of the theory of A.

II. If K satisfies (i-iv) then a suitable structure A may be manufactured as the union of a sequence of elements of K, by iterated application of properties (iii) and (iv). Property (iii) is used to ensure that $K \subseteq$ Sub(A). Property (iv) is used to obtain QE.

We turn to the construction of QE nilrings. We fix a prime number p throughout.

DEFINITION 3 - A ring of characteristic p is restricted iff it satisfies the two conditions :

1. $\forall x,y,z \quad xyz = 0$
2. $\forall x,y \quad x^2 = 0 \Rightarrow xy = yx = 0.$

We know that all QE nilrings are restricted. It is not hard to see that the class K of all finite restricted rings of characteristic p satisfies the conditions of the foregoing theorem, and corresponds to a theory of QE nilrings, namely the model companion of the theory of restricted rings of characteristic p. A similar example is obtained by taking K to be the class of commutative rings in K.

Our objective is of course to identify $2^{\aleph_0}$ classes K of restricted rings satisfying conditions (i-iv) of theorem 2. Condition (ii) is vacuous in this context and condition (iv) implies condition (iii) since the ring (0) is common to all rings.

We require a preliminary discussion of the structure of restricted rings.

NOTATION 4 - Let A be a restricted ring. Set :

$$V = V(A) = \{a \in A : a^2 = 0\}.$$

Then V is an $\mathbf{F}_p$-vector space satisfying VA = AV = (0). Set $\overline{A} = A/V$. Then multiplication induces a bilinear map :

$$\overline{A} \times \overline{A} \to V$$

NOTATION 5 - Let $A \subseteq B_1, B_2$ be restricted rings. Then the free product of B_1 with B_2 over A is denoted $B_1 *_A B_2$. It is easily described via generators and relations, but we require the more explicit description :

Set $V = V(A)$, $V_i = V(B_i)$, and choose vector space decompositions :

$$A + V_i = A \oplus V_i' \quad , \quad V_i' \subseteq V_i$$
$$B_i = (A \oplus V_i') \oplus B_i' \quad .$$

Let $B_i^* = V_i' \oplus B_i'$. Set :

$$C = A \oplus B_1^* \oplus B_2^* \oplus (B_1' \otimes B_2') \oplus (B_2' \otimes B_1') \; .$$

There is an obvious way to define multiplication on C so that $C \simeq B_1 *_A B_2$. This description of the free product will be applied below.
The main ingredient in the construction below is the analysis of the following rather special restricted rings.

DEFINITION 6 - The ring A_n is defined to be the restricted ring of characteristic p which has the following presentation :

generators : $x_1,\ldots,x_n$
relations : $x_i x_j = 0$ for $i < j$
$\sum x_i^2 = 0$

The following two facts are essential.

LEMMA 7 - There is no embedding of A_m into A_n for $m \neq n$.

LEMMA 8 - Fix n. If $A \subseteq B,C$ are restricted rings such that both B and C contain no subring isomorphic to A_n, then $B *_A C$ also contains no subring isomorphic to A_n.

We will now verify these facts by direct calculation, after which it will be a trivial matter to complete the construction.

Proof of Lemma 7 :

Suppose that $x_1,\dots,x_n$ are the generators of A_n. It is then easy to see that $V = V(A_n)$ has a linear basis of the form : $x_i x_j$ $(i > j)$; x_i^2 $(i < n)$.
Also $x_1/V,\dots,x_n/V$ is a basis for $\overline{A}_n = A_n/V$. Suppose now that $y_1,\dots,y_m \in A_n$ satisfy :

$$y_i y_j = 0 \text{ for } i < j \;;\; y_i^2 \neq 0 \text{ for all } i \;;\; \sum y_i^2 = 0 .$$

We can write :

$$y_i = \sum c_{ik}\, x_k + v_i \quad \text{with} \quad v_i \in V .$$

Define supp $y_i = \{k : c_{ik} \neq 0\}$.
We claim :
(*) If $i < j$, $k \in$ supp y_i, $\ell \in$ supp y_j then $k < \ell$.
Indeed for $i < j$ we have :

$$(1) \qquad 0 = y_i y_j = \sum_{k>\ell} c_{ik}\, c_{j\ell}\, x_k\, x_\ell + \sum_k c_{ik}\, c_{jk}\, x_k^2 .$$

Hence $c_{ik}\, c_{j\ell} = 0$ for $k > \ell$, so that :

$$\max(\text{supp } y_i) \le \min(\text{supp } y_j) .$$

However from (1) we also have :

$$\sum_k c_{ik}\, c_{jk}\, x_k^2 = 0$$

and it follows easily that (*) holds.

Now we exploit the relation :

$$(2) \qquad \sum y_i^2 = 0$$

If supp y_i has more than one element this yields an immediate contradiction, since upon expanding y_i^2 we find a nonzero term of the form $c_{ik}\, c_{i\ell}\, x_k\, x_\ell$, and by (*) there is only one such term. Thus we may write :

$$y_i = c_i\, x_{k(i)} + v_i$$

By (2) we see easily that $k(i) = i$, hence $m = n$ as asserted.

Proof of Lemma 8 :

Suppose $x_1,\dots,x_n \in B *_A C$ and $\langle x_1,\dots,x_n\rangle \simeq A_n$.
We may write :

$$x_i = a_i + b_i + c_i + v_i, \quad v_i \in V = V(B *_A C), \quad a_i \in A,\ b_i \in B',\ c_i \in C'.$$

Set $y_i = a_i + b_i + c_i$. Observe that also :

$$\langle y_1,\dots,y_n\rangle \simeq A_n$$

Now use the relation $y_iy_j = 0$ for $i < j$ to conclude :

$$b_i \otimes c_j = c_i \otimes b_j = 0 \text{ ; that is ,}$$

$$b_i = 0 \text{ or } c_j = 0$$

$$b_j = 0 \text{ or } c_i = 0 \qquad (\text{for } i < j).$$

Fix i minimal that b_i or c_i is nonzero. We may suppose that b_i is nonzero. Then $c_j = 0$ for $j \neq i$, so that $y_j \in B$ for $j \neq i$. The relation $\sum y_k^2 = 0$ then implies that $y_i^2 \in B$, hence $b_i \otimes c_i + c_i \otimes b_i = 0$, and since b_i is nonzero therefore $c_i = 0$. Hence $y_1,\dots,y_n \in B$, as claimed.

Application. For $X \subseteq \{2,3,4,\dots\}$ let $K(X)$ be the class of finite restricted rings which do not contain any subring isomorphic to A_n for $n \in X$. Then :

I. K(X) satisfies properties i-iv of theorem 2.

II. If $X \neq Y$ then $K(X) \neq K(Y)$ (Indeed if $n \in X-Y$ then $A_n \in K(Y) - K(X)$).

By Theorem 2 each of the classes K(X) gives rise to a uniformly locally finite QE nilring A(X) satisfying :

$$K(X) = \text{Sub } A(X) .$$

Thus we have proved :

THEOREM 9 - For a fixed prime p, there are $2^{\aleph_0}$ elementarily inequivalent QE nilrings of characteristic p.

Note also :

PROPOSITION 10 - For a QE structure A in a finite language, the following are equivalent :

1. A is uniformly locally finite
2. A is $\aleph_0$-categorical.

COROLLARY 11 -

1. There are $2^{\aleph_0}$ $\aleph_0$-categorical commutative rings of any prime characteristic p.
2. There are $2^{\aleph_0}$ $\aleph_0$-categorical nilpotent groups of class 2.

(These results are obtained by introducing structures bi-interpretable with the $2^{\aleph_0}$ $\aleph_0$-categorical QE nilrings constructed above. For result 2, this involves the Mal'cev correspondence. The correspondence appropriate to case 1 was introduced in [3].

PROBLEM - Construct $2^{\aleph_0}$ commutative QE nilrings.

BIBLIOGRAPHY

[1] Ch. BERLINE, G. CHERLIN, QE nilrings of prime characteristic, to appear in the Proceedings of the Logic Meeting in Brussels and Mons (1980).

[2] M. BOFFA, A. MACINTYRE, F. POINT, The quantifier elimination problem for rings without nilpotent elements and for semi-simple rings, to appear in the Proceedings of the Karpacz conference (1979).

[3] G. CHERLIN, On $\aleph_0$-categorical nilrings, Algebra Universalis 10 (1980), p. 27-30.

[4] C. CHEVALLEY, Démonstration d'une hypothèse de M. Artin, Abhandlungen aus dem Mathematischen Seminar der Hansischen Universität, vol. 11, Leipzig (1936), p. 73-75.

[5] R. FRAÏSSE, Sur certaines relations qui généralisent l'ordre des nombres rationnels, C.R. Acad. Sci. Paris, 237 (1953), p. 540-542.

[6] J. SCHMERL, Countable homogeneous partially ordered sets, Algebra Universalis 9 (1979) p. 317-321.

HIERARCHIES OF SETS AND DEGREES BELOW 0'

by

Richard L. Epstein, Richard Haas, and Richard L. Kramer*

We examine two hierarchies of sets below 0' based on the number of changes a recursive approximation to a set needs to make. Both are generalizations of the notion of being r.e. The first classifies sets by asking what functions dominate the number of changes, as previously set out in Epstein [4]. This extends the ideas of Putnam on trial and error predicates [8]. The second views the changes as dominated by a constructive ordinal, as first suggested by Addison [1], and developed by Ershov [6]. We provide a translation between them and relate these hierarchies to the degrees of unsolvability $\leq$ 0'.

*   *   *

We first review some facts about sets $\leq_T 0'$. All notation comes from Epstein [4].

The reader should be aware that there is another hierarchy of degrees $\leq \underline{0}'$ which is based on the jump operator and is due primarily to Cooper. See Epstein [4], Chapter XI for that.

We indicate the end of a proof by ■, and the end of a subproof by □.

*We are grateful to H. Hodes for correcting a number of errors in an earlier version of this paper. Roger Maddux also aided us with suggestions concerning diagrams.

We notate $0' = \{\langle x,y\rangle : \varphi(y)\downarrow\}$. Recall that $0' \equiv_T K = \{x : \varphi_x(x)\downarrow\}$.

First we note the Quantifier Characterization of Sets Below 0':

$A \leq_T 0'$ iff there are two recursive predicates R,S such that $w \in A$ iff $\forall x \exists y\, R(x,y,w)$

iff $\exists x \forall y\, S(x,y,w)$

Proof: $\Leftarrow|$ $\forall y\, S(x,y,w)$ is recursive in 0' since its the complement of an r.e. predicate. Therefore $\exists x \forall w\, S(x,y,w)$ is r.e. in 0'. Similarly, $\exists x \forall y\, \neg R(x,y,w)$ is r.e. in 0'. Thus A and $\overline{A}$ are r.e. in 0' so A is recursive in 0'.

$\Rightarrow|$ Let f enumerate 0' and let $0'_s = \{x : f(y) = x \leq s$, some $y \leq s\}$ be 0' enumerated to level s. For some e, $A = \Phi_e(0')$ the eth function partial recursive in 0'. Define $A_s(x) = \Phi_{e,s}(0'_s)(x)$ (where $\Phi_{e,s}$ is Φ_e with calculations truncated by s). Then

$$x \in A \text{ iff } \exists t \forall s > t \quad \Phi_{e,t}(0')(x) = 1$$
$$\text{iff } \forall t \exists s > t \quad \Phi_{e,t}(0')(x) = 1 \qquad \boxtimes$$

As a corollary to the proof we have

The Limit Lemma (Shoenfield [11]): $f \leq_T 0'$ iff there is some recursive function g such that

$$f(x) = \lim_s g(s,x).$$

Here $\lim_s g(s,x) = f(x)$ means for all sufficiently large s, $g(s,x) = a = f(x)$. That is, $\exists t \forall s \geq t\ g(s,x) = a$.

When we have $g(s,x)$ as in the Limit Lemma we call $g(s,x) = f_s(x)$ and call that a recursive approximation to f. If f is 0-1 valued we'll view it as a set and say $g(s,x) = A_s(x)$.

If A is r.e. then A has an approximation A_s such that $A_s(x)$ "changes its mind at most once." That is, there is at most one s (if $x \in A$ there is one) such that $A_s(x) = 0$ and $A_{s+1}(x) = 1$ (if $x \notin A$, $A_s(x) = 0$ always).

The r.e. sets have always been considered a distinguished class of sets below 0'. This is partly because the special property just described is easy to utilize, and partly because it reflects the relation in logic between axioms (a recursive set) and theorems (an r.e. set), as "proving" is just a recursive enumeration.

We wish to generalize the notion of r.e.'ness so that r.e. sets are seen as part of a continuum.

Let us classify A by measuring how often an approximation to A changes before it settles down.

<u>Definition</u>: A is n-r.e. iff there is a recursive approximation A_s to A such that for all x,

$A_0(x) = 0$ and

$|\{s : A_s(x) \neq A_{s+1}(x)\}| \leq n.$

Note that this definition can be extended in an obvious way to apply to functions, too. (This is what Putnam [8] calls an n-trial predicate. Similar ideas are presented in Gold [7], but note well that his definition of 2-r.e. is quite different from ours.)

The only 0-r.e. set is $\emptyset$ and the 1-r.e. sets are the usual r.e. sets. The n-r.e. sets are those that arise after n steps in generating the Boolean algebra of r.e. sets. Rogers [9], p. 317 shows that A is in the Boolean algebra generated by the r.e. sets iff $A \leq_{btt} 0'$ ($\leq_{btt}$ means bounded-truth-table reducible). See also Putnam [8].

If $\mathcal{C} \subset 2^{\omega}$ denote by $\mathcal{C}^T$ the set $\{deg(A) : A \in \mathcal{C}\}$. By degree we always mean Turing degree unless otherwise specified. If $\mathcal{C}$ is a class of partial functions from $\mathbb{N}$ to $\mathbb{N}$ define $\mathcal{C}^T = (\mathcal{C} \cap 2^{\omega})^T$.

<u>Theorem 1</u>: $\{A : A \text{ is n-r.e.}\}^T \subsetneqq \{A : A \text{ is n+1-r.e.}\}^T$.

To prove this we construct, for e.g., n = 1, a 2-r.e. set A which is $\neq_T$ any r.e. set. We use the fact that A may change <u>once more</u> than any r.e. set to execute the diagonalization. That's all the freedom we need. This proof is given in Epstein [4], Appendix 2, and is due to R. W. Robinson. To the best of our knowledge, Cooper was the first to prove this, in [2].

Recalling that $X, \overline{X}$ r.e. implies that X is recursive we ask whether $X, \overline{X}$ n+1-r.e. implies that X is n-r.e. To answer this we'll modify the question a little. Say that

A <u>is weakly n-r.e.</u> if $A = \lim_s A_s(x)$ and
$|\{s : A_s(x) \neq A_{s+1}(x)\}| \leq n$.

That is $A_0(x) = 0$ is no longer required. A picture will help:

Diagram 1

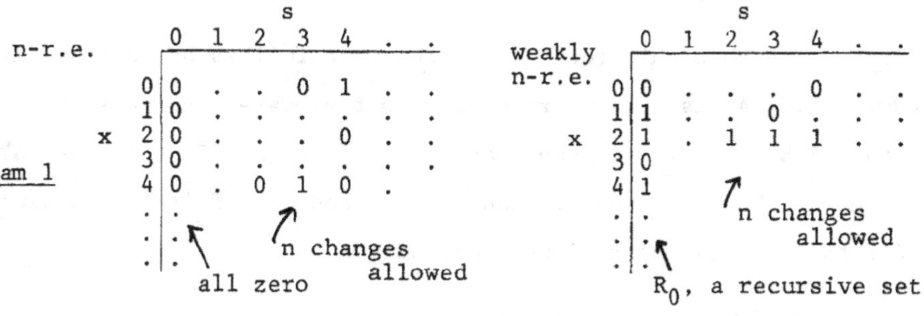

Note that the weakly n-r.e. sets are closed under complementation: just reverse the 0' and 1's.

The weakly 0-r.e. sets are the recursive sets.

<u>Theorem 2</u>: $A, \overline{A}$ are n+1-r.e. iff they are weakly n-r.e.

<u>Proof</u>: $\Leftarrow$| Easy.

$\Rightarrow$| Suppose $A = \lim_s A_s$, $\overline{A} = \lim_s B_s$ are n+1-r.e. approximations. To obtain a weakly n-r.e. approximation C_s to A, we will go to the first stage at which $A_s(x) \neq B_s(x)$ and set $C_0(x) = A_s(x)$. Then we'll only allow $C_s(x) \neq C_{s+1}(x)$ if we see $A_t(x) \neq A_{t+1}(x) \neq B_{t+1}(x)$ since we know that in the end $A(x) \neq B(x)$ all x. Then there will be at most n changes.

Formally: let $s_0 = \mu s\ (A_s(x) \neq B_s(x))$, and, for $m < n+1$,

$$s_{m+1} = \mu s\ (s > s_m \text{ and } A_{s_m}(x) \neq A_s(x) \text{ and } A_s(x) \neq B_s(x)).$$

Then define $C_s(x) = A_{s_0}(x)$ all $s \leq s_0$.

$$C_s(x) = \begin{cases} A_s(x) & \text{if} \quad s_m \leq s < s_{m+1} \\ A_{s_{m+1}}(x) & \text{if} \quad s_{m+1} \leq s \end{cases}$$

We leave to you that $\lim_s C_s(x) = A(x)$. Clearly its n-r.e. ∎

An especially important fact about the r.e. degrees is that they are dense (see Sacks [10]). What can we say about the n-r.e. degrees?

Theorem 3: Given A n+1-r.e. and not recursive, there is some non-recursive r.e. $C \leq_T A$.

Proof: We show this for $n = 1$; the rest follows by an inductive procedure. If A is r.e. we're done. So there are r.e. sets E,F such that $F-E = A$. (F is the numbers put into A, E the ones taken out.) Let f be a 1-1 enumeration of F, and $C = f^{-1}(E)$. Convince yourself that C is the required set. ∎

Corollary: No n-r.e. degree is minimal.

Cooper (unpublished) was the first to prove the Corollary, but by quite different means.

By relativizing the proof of Theorem 3 we can get that if $\underline{0} \leq \underline{d} < \underline{a}$ and both are n+1-r.e. then there is some $\underline{c}$, $\underline{d} < \underline{c} < \underline{a}$ and $\underline{c}$ is r.e. in $\underline{d}$. This does not solve for us the question of whether the n-r.e. degrees are dense: that's still open. But it makes us ask where in our classification schema such a $\underline{c}$ will lie.[1]

We can extend our hierarchy by raising the bound on the number of changes allowed.

Definition: Given any f which is total

A is f-r.e. iff there is some recursive approximation A_s to A such that $|\{s : A_s(x) \neq A_{s+1}(x)\}| \leq f(x)$.

Similarly we may define what it means for $h \leq_T 0'$ to be f-r.e.

... continued

We say a degree $\underline{a}$ is f-r.e. if some $A \in \underline{a}$ is f-r.e.

Note that this extends the weakly n-r.e. definition.

What do we know? Certainly every $A \leq_T 0'$ is f-r.e for some f. Indeed A is f-r.e. for some $f \leq_T 0'$: we can spot recursively in 0' the last place A_s changes its mind.

Let's look at a simple f-r.e. class. Abbreviate identity-function-r.e. as <u>id-r.e.</u>

<u>Theorem 4</u>. $\{A : A \text{ is id-r.e.}\}^T$ is not dense.

If $\{A : A \text{ is id-r.e.}\}^T \subseteq \mathbb{A} \subseteq \mathbb{D}(\leq \underline{0}')$ then $Th(\langle \mathbb{A}; \leq \rangle)$ is undecidable.

<u>Proof</u>: A permitting argument is the construction of a set B which is $<_T C$, a given r.e. set, by allowing $B_s(x) \neq B_{s+1}(x)$ only if $C_s(y) \neq C_{s+1}(y)$ for some $y \leq x$. Any permitting argument produces an identity-r.e. degree. Hence the construction of a minimal degree below a given r.e. degree produces an id-r.e. minimal degree (see e.g., Epstein [3]).

All the degrees used in the proof in Epstein [4] and [5] that $Th(\langle \mathbb{D}(\leq \underline{0}'), \leq \rangle)$ is undecidable are constructed by permitting arguments. Hence they are id-r.e., and the translation of fragments of arithmetic goes through as for $Th(\langle \mathbb{D}(\leq \underline{0}'); \leq \rangle)$. ☒

It is open whether $Th(\langle \{A : A \text{ is r.e.}\}^T, \leq \rangle)$ is undecidable (see Soare [12] for a discussion). We can also ask whether $Th(\langle \{A : A \text{ is n-r.e.}\}^T, \leq \rangle)$ is the same as $Th(\langle \{A : A \text{ is n+1-r.e.}\}^T, \leq \rangle)$.

Let us look at classes of f-r.e. sets. We say A is $\mathfrak{F}$-r.e. if some $f \in \mathfrak{F}$, A is f-r.e.

<u>Theorem 5</u>: Let $\mathfrak{F}$ be a class of total recursive functions such that we can recursively enumerate an index for every function in $\mathfrak{F}$. Let g be a function which dominates every func-

tion in $\mathfrak{F}$. Then there is an A which is g-r.e. but is not $\mathfrak{F}$-r.e.

The idea of the proof is just a modification of the proof that there is a 2-r.e. degree which is not 1-r.e. Here we know that given an f-r.e. set B, the set A which we are constructing is allowed, for sufficiently large x, to make one more change than B does. The proof is in Epstein [4], Appendix 2.

Corollary: Any of the usual hierarchies of recursive functions induces a hierarchy on the recursive-r.e. degrees.

But the recursive-r.e. degrees aren't the whole story.

Theorem 6: There is an $A \leq_T 0'$ which is not f-r.e. for any recursive f.

The essential step in the proof is to show that we can "enumerate" all the recursive-r.e. approximations. The proof appears in Epstein [4], Appendix 2 and is due to Posner (private communication 1973).[2]

* * *

How can we extend this hierarchy? We take another tack as in Ershov [6]. Instead of bounding the number of changes by functions when we pass from the n-r.e. case, we'll bound them by ordinals.

Consider that A being weakly n-r.e. can be viewed as A being given by a collection of n partial recursive functions, $\psi_0, \ldots, \psi_n$. Here

$A(x) = \psi_k(x)$ where k is the largest t such that $\psi_t(x)\downarrow$.

Of course A need not be 0-1 valued. For any total G we could say that G is weakly n-approximable if there are partial recursive functions $\psi_0, \ldots, \psi_n$ such that $G(x) = \psi_k(x)$ where $k = \max t(\psi_t(x)\downarrow)$. How could we extend this to ω? In that case there needn't be a largest t for which $\psi_t(x)\downarrow$. But there is a smallest one! And surely

$\{G : G$ is weakly n-approximable$\} = \{G : \exists\psi_0, \ldots, \psi_n$ partial recursive such that $G(x) = \psi_k(x)$, $k = \mu t(\psi_t(x)\downarrow)\}$. Let's clean this up. After all we may collect these ψ_i's as $\psi(\langle i,x\rangle)$.

Definition: $\psi \xrightarrow[(n)]{} G$, ψ partial recursive if for all x,

$$G(x) = \psi(\langle k,x\rangle) \text{ where } k = \mu t < n(\psi(\langle t,x\rangle)\downarrow).$$

Now it's easy to define an ω-r.e. set. Let ψ be partial recursive; then $\psi \xrightarrow[(\omega)]{} G$ if $G(x) = \psi(\langle k,x\rangle)$ such that $k = \mu t(\psi(\langle t,x\rangle)\downarrow)$.

From this definition it follows that A is ω-r.e. iff $A \leq_{tt} 0'$ ($\leq_{tt}$ means truth-table reducible).

Let us define

$$\nabla_n = \{A : \text{some partial recursive } \psi,\ \psi \xrightarrow[(n)]{} A\}$$

$$\nabla_\omega = \{A : \text{some partial recursive } \psi,\ \psi \xrightarrow[(\omega)]{} A\}.$$

Theorem 7: 1. $A \in \nabla_{n+1}$ iff A is weakly n-r.e.

2. $A \in \nabla_\omega$ iff A is f-r.e. for some recursive f.

Proof: We outlined 1. above. Note that $A \in \nabla_n$ iff for every x we guess at $A(x)$ at most n times, which is the same as changing our guess n-1 times.

2. Suppose A is f-r.e. for some recursive f. That is $A = \lim_s A_s(x)$ where the changes for each x are bounded by $f(x)$. The idea is simple: we reverse the order of the labels on the changes since we know that we can label our first guess by $f(x)$. Formally, define

$$\psi(\langle r,x\rangle) = \begin{cases} \text{undefined if } r > f(x) \\ A_0(x) \quad \text{if} \quad r = f(x) \\ A_y(x) \quad \text{if} \quad r < f(x) \text{ and } \psi(\langle r+1,x\rangle) \\ \text{was chosen to be } A_z(x) \text{ and} \\ y = \mu t > z \quad (A_t(x) \neq A_z(x)). \end{cases}$$ □

Now let us suppose that we are given $\psi \xrightarrow[(\omega)]{} A$. Let ψ_s be ψ with all computations truncated at stage s. Define

$$A_s(x) = \begin{cases} 0 \text{ if } s = 0 \\ \psi(\langle r,x\rangle) \quad \text{where } r = \mu t \leq s\ \psi_s(\langle t,x\rangle)\downarrow \\ \qquad\qquad \text{if } s > 0 \text{ and there is such an } r \\ 0 \text{ otherwise} \end{cases}$$

Then $\lim_s A_s(x) = A(x)$. And the number of changes that $A_s(x)$ makes can't be more than r, where for the first time $\psi_s(q,x)\downarrow$ for any q, $r = \mu t \leq s\ (\psi_s(\langle t,x\rangle)\downarrow)$. Calling that $r = f(x)$ we note that f must be total as for all x, $\exists t\ \psi(\langle t,x\rangle)\downarrow$. ⊠

Corollary: $\{A : A \leq_{btt} 0'\}^T \subsetneq \{A : A \leq_{tt} 0'\}^T$.

Proof: Theorem 5 and 7, plus the observation that the former class of sets is the Boolean algebra generated by the r.e. sets, and the latter class of sets is ∇_ω, yield the corollary. ⊠

There is no reason to stop this hierarchy with ω. Let α be any constructive ordinal (see Rogers [9], Chapter 11). Denote by α-S, α-T, α-U, ... the list of recursively related univalent systems of notation for α. Thus given α-S we have a recursive ordering on a recursive set of numbers, where we can recursively distinguish 0, successors and limit ordinals, and the predecessor and successor functions are recursive. By $z = (\gamma)_S$ we will mean that z denotes γ in the system S.

Definition: Given α-S, define

$\psi \xrightarrow[(\alpha-S)]{} \varphi$ iff $\varphi(x) = \psi(\langle(\gamma)_S,x\rangle)$ where γ is the least ordinal $< \alpha$ such that $\psi(\langle(\gamma)_S,x\rangle)\downarrow$.

We read $\psi \xrightarrow[(\alpha-S)]{} \varphi$ as "φ is the α-limit of ψ in notation S" or "ψ α-approximates φ in notation S." When working with a fixed notation we will often delete reference to "notation S."

We need not suppose that φ is total in this definition.

Lastly define

$$\nabla_{\alpha\text{-}S} = \{\varphi : \text{some partial recursive } \psi, \ \psi \xrightarrow[(\alpha\text{-}S)]{} \varphi\}$$

As the next theorem will demonstrate, some of these classes will be notation-dependent. We leave to the reader, however, that ∇_n, $n \geq 0$ and ∇_ω are not. That is, for any system S, $\nabla_{n\text{-}S} = \nabla_n$, and $\nabla_{\omega\text{-}S} = \nabla_\omega$ as previously defined. Immediately we can conclude

Corollary (to Theorem 7): $Th(\langle \nabla^T_{(\alpha\text{-}S)}; \leq \rangle)$ is undecidable if $\alpha \geq \omega$.

Proof: See Theorem 4.

More importantly, we really do have all the functions below 0' now.

Theorem 8: If $f \in \nabla_{\alpha\text{-}S}$ then $f \leq_T 0'$.

If $f \leq_T 0'$ then for some notation S, $f \in \nabla_{\omega^2\text{-}S}$.

Moreover, given any α-S, we may find β-U, $\beta \geq \alpha$ such that we can pass from α-S to β-U recursively and $f \in \nabla_{\beta\text{-}U}$.

Proof:[3] Suppose that for some α-S, $f \in \nabla_{\alpha\text{-}S}$. We will give an informal proof that $f \leq_T 0'$. As S is fixed we will suppress reference to it throughout.

Diagram 2

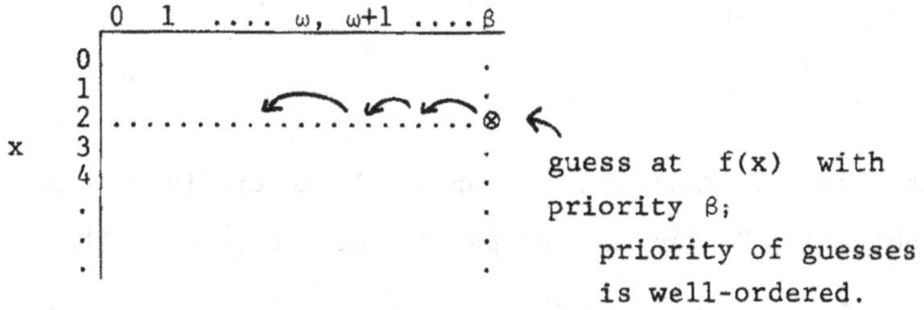

To calculate f(x) look for some β with $\psi(\langle \beta,x \rangle)\downarrow$. Then ask if there is a $\zeta < \beta$ such that $\psi(\langle \zeta,x \rangle)\downarrow$, a question recursive in 0'. If not, then $\psi(\langle \beta,x \rangle) = f(x)$. If there is such a ζ go to it and ask the same question.

Since α is well-ordered we can be sent to a smaller ξ with

$\psi(\langle\xi,x\rangle)\downarrow$ only a finite number of times. The least one we reach is then $f(x)$. This procedure for finding $f(x)$ is clearly recursive in $0'$.

Note that if f is partial the same proof works. Then for each x, $f(x)\not\downarrow$ iff we simply never make a guess at $f(x)$. Thus if $f \in \nabla_\alpha$ and is partial or total then $f \leq_T 0'$. □

Now assume $f \leq_T 0'$. We will obtain a system of notation S such that $f \in \nabla_{\omega^2\text{-}S}$.

Since $f \leq_T 0'$, $f(x) = \lim_s f(s,x)$. Let $X = \{\langle x,s\rangle : s = 0$ or $f(s,x) \neq f(s+1,x)\}$. Consider the recursive well-ordering R on X given by $\langle x,s\rangle R \langle y,t\rangle$ iff $x \leq y$ or $(x = y$ and $s \geq t)$. This has order type ω.

Diagram 3

e.g.

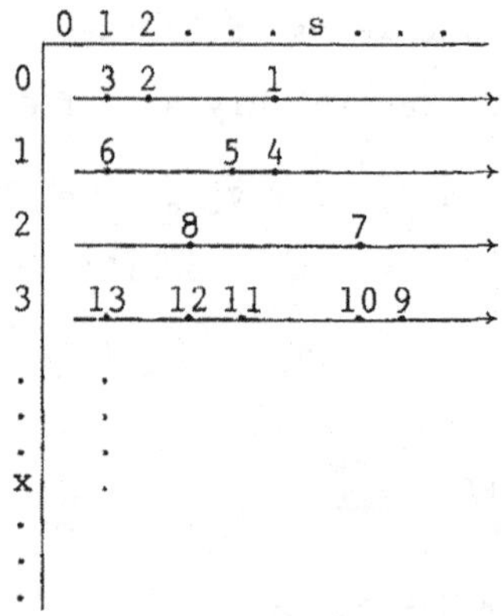

As in the proof of Theorem XX, Chapter 11, Rogers [9] we obtain a system of notation S for ω^2 by padding out $R|_X$ as follows:

$$xSy \text{ iff } x = \langle v,n\rangle \text{ and } y = \langle u,m\rangle \text{ and } \langle v,u\rangle \in X$$
$$\text{and if } v \neq u,\ vRu; \text{ otherwise if } v = u,\ n < m.$$

Now define

$$\psi(\langle\langle x,s\rangle,x\rangle) = \begin{cases} f(s,x) & \text{if } s = 0 \text{ or } f(s,x) \neq f(s-1,x) \\ \not\downarrow & \text{otherwise} \end{cases}$$

Clearly $\psi \xrightarrow[\omega^2-S]{} f$. □

Lastly, given any α-S, we can obtain a notation U for $\alpha+\omega^2$ in the same manner as described above, for which $f \in \nabla_{\alpha+\omega^2-U}$. ■

Note that if we were to define classes $\nabla'_{\alpha-R}$ where we require only that α-R is a recursive well-ordering on a recursive field, then given any $f \leq_T 0'$, there is some R such that $f \in \nabla_{\omega-R}$.

<u>Theorem 9</u>: Given a notation α-S.

The classes $\nabla^T_{\beta-S}$ for $\beta \leq \alpha$, form a hierarchy. That is,

$$\left(\bigcup_{\zeta<\beta} \nabla^T_{\zeta-S}\right) \subsetneq \nabla^T_{\beta-S} .$$

For $\alpha = \beta+1$ this is a modification of the proof of Theorem 1 that there is a 2-r.e. set which $\not\equiv_T$ any 1-r.e. set. After all, we're allowed for each variable x one more change than before. The changes in the proof are really only to accomodate the ordinal notation. Now consider α a limit ordinal. Essentially in this proof we're allowed to construct an A with an approximation whose numbers of changes, for each x, dominates the number of changes to any given $Y \in \nabla_\xi$, for $\xi < \alpha$. Thus the proof is, up to modification to accommodate ordinal notation (which is not simple), the proof of Theorem 5.

Note that for many-one degrees the hierarchy is better behaved: if $A \leq_m B \in \nabla_{\alpha-S}$ then $A \in \nabla_{\alpha-S}$. This fails for Turing degrees: by the Corollary to Theorem 3 any $A \leq_T 0'$ which has minimal degree provides a counterexample for ∇_1.[4]

<u>Theorem 10</u>: Let S be a notation for $\omega \cdot n$.

If $A \in \nabla_{\omega \cdot n-S}$ we have A is ∇_n-r.e.

Conversely, if A is ∇_n-r.e. then $A \in \nabla_{\omega \cdot n-W}$ where W is the notation for $\omega \cdot n$ given by the canonical ordering on $\omega \times n$ (namely $\langle x,r\rangle < \langle y,t\rangle$ iff $x < y$ or if $x = y$ then $r < t$).

We will denote by $\nabla_{\omega\cdot n}$ the class of ∇_n-r.e. functions.

Proof: Let $A \in \nabla_{\omega\cdot n-S}$. We will approximate A via the information given by ψ where $\psi \xrightarrow[(\omega\cdot n-S)]{} A$. And we'll construct a total function $f \in \nabla_n$ which dominates the number of changes that approximate makes. We'll show that $f \in \nabla_n$ by showing that we need to guess recursively at most n times at $f(x)$ until we are correct. We present an informal proof and leave the details to the reader.

First note that given any q,z which notate ordinals $\leq \omega\cdot n$ we can recursively determine s and t such that $(q)_S = \omega\cdot k+t$ and $(z)_S = \omega\cdot m+s$, and whether $k < m$ or $m < k$ or $m = k$. We proceed with the proof, deleting subscripts for legibility.

Diagram 4

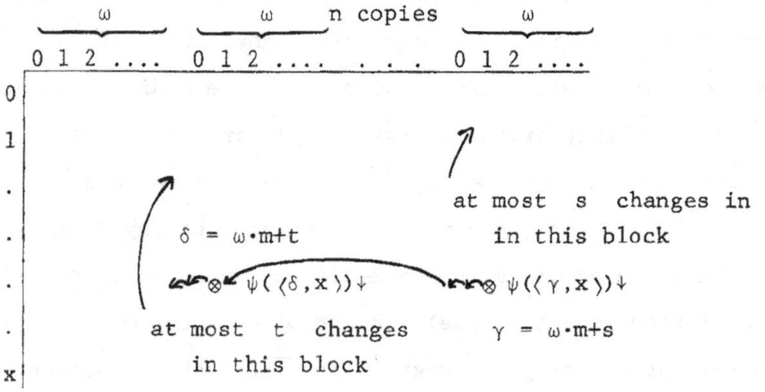

Look at the first $\psi(\langle\gamma,x\rangle)$ which is $\downarrow$ (in a computation search). We have $\gamma = \omega\cdot m+s$ for some $m < n$. As long as we stay in the $m^{\underline{th}}$ ω-block we know that ψ can change its guess at most s times. So $f(0,x) = s$. If we later have $\psi(\langle\delta,x\rangle)\downarrow$ and $\delta = \omega\cdot k+t$ and $k < m$ we are in another block, but now ψ can change its mind at most t times and stay in this block. So $f(1,x) = s+t$.

We will have to change our guess at $f(x)$ at most n times, once for each time ψ goes to another ω-block. The final time ψ

shifts is, say $\psi(\epsilon,x)\downarrow$ and $\epsilon = \omega\cdot r+u$, we have $f(r,x) = f(r-1,x) + u = f(x)$. □

Now suppose that $A = \lim_s A_s(x)$ where the number of changes to our guess at $A(x)$ is dominated by $f \in \nabla_n$.

A schematic presentation will suffice to show that $A \in \nabla_{\omega\cdot n}$. Let $f(0,x), \ldots, f(m,x)$ be the m guesses we make at $f(x)$, $m \leq n-1$.

$$A(x) = \begin{array}{ll} \left.\begin{array}{l} \psi_0^1(x) \\ \vdots \\ \psi_{k_1}^1(x) \end{array}\right\} & \text{the first } k_1\text{-changes of } A_s(x) \text{ where } f(0,x) = k_1. \\ \hline \left.\begin{array}{l} \psi_1^2(x) \\ \vdots \\ \psi_{k_2}^2(x) \end{array}\right\} & \text{the next } k_2\text{-changes of } A_s(x) \text{ where } f(1,x) = k_2. \\ \hline \vdots & \\ \hline \left.\begin{array}{l} \psi_1^n(x) \\ \vdots \\ \psi_{k_m}^n(x) \end{array}\right\} & \text{the } m^{th}\ k_m\text{-changes where } f(m-1,x) = k_m \end{array}$$

That is, we have for $t \leq k_r$, $\psi(\langle\langle r,t\rangle,x\rangle) = \psi_{k_{r-t}}^{m-r}(x)$. ⊠

Similarly we may prove

<u>Theorem 11</u>: If $A \in \nabla_{\omega\cdot\alpha-S}$ then A is $\nabla_{\alpha-S}$-r.e.

And if A is $\nabla_{\alpha-S}$-r.e. then $A \in \nabla_{\omega\cdot\alpha-W}$ where W notates $\omega\cdot\alpha$ via the canonical ordering given by S on $\omega\times\alpha$ (namely $\langle x,r\rangle < \langle y,t\rangle$ iff $x < y$ or, if $x = y$, $(r)_S < (t)_S$).

<u>Corollary</u>: For each α-S the $\nabla_{\alpha\text{-}S}$-r.e. sets do not exhaust the sets $\leq_T 0'$. That is, given $\nabla_{\alpha\text{-}S}$ we can find some $A \leq_T 0'$ such that A is not f-r.e. for any $f \in \nabla_{\alpha\text{-}S}$.

<u>Proof</u>: Given α-S consider a notation S' for $\alpha+1$ obtained recursively from S (see Rogers [9], p. 206). By Theorem 9 we have an $A \in \nabla_{\omega\cdot(\alpha+1)\text{-}S''} - \nabla_{\omega\cdot\alpha\text{-}S''}$, S'' obtained from S' as above. Then A is $\nabla_{\alpha+1\text{-}S''}$ r.e. but not $\nabla_{\alpha\text{-}S''}$ r.e., hence not $\nabla_{\alpha\text{-}S}$ r.e. ∎

The classes ∇_α for $\alpha \leq \omega$ are given canonically. Using Theorem 10 we can give canonical classes ∇_α for $\alpha = \omega\cdot n$ any n. Using Theorem 11 we may now pick out canonical classes for any $\alpha = \omega\cdot\beta$ where $\alpha \leq \omega^\omega$. A picture will clarify this.

<u>f-r.e.</u>	∇_1-r.e.	...	∇_n-r.e.	...	∇_ω-r.e.	...	$\nabla_{\omega\cdot n}$-r.e.	...
	"		"		"		"	
<u>∇_α</u>	∇_ω	...	$\nabla_{\omega\cdot n}$	...	∇_{ω^2}	...	$\nabla_{\omega^2\cdot n}$	...

<u>f-r.e.</u>	∇_{ω^2}-r.e.	...	∇_{ω^ω}-r.e.
	"		"
<u>∇_α</u>	∇_{ω^3}	...	∇_{ω^ω}

* * *

M. Lerman has recently communicated to us that he and L. Hay have proved:

> for all $n \geq 1$ there are two n+1-r.e. degrees $\underline{c} < \underline{d}$ such that the interval $\{\underline{b} : \underline{c} \leq \underline{b} \leq \underline{d}\}$ has no n-r.e. degrees.

He suggests that the proof is a not difficult modification of the proof of Theorem 1.

Also R. Shore and L. Hay have communicated to us that they have shown that there is a 2 r.e. degree $\underline{a} < \underline{0}'$ such that there is no r.e. degree $\underline{b}$, $\underline{a} < \underline{b} < \underline{0}'$.

1. Hay and Lerman have observed that by a permitting argument one can prove that if $\underline{a} > \underline{0}$ is r.e. then for $\forall m+1$ there is some $\underline{b}$ which is $m+1$-r.e. and not m-r.e., $\underline{a} \geq \underline{b} > \underline{0}$. A corollary to Theorem 3 is that the same is true if $\underline{a}$ is n-r.e. for any $n \geq 1$.
2. L. Hay has pointed out that Theorem 6 follows from the fact that there is an $A \leq_T 0'$ but $A \not\leq_{tt} 0'$. And that $A \leq_{tt} 0'$ iff A is f-r.e. for some recursive f appears as Theorem 2,3 in "Δ_2^0-Mengen" by H. G. Carstens, <u>Arch. Math Logik</u> <u>18</u>(1976), 55-65.
3. This proof is essentially the same as Theorem 5 and 6(2) of Ershov [6] (part II).
4. Actually it fails for every ∇_n by Theorem 3 and the fact that for every r.e. $\underline{a}$ there is some $\underline{m} < \underline{a}$ of minimal degree (see Epstein [5]).

Bibliography

[1] Addison, J., The method of alternating chains, in *Theory of Models*, North-Holland, Amsterdam, 1965 (p. 1-16).

[2] Cooper, S. B., Doctoral Dissertation, University of Leicester, 1971.

[3] Epstein, Richard L., *Minimal Degrees of Unsolvability and the Full Approximation Construction*, Memiors of the A.M.S., no. 162, 1975.

[4] Epstein, Richard L., *Degrees of Unsolvability: Structure and Theory*, Lecture Notes in Mathematics no. 759, Springer-Verlag, New York.

[5] Epstein, Richard L., *Initial Segments of Degrees* $\leq 0'$, to appear.

[6] Ershov, A. Hierarchy of Sets I, II, III, *Algebra and Logic*, vol. 7, no. 1, no. 4 (1968) and vol. 9, no. 1 (1970). (English translation, Consultants Bureau, N.Y.).

[7] Gold, Limiting recursion, *Journal of Symbolic Logic*, vol. 30, no. 1, p. 28-48, 1965.

[8] Putnam, H., Trial and error predicates and the solution to a problem of Mostowski, *Journal of Symbolic Logic*, vol. 30, no. 1, p. 49-57, 1965.

[9] Rogers, Hartley, *Theory of Recursive Functions and Effective Computation*, McGraw-Hill, New York.

[10] Sacks, Gerald, *Degrees of Unsolvability*, Annals of Math. Studies, no. 55, Princeton, New Jersey, revised edition, 1965.

[11] Shoenfield, J. R., On the degrees of unsolvability, *Annals of Mathematics*, vol. 69, p. 644-653, 1959.

[12] Soare, Robert, Recursively enumerable sets and degrees, *Bull. A.M.S.*, vol. 84, no. 6 (1978), p. 1149.

Iowa State University
Ames, Iowa 50011

University of California
Berkeley, California 94720

Iowa State University
Ames, Iowa 50011

THE PLUS-CUPPING THEOREM FOR THE RECURSIVELY ENUMERABLE DEGREES[1]

Peter A. Fejer
Department of Mathematics
Cornell University
Ithaca, NY 14853/USA

Robert I. Soare
Department of Mathematics
University of Chicago
Chicago, IL 60637/USA

§1. Introduction.

Let $\mathbf{R} = (\mathbf{R}, \leq, \cup, \mathbf{0}, \mathbf{0}')$ denote the upper semi-lattice of recursively enumerable (r.e.) degrees where $\leq$ is the ordering induced by Turing reducibility $\leq_T$, $\cup$ denotes the least upper bound, and $\mathbf{0}$ and $\mathbf{0}'$ denote the least and greatest elements respectively of $\mathbf{R}$. Warning: All sets and degrees considered here will be r.e. The former will be denoted by $A, B, C, D, \ldots$ and the latter by $\mathbf{a}, \mathbf{b}, \mathbf{c}, \mathbf{d}, \ldots$. The infimum of degrees $\mathbf{a}, \mathbf{b} \in \mathbf{R}$ does not always exist but when it does it is written $\mathbf{a} \cap \mathbf{b}$. We use $\mathbf{a} \leq \mathbf{b}, \mathbf{c}$ to abbreviate $\mathbf{a} \leq \mathbf{b}$ and $\mathbf{a} \leq \mathbf{c}$. A degree $\mathbf{a}$, $\mathbf{0} < \mathbf{a} < \mathbf{0}'$, cups (caps) if there is a degree $\mathbf{b}$, $\mathbf{0} < \mathbf{b} < \mathbf{0}'$, such that $\mathbf{a} \cup \mathbf{b} = \mathbf{0}'$ ($\mathbf{a} \cap \mathbf{b} = \mathbf{0}$).

One of the most elegant and pleasing results on the r.e. degrees is the Sacks density theorem [11] which asserts that if $\mathbf{a} < \mathbf{b}$ then there exists $\mathbf{c}$, $\mathbf{a} < \mathbf{c} < \mathbf{b}$. This led Shoenfield to formulate a conjecture [12] that $\mathbf{R}$ is a dense structure as an upper semi-lattice analogously as the rationals are a dense structure as a linearly ordered set. (The conjecture asserts that if $\vec{\mathbf{a}} \in \mathbf{R}$ satisfies the diagram $D(\vec{x})$ and $D_1(\vec{x}, y)$ is any consistent diagram in $L(\leq, \cup, \mathbf{0}, \mathbf{0}')$ extending $\mathbf{D}$, then there exists $\mathbf{b} \in \mathbf{R}$ such that $D_1(\vec{\mathbf{a}}, \mathbf{b})$.) Shoenfield listed two consequences of his conjecture:

1 This paper is based on a talk given by the second author in the logic seminar at the University of Connecticut during the fall of 1979. The simplified proof of the main theorem was discovered by the authors through a series of refinements beginning with a study of Harrington's original notes [3] during the recursion theory seminar at the University of Chicago in 1978. Further improvements were made at the University of Connecticut. This work was partially supported by NSF Grant MCS 7905782. Both authors were also supported by the University of Connecticut where they were visitors for the Logic Year during the fall of 1979.

(1.1) If $\mathbf{a}, \mathbf{b} \in \mathbf{R}$ are incomparable then they have no greatest lower bound in $\mathbf{R}$;

(1.2) Given degrees $\mathbf{0} < \mathbf{b} < \mathbf{a}$, there exists $\mathbf{c} < \mathbf{a}$ such that $\mathbf{a} = \mathbf{b} \cup \mathbf{c}$.

Shoenfield's conjecture was first disproved when Yates [16] and Lachlan [5] refuted (1.1) by constructing a minimal pair, namely $\mathbf{a}, \mathbf{b} \in R$, both $> \mathbf{0}$, such that $\mathbf{a} \cap \mathbf{b} = \mathbf{0}$. Lachlan [4] also refuted (1.2) and Cooper [1] and Yates proved that (1.2) is even false for $\mathbf{a} = \mathbf{0}'$. The latter is called the "anti-cupping theorem" because it asserts that there exists $\mathbf{b} < \mathbf{0}'$ which does not cup. Harrington [2] then improved the anti-cupping theorem by showing that $\mathbf{0}'$ could be replaced by any high degree $\mathbf{a}$ (i.e., $\mathbf{a} \in \mathbf{R}$ such that $\mathbf{a}' = \mathbf{0}''$), see [8].

Of course it is easy to see that some degrees <u>do</u> cup since the Sacks splitting theorem [10] asserts that for any $\mathbf{a} > \mathbf{0}$ there are incomparable degrees $\mathbf{b}, \mathbf{c} < \mathbf{a}$ such that $\mathbf{a} = \mathbf{b} \cup \mathbf{c}$. The purpose of this paper is to give a fairly easy proof of the plus-cupping theorem which asserts that there exists $\mathbf{a} > \mathbf{0}$ such that $\mathbf{b}$ cups for every $\mathbf{b}$ in the entire <u>interval</u> $(\mathbf{0}, \mathbf{a}] = \{\mathbf{b} : \mathbf{0} < \mathbf{b} \leq \mathbf{a}\}$. Immediate corollaries are the cup and cap theorem (which asserts that there is a degree which both cups and caps), and the cup or cap theorem (which asserts that <u>every</u> degree $\mathbf{a} \neq \mathbf{0}, \mathbf{0}'$ either cups or caps). These three results were originally proved separately by Harrington using Lachlan's rather complicated "monstrous injury" priority method [6]. Then M. Stob and others observed that the first result implies the next two. Later we discovered that the first result can be proved using a variation of the nested strategies method used to prove the existence of a minimal pair [15, §4]. This method is much simpler than the monstrous injury method and lies somewhere between finite and infinite injury.

Harrington [3] indeed proved a stronger version of the plus-cupping theorem, namely

(1.3) $$(\exists\, \mathbf{a} > \mathbf{0})(\forall \mathbf{b})_{\mathbf{0} < \mathbf{b} \leq \mathbf{a}}(\forall \mathbf{d})_{\mathbf{a} \leq \mathbf{d}}(\exists \mathbf{c})_{\mathbf{c} < \mathbf{d}}[\mathbf{b} \cup \mathbf{c} = \mathbf{d}] .$$

We do not give a proof of (1.3) which seems to require the more complicated monstrous injury style proof. The latter method was invented by Lachlan [6] to prove that the Sacks density theorem and splitting theorem could not be combined. Namely, Lachlan proved

(1.4) $$(\exists \mathbf{a})(\exists \mathbf{b})[\mathbf{b} < \mathbf{a} \mathbin{\&} \neg(\exists \mathbf{c})(\exists \mathbf{d})[\mathbf{b} < \mathbf{c}, \mathbf{d} < \mathbf{a} \mathbin{\&} \mathbf{c} \cup \mathbf{d} = \mathbf{a}] .$$

The method derived its name from the extreme degree of technical complication in Lachlan's proof. Harrington recently improved this by showing that $\underset{\sim}{a}$ could be chosen to be $\underset{\sim}{0}'$.

Harrington and Shelah have announced that the monstrous injury method can be combined with the plus-cupping and anti-cupping ideas to prove undecidability of the first order theory of $\underset{\sim}{R}$ as follows. They first construct degrees $\underset{\sim}{a}$, $\underset{\sim}{b}$, and $\underset{\sim}{m}$ satisfying the formula

$$\Phi(\underset{\sim}{a},\underset{\sim}{b},\underset{\sim}{m}): \underset{\sim}{b} \cup \underset{\sim}{a} \not\geq \underset{\sim}{m} \;\&\; (\forall \underset{\sim}{d} \leq \underset{\sim}{a})[\underset{\sim}{b} \cup \underset{\sim}{d} \geq \underset{\sim}{a} \vee \underset{\sim}{d} \leq \underset{\sim}{m}].$$

Next given a countable partial ordering $(P,\leq_P)$ where $\leq_P$ is $\underset{\sim}{0}'$-presented, they construct degrees $\underset{\sim}{a}$, $\underset{\sim}{b}$, $\underset{\sim}{c}$ such that $(P,\leq_P)$ is isomorphic to the following set of degrees under $\leq$,

$$\{\underset{\sim}{d}: (\exists \underset{\sim}{m})[\Phi(\underset{\sim}{a},\underset{\sim}{b},\underset{\sim}{m}) \;\&\; \Phi(\underset{\sim}{m},\underset{\sim}{c},\underset{\sim}{d})]\}.$$

Therefore, since the theory of partial orderings is undecidable, so is the theory of $(\underset{\sim}{R},\leq)$.

In a sense this paper may be viewed as a first step towards understanding the monstrous injury method and the results of Lachlan and Harrington-Shelah mentioned above, as well as the recent result of Lachlan [7] that there is a degree $\underset{\sim}{a} > \underset{\sim}{0}$ which does not bound a minimal pair. In the proof of each of these results the requirements R are of an unusual nature and may be satisfied in one of two ways according to whether a sufficiently small number is ever enumerated in a given set B during one of infinitely many periods in the construction called "gaps". If so then we can preserve a certain corresponding C-computation $\Theta(C;x) = y$ forever, and satisfy requirement R in the first way. If not then at the end of each gap we enumerate a number into C (thereby possibly "capriciously destroying" the C-computation we are trying to preserve), and we wait for the next gap to begin. If no sufficiently small element is enumerated in B during any of the gaps, then we will show that B is recursive, and will satisfy requirement R in the second way.

Each of the monstrous injury proofs of the theorems mentioned above has requirements of this kind. The argument presented here is simpler than the others for two reasons. First each requirement R will contribute at worst an infinite set which is recursive (rather than r.e. and nonrecursive). Secondly, combining the requirements here requires only a linear nesting of strategies (as in the simplified proof [15, §4] of the minimal pair) while the usual monstrous injury method requires a much more complicated tree of strategies.

Background information, definitions and unexplained notation can be found in Rogers [9]. In addition we let $\varphi(x)\downarrow = y$ denote that the partial recursive function φ is defined on x and equal to y, and $\varphi(x)\uparrow$ denote that $\varphi(x)$ diverges. We identify a set A with its characteristic function and let $f \upharpoonright z$ denote the restriction of f to arguments $< z$.

§2. The corollaries and the strategy for meeting a single requirement.

After deriving the corollaries, we list the requirements which we must meet to prove the theorem and we describe the strategy for meeting a single requirement.

Theorem 2.1. (Plus-Cupping Theorem-Harrington)

$$(\exists \underset{\sim}{a} > \underset{\sim}{0})(\forall \underset{\sim}{b})_{\underset{\sim}{0} < \underset{\sim}{b} \leq \underset{\sim}{a}}(\exists \underset{\sim}{c})_{\underset{\sim}{c} < \underset{\sim}{0}'}[\underset{\sim}{b} \cup \underset{\sim}{c} = \underset{\sim}{0}'] .$$

Lemma 2.2. If $\underset{\sim}{a}$ is as in Theorem 2.1, then $\underset{\sim}{a}$ caps.

Proof. Suppose that $\underset{\sim}{a}$ does not cap. Let $\underset{\sim}{b} > \underset{\sim}{0}$. Since $\underset{\sim}{a}$ does not cap with $\underset{\sim}{b}$ there is a $\underset{\sim}{c} > \underset{\sim}{0}$, $\underset{\sim}{c} \leq \underset{\sim}{a}, \underset{\sim}{b}$. Since $\underset{\sim}{0} < \underset{\sim}{c} \leq \underset{\sim}{a}$, $\underset{\sim}{c}$ cups, so since $\underset{\sim}{c} \leq \underset{\sim}{b}$, $\underset{\sim}{b}$ cups. Thus every $\underset{\sim}{b} > \underset{\sim}{0}$ cups, contradicting the anti-cupping theorem [1]. □

Corollary 2.3. (Cup and Cap Theorem-Harrington)

$$(\exists \underset{\sim}{a} > \underset{\sim}{0})(\forall \underset{\sim}{b})_{\underset{\sim}{0} < \underset{\sim}{b} \leq \underset{\sim}{a}}(\exists \underset{\sim}{c} < \underset{\sim}{0}')(\exists \underset{\sim}{d} > \underset{\sim}{0})[\underset{\sim}{b} \cup \underset{\sim}{c} = 0' \ \& \ \underset{\sim}{b} \cap \underset{\sim}{d} = \underset{\sim}{0}].$$

Proof. Take $\underset{\sim}{a}$ as in Theorem 2.1. Then every nonzero degree $\leq \underset{\sim}{a}$ cups and by Lemma 2.2 $\underset{\sim}{a}$ caps, so every nonzero degree $\leq \underset{\sim}{a}$ caps. □

Harrington originally proved that there is a degree which both cups and caps. Lemma 2.2, which gives the strengthened version of the Cup and Cap Theorem given in Corollary 2.3, was pointed out by K. Ambos.

(Jockusch has shown (unpublished) that the cup and cap theorem is false for wtt-degrees [9, p.158] in place of Turing degrees.) The next result asserts not only that any $\underset{\sim}{d} \neq \underset{\sim}{0}, \underset{\sim}{0}'$ either cups or caps but indeed that $\underset{\sim}{d}$ either cups or else caps with the fixed degree $\underset{\sim}{a}$ of Theorem 2.1.

Corollary 2.4. (Cup or Cap Theorem-Harrington)

$$(\forall \underset{\sim}{d})_{\underset{\sim}{0} < \underset{\sim}{d} < \underset{\sim}{0}'}[\underset{\sim}{d} \cap \underset{\sim}{a} = \underset{\sim}{0} \vee d \text{ cups}].$$

Proof. Suppose $\mathbf{d} \cap \mathbf{a} = \mathbf{0}$ is false. Choose $\mathbf{b} \leq \mathbf{a}, \mathbf{d}, \mathbf{b} > \mathbf{0}$. Now by Theorem 2.1, there exists $\mathbf{c} < \mathbf{0}'$ such that $\mathbf{b} \cup \mathbf{c} = \mathbf{0}'$. But then $\mathbf{d} \cup \mathbf{c} = \mathbf{0}'$, because $\mathbf{b} \leq \mathbf{d}$. □

(K. Ambos has shown that the cup or cap theorem is false for wtt-degrees.)

Definition 2.5. A degree $\mathbf{a}$ has the strong anti-cupping property via witness $\mathbf{b} < \mathbf{a}$ if $b > 0$ and

$$(\forall \mathbf{c})[\mathbf{c} \not\geq \mathbf{a} \Longrightarrow \mathbf{c} \cup \mathbf{b} \not\geq \mathbf{a}].$$

For every high degree $\mathbf{a}$ such a witness $\mathbf{b} < \mathbf{a}$ is constructed in [8]. The proof of Theorem 2.1 nowhere uses the completeness of $\mathbf{0}'$ but merely the fact that $\mathbf{0}'$ is nonrecursive. Namely, we will actually prove the following stronger version of Theorem 2.1.

Theorem 2.6. $(\forall \mathbf{d} > \mathbf{0})(\exists \mathbf{a} > \mathbf{0})(\forall \mathbf{b})_{\mathbf{0} < \mathbf{b} \leq \mathbf{a}}(\exists \mathbf{c})_{\mathbf{c} \not\geq \mathbf{d}}[\mathbf{b} \cup \mathbf{c} \geq \mathbf{d}]$.

Corollary 2.7. (K. Ambos) If $\mathbf{d}$ has the strong anticupping property via $\mathbf{e}$, $\mathbf{0} < \mathbf{e} < \mathbf{d}$, then $\mathbf{e}$ is half of a minimal pair and indeed caps with the degree $\mathbf{a}$ of Theorem 2.6.

Proof. Let $\mathbf{a}$ be as in Theorem 2.6. If $\mathbf{e}$ fails to cap with $\mathbf{a}$ then choose $\mathbf{b} > \mathbf{0}$, such that $\mathbf{b} < \mathbf{a}, \mathbf{e}$. But then by Theorem 2.6, $\mathbf{b} \cup \mathbf{c} \geq \mathbf{d}$ for some $\mathbf{c} \not\geq \mathbf{d}$, so $\mathbf{e} \cup \mathbf{c} \geq \mathbf{d}$, contrary to the hypothesis on $\mathbf{e}$. □

We now list the requirements necessary to prove Theorem 2.6 and explain the basic strategy for a single requirement, using the terminology of "gaps" introduced by Lachlan [7]. Let $\{\Theta_e\}_{e\in\omega}$ be a standard listing of all partial recursive (p.r.) functionals and $\{\Phi_e, B_e\}$, a standard listing of all pairs such that Φ_e is a p.r. functional and B_e is an (r.e.) set. Let D be any nonrecursive (r.e.) set. For a set X or p.r. functional Ψ, let X_s and Ψ_s denote the result after s steps in the enumeration. Let θ_i and φ_i be the (p.r.) use functions for Θ_i and Φ_i (namely $\theta_i(X_s; x, s)$ is the greatest element used in the computation $\Theta_{i,s}(X_s; x)$ if defined and is 0 otherwise). We assume $\theta_i(X; x, s) < s$ and likewise for φ_i.

To prove Theorem 2.6 it suffices to construct A coinfinite to meet for all e the requirements

P_e: W_e infinite $\Longrightarrow W_e \cap A \neq \emptyset$; and

R_e: $\Phi_e(A) = B_e \Longrightarrow [B_e$ is recursive $\vee (\exists C_e)[D \leq_T B_e \oplus C_e \;\&\; D \not\leq_T C_e]]$.

Fix e. Thus we are given D and B_e and we are enumerating A and C_e. We may assume that the enumeration of B_e satisfies

$$(2.1) \qquad x \in B_{e,s+1} - B_{e,s} \implies \Phi_{e,s}(A_s; x) = 1,$$

because we are only interested in those e such that $\Phi_e(A) = B_e$, so we may withhold an element from B_e until $\Phi_{e,s}(A_s; x) = 1$. (This gives us considerable power over B_e by restraining elements from entering A.)

To measure whether $\Phi_e(A) = B_e$ we define the recursive functions,

$$\ell^A(e,s) = \max\{x: (\forall y < x)[\Phi_{e,s}(A_s; y) = B_{e,s}(y)]\}, \text{ and}$$

$$m^A(e,s) = \max\{\ell^A(e,t): t < s\}.$$

Call s an <u>e-expansion stage</u> if $\ell^A(e,s) > m^A(e,s)$. (Thus, if $\Phi_e(A) = B$ there are infinitely many e-expansion stages.)

To attempt to arrange $D \leq_T B_e \oplus C_e$ we have a list of "coding markers" $\{\Gamma_m\}_{m\in\omega}$. Let Γ_m^s denote the position of Γ_m at the end of stage s. We will arrange that if $\Phi_e(A) = B$ then for all m and s

(2.2) if m enters D at stage v+1 then Γ_m^v is not the final position of Γ_m;

(2.3) $\Gamma_m^{s+1} \geq \Gamma_m^s$; and

(2.4) $\Gamma_m^{s+1} > \Gamma_m^s \implies (\exists x \leq \Gamma_m^s)[x \in C_{e,s+1} - C_{e,s}]$

$$\vee [B_{e,s} \upharpoonright \Gamma_m^s \neq B_{e,v} \upharpoonright \Gamma_m^s \text{ where v is the last e-expansion stage} < s].$$

Thus, if $\Phi_e(A) = B_e$ and all the markers come to rest, say $f(m) = \lim_s \Gamma_m^s$, then $f \leq_T B_e \oplus C_e$ by (2.3) and (2.4), and hence $D \leq_T B_e \oplus C_e$ by (2.2). (The trick in meeting requirement R_e is to insure that if some marker Γ_m moves infinitely often then B_e is recursive.)

To arrange $D \not\leq_T C_e$ we attempt to satisfy for all i the requirement

$$N_{\langle e,i\rangle}: \Theta_i(C_e) \neq D.$$

Fix i. We first try to meet $N_{\langle e,i\rangle}$ by preserving agreements between $\Theta_{i,s}(C_{e,s}; x, s)$ and $D_s(x)$ as in the usual Sacks preservation method [13, §2]. Define the recursive function,

$$\ell(e,i,s) = \max\{x: (\forall y < x)[\Theta_{i,s}(C_{e,s}; y, s) = D_s(y)]\}.$$

However, unlike [13, §2], here we have no formal restraint on C_e (except that a number z enters C_e at stage $t+1$ only if t is e-expansionary and z is a marker position). Instead, if $\theta = \theta_i(C_{e,s}; x, s)$ for some $x \leq \ell(e,i,s)$ then $N_{\langle e,i \rangle}$ attempts to clear $C_e \upharpoonright \theta$ of all markers Γ_m, $m \geq i$, (since these marker positions might later enter C_e destroying the C_e-computation.) For $N_{\langle e,i \rangle}$ we define an A-restraint function $r: N \to N$ such that:

(2.5) either r has infinitely many zeros or else r is eventually constant; and

(2.6) if A eventually obeys r (i.e., there are only finitely many s such that $(\exists x)[x \leq r(s+1) \ \&\ x \in A_{s+1} - A_s])$ and r is not eventually constant then B is recursive.

We define r and attempt to construct a series of "gaps" as follows. To begin a new gap we wait for a stage $s+1$ such that

(2.7) $z < \ell^A(e,s)$; and

(2.8) $(\exists x)[x \leq \ell(e,i,s) \ \&\ z \leq \theta_i(C_{e,s}; x, s)]$,

where $z = \Gamma_i^s$. The gap is closed at stage $t+1$ where t is the next e-expansion stage $\geq s+1$, and we set $r(v) = 0$ for all v, $s+1 \leq v \leq t$. If there is no such t then the gap is never closed and $r(v) = 0$ for all $v \geq s+1$. Initially we set $r(v) = 0$ for all $v \leq s$ where $s+1$ is the beginning of the first gap. (If the enumeration of A ever violates the restraint function r then we begin constructing the series of gaps all over again forgetting what was previously done.) During the gap the positive requirements P_j are free to contribute elements to A, and so various A-computations $\Phi_{e,s}(A_s; x) = B_{e,s}(x)$ may be destroyed, thereby allowing $B_e(x)$ to change in value. However, at the next e-expansion stage t, $\ell^A(e,t) > m^A(e,t) \geq \ell^A(e,s) \geq z$ so $\Phi_{e,t}(A_t; x) = B_{e,t}(x)$ for all $x \leq z$. In closing the gap at stage $t+1$ we perform the following four steps.

<u>Step 1</u>. <u>Case 1</u>. If $B_{e,t} \upharpoonright z \neq B_{e,s} \upharpoonright z$, where $z = \Gamma_i^s$, go to Step 2.

<u>Case 2</u>. If $B_{e,t} \upharpoonright z = B_{e,s} \upharpoonright z$, enumerate z into C_e.

<u>Step 2</u>. Move markers Γ_m, $m \geq i$, (maintaining their order) to elements greater than both t and their present positions.

(Thus, if Case 1 holds, no element is enumerated in C_e at Step 1 and the computations $\Theta_{i,s}(C_{e,s}; x) = D_s(x)$ for $x < \ell(e,i,s)$ are cleared at Step 2 of any marker Γ_m, $m \geq i$, since $\theta_i(C_{e,s}; x, s) < s$. Note that no elements enter C_e during a gap so the C_e-computations at stage s are still valid at stage t.)

Step 3. If $m = \mu m'[m' \in D_t - D_v]$ where v is the last e-expansion stage $< s$, put Γ_m into C_e and move all $\Gamma_{m'}$, $m' \geq m$.

Step 4. Set $r(t+1) = t$.

This will protect the A-computations $\Phi_{e,t}(A;x) = B_{e,t}(x)$ for $x < \ell^A(e,t)$ until the start of the next gap, say at stage $s_1 > t+1$, thereby insuring that

$$(2.9) \qquad B_{e,s_1} \restriction z = B_{e,t} \restriction z,$$

by our convention (2.1).

If there are infinitely many gaps then r has infinitely many zeros so (2.5) is met. To see that (2.6) holds, assume that $\Phi_e(A) = B$, A obeys r, and r is not eventually constant. Then we must open infinitely many gaps. Hence, $\lim_s \Gamma_i^s = \infty$ because Γ_i moves whenever a gap is closed. We will show that Case 2 of Step 1 holds at the closing of all gaps begun after some stage s_0. Thus B_e is recursive because $B_e \restriction \Gamma_i$ does not change during the gaps, but by (2.9) $B_e \restriction \Gamma_i$ does not change during the intervals between gaps. Hence, for any x, find $s > s_0$ such that $x < \Gamma_i^s$ and a gap is opened at stage $s+1$. Then $x \in B_e$ iff $x \in B_{e,s}$.

To see that s_0 exists first note that C_e is recursive. For each x choose s such that $x < \Gamma_i^s$. Now (ignoring the position of Γ_m, $m < i$, which we assume come to rest) $x \in C_e$ iff $x \in C_{e,s}$. Thus $\Theta_i(C_e) \neq D$, since D is nonrecursive. Let $x_0 = (\mu x)[\Theta_i(C_e;x) \neq D(x)]$. Now $\Theta_i(C_e;x_0)$ must diverge because otherwise we would not open infinitely many gaps. Thus there exists s_0 such that at each gap begun at a stage $s > s_0$ we attempt to clear a computation $\Theta_{i,s}(C_{e,s};x_0)$ of markers. If Case 1 of Step 1 holds at the closing of this gap then the computation $\Theta_{i,s}(C_{e,s};x_0)$ would be preserved forever, contrary to the divergence on x_0.

§3. The proof of Theorem 2.6.

The strategy just given for meeting a single requirement $N_{\langle e,i\rangle}$, say N_0, produces an A-restraint function $r(0,s)$ such that $\liminf_s r(0,s) < \infty$. As in the minimal pair construction [15, §4] we must modify the strategy σ_1 for N_1 so that the two restraint functions drop back simultaneously. To do this N_1 must guess the value $k = \liminf_s r(0,s)$, and must simultaneously play infinitely many strategies σ_1^k, $k \in \omega$, one for each possible value of k. Each strategy

σ_1^k is played like σ_0 but with $S^k = \{s : r(0,s) = k\}$ in place of ω as the set of stages on which it is active. Strategy σ_1^k still succeeds if any restraint it imposes is maintained during intermediate stages $s \notin S^k$ while σ_1^k is dormant. Thus, at stage s if $k = r(0,s)$, we play σ_1^k, maintain the restraints imposed previously by the dormant σ_1^i, $i < k$, and discard restraints imposed by σ_1^j, $j > k$. Therefore, if $k = \liminf_s r(0,s)$, then: (1) the strategy σ_1^k succeeds in meeting N_1; (2) the strategies σ_1^i, $i < k$, impose finitely much restraint over the whole construction; and (3) the strategies σ_1^j, $j > k$, drop all restraint at each stage $s \in S^k$. Thus, the entire restraint $r(1,s)$ imposed by N_0 and N_1 <u>together</u> has $\liminf_s r(1,s) < \infty$.

In addition we need a new trick here not found in the proof of the minimal pair. Namely, σ_1^k is allowed to <u>open</u> an N_1^k-gap (and drop its A-restraint) only at a stage $s \in S^k$. However σ_1^k is allowed to <u>close</u> that gap (thereby reimposing A-restraint) at a stage $t \notin S^k$ (providing t is an e-expansion stage for the corresponding e). This allows us to impose a sufficiently small amount of restraint so that $\liminf_s r(n,s) < \infty$ for all n, and yet close the gap often enough to allow enumeration in C_e and meet Lemma 3 below.

In the full construction we have for each e coding markers $\{\Gamma_{e,m}\}_{m\in\omega}$. We sometimes write $N^k_{\langle e,i\rangle}$ for $\sigma^k_{\langle e,i\rangle}$ and refer to gaps opened by $\sigma^k_{\langle e,i\rangle}$ as $N^k_{\langle e,i\rangle}$-gaps. These gaps will not be opened to clear marker $\Gamma_{e,i}$ as in our sketch but rather to clear $\Gamma_{e,m}$ where $m = \langle i,k,p\rangle$ and

$$p = p(n,k,s) = \text{card}\{t \leq s : r(n-1,t) < k\} \quad \text{for} \quad n = \langle e,i\rangle,$$

so that no marker will be moved infinitely often by N_n^k if $k > \liminf_s r(n-1,s)$.

<u>Construction of A and C_e.</u>

<u>Stage $s = 0$.</u> Do nothing.

<u>Stage $s+1$.</u>

<u>Step 1.</u> (Closing gaps.) In increasing order examine all $e \leq s$ such that s is an e-expansion stage. See whether there exists $m \leq s$, $m = \langle i,k,p\rangle$ such that some $N^k_{\langle e,i\rangle}$-gap was opened at some stage $v \leq s$ via marker $\Gamma_{e,m}$ and that gap has not been closed or cancelled. If not go to the next e or to Step 2. If so fix the least such m and let v correspond to m. Let $z = \Gamma^v_{e,m}$.

<u>Substep (a).</u> If $B_{e,v} \upharpoonright z = B_{e,s} \upharpoonright z$ enumerate z in C_e.

<u>Substep (b).</u> Move $\Gamma_{e,m'}$, for all $m' \geq m$, in order, to

integers greater than both s and their present positions.

Substep (c). For all i and k if there is an $N^k_{\langle e,i\rangle}$ gap open, declare the gap to be closed and let $N^k_{\langle e,i\rangle}$ assign s as A-restraint.

Step 2. (Coding D into $B_e \oplus C_e$.) Examine each $e \leq s$ such that s is an e-expansion stage. Let s' be the greatest e-expansion stage $< s$ if one exists and 0 otherwise. Let $m = \mu x\,[x \in D_s - D_{s'}]$. Enumerate the current position of $\Gamma_{e,m}$ into C_e and move markers $\Gamma_{e,m'}$, for $m' \geq m$, in order to new larger positions. (If m fails to exist go to Step 3.)

Step 3. (Opening gaps.) Perform for each $n = \langle e,i\rangle \leq s$ in increasing order the following procedure. Let $k = r(n-1, s+1)$, where $r(-1,t) = 0$ for all t. For every $j > k$, cancel any gap or restraint previously imposed for N^j_n. Let z be the current position of $\Gamma_{e,m}$ where $m = \langle i,k,p(n,k,s)\rangle$, and $p(n,k,s) = \text{card}\{t \leq s : r(n-1,t) < k\}$. Now we open an $N^k_{\langle e,i\rangle}$-gap via $\Gamma_{e,m}$ if

(1) there is not now an open $N^k_{\langle e,i\rangle}$ gap;

(2) $z < \ell^A(e,s)$; and

(3) $(\exists x)[x \leq \ell(e,i,s)\ \&\ z \leq \theta_i(C_{e,s}; x, s)\ \&$ computations still exist].

If so we cancel all restraint of priority N^k_n. For each $j \leq k$, let $r(n,j,s+1)$ be the maximum restraint of priority N^j_n previously assigned and not yet cancelled (including any restraint assigned at Step 1). Define $r(n,s+1) = \max(\{k\} \cup \{r(n,j,s+1) : j \leq k\})$.

Step 4. (Making A simple.) For each $j \leq s$, if $W_{j,s} \cap A_s = \emptyset$ and

$$(\exists y)[y \in W_{j,s}\ \&\ y > 2j\ \&\ y > r(j,s+1)],$$

choose the least such y and enumerate y in A.

This completes the construction.

Lemma 1. $(\forall n)[\liminf_s r(n,s) < \infty]$.

Proof. Recall that $r(-1,s) = 0$ for all s. Fix n and assume the lemma for $n-1$. Let $k = \liminf_s r(n-1,s)$, and $S = \{s : r(n-1,s) = k\}$. Choose s_0 such that $r(n-1,s) \geq k$ for all $s \geq s_0$. Now for $j < k$ no new N^j_n-gap can be opened after stage s_0, and each existing gap will be closed at most once during which time N^j_n may increase its restraint. Let r_0 be the maximum restraint ever imposed by N^j_n for $j < k$. Now for each sufficiently large $s \in S$, $r(n,s) = \max\{k, r_0, r(n,k,s)\}$, since any restraint for N^j_n, $j > k$, is cancelled at such a stage s.

Now either $r(n,k,s)$ is eventually constant or else we open infinitely many N_n^k-gaps, say at stages $s_1 < s_2 < s_3 < \dots$, where $s_i \in S$ and $r(n,k,s_i) = 0$ for each i. In either case $\liminf_{s \in S} r(n,s) < \infty$ so $\liminf_s r(n,s) < \infty$. □

Lemma 2. $(\forall j)[W_j \text{ infinite} \Longrightarrow W_j \cap A \neq \emptyset]$.

Proof. Let $r = \liminf_s r(j,s)$. If W_j is infinite choose $y > r, 2j$, and choose s such that $y \in W_{j,s}$ and $r(j,s+1) = r$. Now if $W_{j,s} \cap A_s = \emptyset$ then y (or some smaller $y' \in W_{j,s}$) enters A_{s+1} insuring $W_{j,s} \cap A_{s+1} \neq \emptyset$. □

Lemma 3. If $\Phi_e(A) = B_e$ then

(a) $(\forall i)(\forall k)$[any $N^k_{\langle e,i\rangle}$-gap is eventually closed or cancelled].

(b) $(\forall m)$[if $m \in D_{s+1} - D_s$ then $\Gamma^s_{e,m}$ is not the final position of $\Gamma_{e,m}$].

Proof. Suppose that either there is an open $N^k_{\langle e,i\rangle}$-gap at the end of stage $s+1$ or $m \in D_{s+1} - D_s$. Let t be the least e-expansion stage $> s$. Then during Step 1 of stage $t+1$ we close the $N^k_{\langle e,i\rangle}$-gap (if it has not already been cancelled), and during Step 2 the position of $\Gamma_{e,m}$ changes. □

Lemma 4. If $\Phi_e(A) = B_e$ then $(\forall i)[\Theta_i(C_e) \neq D]$.

Proof. Assume $\Phi_e(A) = B_e$ and $\Theta_i(C_e) = D$ with i minimal. We claim that $\lim_s \Gamma^s_{e,m} = \infty$ for some m. If so then C_e is recursive and hence D is recursive, contrary to hypothesis. (To see that C_e is recursive choose the minimal m satisfying the claim, and s_0 such that no marker $\Gamma_{e,j}$, $j < m$, contributes an element to C_e after stage s_0. To test whether $x \in C_e$, find $s > s_0$ such that $x < \Gamma^s_{e,m}$. Now $x \in C_e$ iff $x \in C_{e,s}$.)

To prove the claim assume to the contrary that $\lim_s \Gamma^s_{e,m} < \infty$ for all e, m. Let $n = \langle e,i\rangle$, $k = \liminf_s r(n-1,s)$, and $p = \text{card}\{t : r(n-1,t) < k\}$. Let z be the final position of $\Gamma_{e,m}$, for $m = \langle i,k,p\rangle$. Let $\theta(x) = \lim_s \theta_i(C_{e,s}; x, s)$. Now there must exist x such that $z < \theta(x)$ else $D \leq_T C_e \restriction z+1$, and so D is recursive. Choose s_0 such that $\Gamma^{s_0}_{e,m} = z$, and $r(n-1,s) \geq k$ for all $s \geq s_0$. Hence, there exists a stage $s > s_0$ when we open an $N^k_{\langle e,i\rangle}$-gap via $\Gamma_{e,m}$. (This gap is never cancelled since $s \geq s_0$.) But by Lemma 3 this gap must be closed at some stage $t > s$, and $\Gamma_{e,m}$ moves at stage t, a contradiction. □

Lemma 5. If $\Phi_e(A) = B_e$ and B_e is not recursive, then $(\forall m)[\lim_s \Gamma^s_{e,m} < \infty]$.

Proof. Assume $\Phi_e(A) = B_e$. Choose m minimal such that $\lim_s \Gamma^s_{e,m} = \infty$, say $m = \langle i,k,p \rangle$. Let $n = \langle e,i \rangle$. We prove that B_e is recursive. Once the $\Gamma_{e,m'}$ with $m' < m$ have settled, $\Gamma_{e,m}$ can move only if its present position is put into C_e. This can happen at most once at Step 2 of a stage and, afterwards, each time $\Gamma_{e,m}$ moves, it is because an N^k_n gap, opened through $\Gamma_{e,m}$, is closed. If $j < \liminf_s r(n-1,s)$, then there are only finitely many N^j_n gaps opened during the construction. If $j > \liminf_s r(n-1,s)$, then N^j_n can only open finitely many gaps through any given marker $\Gamma_{e,m}$ (since $\lim_s p(n,j,s) = \infty$). Thus $k = \liminf_s r(n-1,s)$.

By Lemma 4, choose $x_0 = \mu x[\Theta_i(C_e,x) \neq D(x)]$. Choose s_0 such that for all $s \geq s_0$,

(3.1) $(\forall j < m)[\Gamma_{e,j}$ does not move at stage $s]$;

(3.2) $(\forall j < n)[P_j$ does not act at stage $s]$;

(3.3) $r(n-1,s) \geq k$;

(3.4) the position of $\Gamma_{e,m}$ is not enumerated into C_e during Step 2 of stage s;

(3.5) $(\forall y < x_0)[\Theta_{i,s}(C_{e,s};y) = \Theta_i(C_e;y)$ and $\theta_i(C_{e,s};y,s) = \lim_s \theta_i(C_{e,s};y,s)]$; and

(3.6) $(\forall y \leq x_0)[y \in D_s \iff y \in D]$.

Now after stage s_0 there must be infinitely many stages $s_1 < t_1+1 \leq s_2 < t_2+1 \leq \ldots$ such that N^k_n opens a gap via $\Gamma_{e,m}$ at stage s_j with $x = x_0$ in (3) of Step 3 and this gap closes at stage t_j+1. Now $\Theta_i(C_e;x_0)$ must diverge or else we would not open infinitely many gaps. But when the gap is closed at t_j+1 we cannot have $B_{e,s_j} \upharpoonright z \neq B_{e,t_j} \upharpoonright z$, where $z = \Gamma^{s_j}_{e,m}$, or else the computation $\Theta_{i,t_j}(C_{e,t_j};x_0)$ would be cleared of all markers $\Gamma_{e,q}$, for $q \geq m$, and would be preserved forever contrary to the divergence on x_0. Hence, for every j, $B_{e,s_j} \upharpoonright z = B_{e,t_j} \upharpoonright z$ where $z = \Gamma^{s_j}_{e,m}$. But $z < \ell^A(e,t_j)$ since t_j is an e-expansion stage, so N^k_n assigns A-restraint to protect the computations $\Phi_{e,t_j}(A_{t_j};y)$, $y < \ell^A(e,t_j)$ until stage s_{j+1}. (By (3.2) and (3.3) this restraint is not injured or cancelled before stage s_{j+1}.) Thus, by (2.1) $B_{e,t_j} \upharpoonright z = B_{e,s_{j+1}} \upharpoonright z$. Hence, B_e is recursive since to compute whether $x \in B_e$, find the least stage $s_j \geq s_1$ such that $x < \Gamma^{s_j}_{e,m}$. Now $x \in B_e$ iff $x \in B_{e,s_j}$. □

Lemma 6. If $\Phi_e(A) = B_e$ and B_e is not recursive then $D \leq_T B_e \oplus C_e$.

Proof. Fix e such that $\Phi_e(A) = B_e$ and B_e is not recursive. Now by Lemma 5, $f(m) =_{dfn} \lim_s \Gamma^s_{e,m} < \infty$ for all m. Furthermore, $f \leq_T B_e \oplus C_e$. To see this find an e-expansion stage s such that no $z \leq \Gamma^{s+1}_{e,m}$ enters B_e or C_e after stage s. Then $\Gamma_{e,m}$ cannot move after stage s+1, and furthermore $m \in D$ iff $m \in D_{s+1}$. □

References

[1] S. B. Cooper, On a theorem of C.E.M.Yates, handwritten notes, 1974.

[2] L. Harrington, On Cooper's proof of a theorem of Yates I, and II, handwritten notes, 1976.

[3] L. Harrington, Plus-cupping in the r.e. degrees, handwritten notes, 1978.

[4] A. H. Lachlan, The impossibility of finding relative complements for recursively enumerable degrees, J. Symbolic Logic 31 (1966), 434-454.

[5] ——————, Lower bounds for pairs of r.e. degrees, Proc. London Math. Soc. 16 (1966), 537-567.

[6] ——————, A recursively enumerable degree which will not split over all lesser ones, Ann. Math. Logic 9 (1975), 307-365.

[7] ——————, Bounding minimal pairs, J. Symbolic Logic 44 (1979), 626-642.

[8] David Miller, High recursively enumerable degrees and the anti-cupping property, this volume.

[9] H. Rogers, Jr., Theory of recursive functions and effective computability, McGraw-Hill, New York, 1967.

[10] G. E. Sacks, On the degrees less than $\underset{\sim}{0}'$, Ann. of Math. (2) 77 (1963), 211-231.

[11] ——————, The recursively enumerable degrees are dense, Ann. of Math. (2) 80 (1964), 300-312.

[12] J. R. Shoenfield, Application of model theory to degrees of unsolvability, Sympos. Theory of Models, North-Holland, Amsterdam, 1965, 359-363.

[13] R. I. Soare, The infinite injury priority method, J. Symbolic Logic 41 (1976), 513-530.

[14] R. I. Soare, _Recursively enumerable sets and degrees_, Bull. A.M.S. 84 (1978), 1149-1181.

[15] ——————, _Fundamental methods for constructing recursively enumerable degrees_, Recursion theory, its generalizations and applications, Logic Colloquium 79, Leeds, England, Cambridge University Press, to appear.

[16] C. E. M. Yates, _A minimal pair of r.e. degrees_, J. Symbolic Logic 31 (1966), 159-168.

NATURAL α-RE DEGREES

Sy D. Friedman
M.I.T.

In this paper we provide an explicit positive solution to Post's Problem in α-recursion theory, for many admissible ordinals α. The Sacks-Simpson Theorem (Sacks-Simpson [72]) yields a positive solution for all admissible α via an α-finite injury argument. By way of contrast, our approach makes no use of the priority method. Instead we find new ways to combine Skolem hulls with the transitive collapse lemma. Thus our proof is really very close to Gödel's proof of the G C H in L (Gödel [39]).

If λ is a limit ordinal, $\lambda < \alpha$, then $\alpha\text{-cof}(\lambda) = \alpha$-cofinality of λ is just the cofinality of λ when evaluated inside L_α; thus, $\alpha\text{-cof}(\lambda)$ = least γ s.t. there is an unbounded $f: \gamma \longrightarrow \lambda$, $f \in L_\alpha$. A set $X \subseteq \alpha$ is <u>low</u> if $X' \leq_\alpha \emptyset'$ where X' is the α-jump of X. And, X is <u>hyperregular</u> if $\langle L_\alpha, X\rangle$ is an admissible structure. We suggest consulting Simpson [74] for further clarification of the basic notions of α-recursion theory.

<u>Theorem 1</u>. Suppose $\alpha > \omega$ is admissible and $L_\alpha \models$ There is no largest cardinal. Then $S(\omega) = \{\lambda < \alpha \mid \alpha\text{-cof}(\lambda) = \omega\}$ is a low, hyperregular α-RE set whose α-degree is strictly between 0 and $0'$.

<u>Proof</u>: An <u>α-cardinal</u> is an ordinal κ such that $L_\alpha \models \kappa$ is a cardinal. If κ is an α-cardinal then we let κ^+ denote the least α-cardinal greater than κ. We show that if $\omega < \kappa$ is a <u>regular</u> α-cardinal (that is, $L_\alpha \models \kappa$ is regular) then $\langle L_{\kappa^+}, S(\omega) \cap L_{\kappa^+}\rangle$ is a Σ_1-elementary substructure of $\langle L_\alpha, S(\omega)\rangle$. Thus any $\Sigma_1\langle L_\alpha, S(\omega)\rangle$ function f with domain $\subseteq\kappa$ (and defining parameter $p \in L_{\kappa^+}$) has range contained in L_{κ^+}. This establishes the admissibility of $\langle L_\alpha, S(\omega)\rangle$. Also note that $X = \{\kappa^+ \mid \omega < \kappa$ is a regular α-cardinal$\}$ is Π_1 over L_α and thus $\leq_\alpha 0'$. Moreover if $\phi(x)$ is a Σ_1 formula defining the complete Σ_1 set C for $\langle L_\alpha, S(\omega)\rangle$ then for $\gamma \in X$:

$$C \cap L_\gamma = \{x \in L_\gamma \mid \langle L_\gamma, S(\omega) \cap L_\gamma\rangle \models \phi(x)\}$$

and therefore $C \leq_\alpha X \vee S(\omega)$, where $\vee$ denotes α-recursive join. But C has α-degree $S(\omega)'$ and $X \vee S(\omega) \leq_\alpha 0'$, so $S(\omega)$ is low.

Suppose then that $\omega < \kappa$ is a regular α-cardinal and $\langle L_\alpha, S(\omega)\rangle \models \exists y \phi(x,y,S(\omega))$ where ϕ is Δ_0 and $x \in L_{\kappa^+}$. We wish to show that $\langle L_{\kappa^+}, S(\omega) \cap L_{\kappa^+}\rangle \models \exists y\, \phi(x,y,S(\omega) \cap L_{\kappa^+})$. Choose $y \in L_\alpha$ so that $\langle L_\alpha, S(\omega)\rangle \models \phi(x,y,S(\omega))$ and let λ be a regular α-cardinal so that $y \in L_\lambda$. Also choose δ between κ and κ^+ so that $x \in L_\delta \prec L_{\kappa^+}$ and α-cof$(\delta) > \omega$. Thus any α-finite ω-sequence from L_δ belongs to L_δ: If $f:\omega \to L_\delta$ is α-finite then $\gamma = \sup(\text{Range}(f)) < \delta$ and $g \circ f \in L_\kappa$ where g is a δ-finite injection of γ into κ. Thus $f = g^{-1} \circ (g \circ f) \in L_\delta$.

Now we let $H = \Sigma_1$ Skolem hull of $L_\delta \cup \{y\}$ inside $\langle L_\lambda, S(\omega) \cap L_\lambda \rangle$. We claim that any α-finite ω-sequence from H belongs to H. For, let h be a Σ_1 Skolem function for L_λ; thus $H = h[\omega \times (L_\delta \cup \{y\})^{<\omega}]$. If $y_0, y_1, \ldots \in H$ is α-finite then we can choose α-finite sequences $\vec{x}_0, \vec{x}_1, \ldots$ (from $(L_\delta)^{<\omega}$) and $n_0, n_1, \ldots$ (from ω) so that $y_i = h(n_i, \vec{x}_i, y)$ for each i. But then $\langle (n_0, \vec{x}_0), (n_1, \vec{x}_1), \ldots \rangle \in L_\delta$ and as the Σ_1 sentence $\exists z\, \forall i (z(i) = h(n_i, \vec{x}_i, y))$ is true in L_λ, it is true in H. So $\langle y_0, y_1, \ldots \rangle \in H$.

Transitively collapse H to L_γ, $\delta \leq \gamma < \kappa^+$. Then <u>$S(\omega) \cap H$ collapses to $S(\omega) \cap L_\gamma$</u>: Call the collapsing map π. If $\beta \in S(\omega) \cap H$ then H contains an ω-sequence $\beta_0, \beta_1, \ldots$ cofinal in β. Then $\pi(\beta_0, \beta_1, \ldots)$ is cofinal in $\pi(\beta)$ so $\pi(\beta) \in S(\omega)$. Conversely if $\pi(\beta) \in S(\omega) \cap L_\gamma$ then there is an ω-sequence $\pi(\beta_0)$, $\pi(\beta_1), \ldots$ cofinal in $\pi(\beta)$; then $\langle \beta_0, \beta_1, \ldots \rangle \in H$ and $\beta_0, \beta_1, \ldots$ must be cofinal in β, else $\sup_i \beta_i < \overline{\beta} < \beta$ for some $\overline{\beta} \in H$ and $\sup_i \pi(\beta_i) < \pi(\overline{\beta}) < \pi(\beta)$ shows that $\pi(\beta_0), \pi(\beta_1), \ldots$ is not cofinal in $\pi(\beta)$.

We now have $\langle L_\lambda, S(\omega) \cap L_\lambda \rangle \models \phi(x, y, S(\omega) \cap L_\lambda) \to \langle H, S(\omega) \cap H \rangle \models \phi(x, y, S(\omega) \cap H) \to \langle L_\gamma, S(\omega) \cap L_\gamma \rangle \models \phi(x, \pi(y), S(\omega) \cap L_\gamma)$. Thus since $\gamma < \kappa^+$ we have $\langle L_{\kappa^+}, S(\omega) \cap L_{\kappa^+} \rangle \models \exists y\ \phi(x, y, S(\omega) \cap L_{\kappa^+})$.

It only remains to show that $S(\omega)$ is not α-recursive. Suppose that $\exists y\, \phi(x, y, p)$ is a Σ_1 formula (with parameter p, ϕ is Δ_0) and we show that this formula does not define the complement of $S(\omega)$. Choose a successor α-cardinal κ^+ so that $p \in L_\kappa$ and $y \in L_\alpha$ so that $\phi(\kappa^+, y, p)$ holds (note that $\kappa^+ \notin S(\omega)$). Choose $\delta < \alpha$ so that $\kappa^+, y \in L_\delta$ and inductively define:

$$H_0 = \Sigma_1 \text{ Skolem hull}(L_\kappa \cup \{\kappa^+, y\}) \text{ inside } L_\delta$$

$$\kappa_0 = H_0 \cap \kappa^+$$

$$H_{n+1} = \Sigma_1 \text{ Skolem hull}(L_\kappa \cup \{\kappa_n, \kappa^+, y\}) \text{ inside } L_\delta$$

$$\kappa_{n+1} = H_{n+1} \cap \kappa^+.$$

Then $H = \bigcup_n H_n$ is the Σ_1 Skolem hull of $L_{\overline{\kappa}} \cup \{\kappa^+, y\}$ inside L_δ, where $\overline{\kappa} = \sup_n \kappa_n$. If $\pi: H \xrightarrow{\sim} L_\lambda$ then $\pi(\kappa^+) = \overline{\kappa}$. Moreover $L_\lambda \models \phi(\overline{\kappa}, \pi(y), p)$. But then $\exists y \phi(\overline{\kappa}, y, p)$ is true since ϕ is Δ_0. This shows that $\exists y \phi\ (x, y, p)$ does not define the complement of $S(\omega)$ as $\overline{\kappa}$ has α-cofinality ω. ⊣

The preceding proof is easily modified as follows: If κ is a regular α-cardinal then let $S(\kappa) = \{\lambda < \alpha \mid \alpha\text{-cof}(\lambda) = \kappa\}$. If there is no largest α-cardinal and $\kappa < \mu$ are regular α-cardinals then $<L_{\mu^+}, S(\kappa) \cap L_{\mu^+}>$ is a Σ_1-elementary substructure of $<L_\alpha, S(\kappa)>$. Moreover the same proof shows that if κ_1, κ_2 are both regular α-cardinals less than a regular α-cardinal μ then $<L_{\mu^+}, S(\kappa_1) \cap L_{\mu^+}, S(\kappa_2) \cap L_{\mu^+}>$ is a Σ_1-elementary substructure of $<L_\alpha, S(\kappa_1), S(\kappa_2)>$. Thus $S(\kappa_1) \vee S(\kappa_2)$ is low, hyperregular and α-RE. The next result shows that $S(\kappa_1)$, $S(\kappa_2)$ represent incomparable α-degrees.

Theorem 2. Suppose there is no largest α-cardinal and κ_1, κ_2 are distinct regular α-cardinals. Then $S(\kappa_1)$ is not α-recursive in $S(\kappa_2)$.

Proof: Suppose that ϕ is Δ_0 and $\exists y \phi(x, y, p, S(\kappa_2))$ defines the complement of $S(\kappa_1)$ over $<L_\alpha, S(\kappa_2)>$. We derive a contradiction by showing that $<L_\alpha, S(\kappa_2)> \models \exists y \phi(x, y, p, S(\kappa_2))$, for some $x \in S(\kappa_1)$. Choose a regular α-cardinal κ so that $p, \kappa_1, \kappa_2 \in L_\kappa$ and let $y \in L_\alpha$ be so that $\phi(\kappa^+, y, p, S(\kappa_2))$. The choice of y is possible as $\kappa^+ \notin S(\kappa_1)$. If λ is a regular α-cardinal such that $y \in L_\lambda$ then inductively define a κ_1-sequence of Skolem hulls:

$$H_0 = \Sigma_1 \text{ Skolem hull of } L_\kappa \cup \{\kappa^+, y\} \text{ in } <L_\lambda, S(\kappa_2) \cap L_\lambda>$$

$$\gamma_0 = H_0 \cap \kappa^+$$

$$H_{\delta+1} = \Sigma_1 \text{ Skolem hull of } L_\kappa \cup \{\kappa^+, y, \gamma_\delta\} \text{ in } <L_\lambda, S(\kappa_2) \cap L_\lambda>$$

$$\gamma_{\delta+1} = H_{\delta+1} \cap \kappa^+$$

$$H_\beta = \bigcup_{\beta' < \beta} H_{\beta'}, \quad \gamma_\beta = \bigcup_{\beta' < \beta} \gamma_{\beta'}, \quad \beta \text{ limit.}$$

Lemma. For $\delta < \kappa_1$, any α-finite κ_2-sequence from $H_{\delta+1}$ belongs to $H_{\delta+1}$.

Proof of Lemma: First note that if $f: \kappa_2 \to H_{\delta+1}$ is $\Sigma_1 <L_\lambda, S(\kappa_2) \cap L_\lambda>$ (with parameter in $H_{\delta+1}$) then $f \in H_{\delta+1}$. For, if $f(x) = y \leftrightarrow <L_\lambda, S(\kappa_2) \cap L_\lambda> \models \exists z \psi(x, y, z)$ where ψ is Δ_0 then $<H_{\delta+1}, S(\kappa_2) \cap H_{\delta+1}> \models \exists b \forall x \in \kappa_2 \exists y, z \in b \psi(x, y, z)$, as this sentence is true in $<L_\lambda, S(\kappa_2) \cap L_\lambda>$. So $f \in H_{\delta+1}$. Now it suffices to show that any α-finite $f: \kappa_2 \to H_{\delta+1}$ is $\Sigma_1 <L_\lambda, S(\kappa_2) \cap L_\lambda>$. But there is a $\Sigma_1 <H_{\delta+1}, S(\kappa_2) \cap H_{\delta+1}>$ injection $g: H_{\delta+1} \to L_\kappa$ and $g \circ f \in L_\kappa$ (since $\kappa_2 < \kappa$ and κ is a regular α-cardinal). So $f = g^{-1} \circ (g \circ f)$ is $\Sigma_1 <H_{\delta+1}, S(\kappa_2) \cap H_{\delta+1}>$ and hence $\Sigma_1 <L_\lambda, S(\kappa_2) \cap L_\lambda>$. $\dashv$

Let $H = \bigcup_{\delta < \kappa_1} H_\delta$ and $\gamma = H \cap \kappa^+$. Then $\gamma \in S(\kappa_1)$ and if $\pi: H \xrightarrow{\sim} L_{\bar{\gamma}}$ then $\pi(\kappa^+) = \gamma$. For our contradiction it suffices to show that $\pi[S(\kappa_2) \cap H] = S(\kappa_2) \cap L_{\bar{\gamma}}$ as this implies $<L_{\bar{\gamma}}, S(\kappa_2) \cap L_{\bar{\gamma}}> \models \phi(\gamma, \pi(y), p, S(\kappa_2) \cap L_{\bar{\gamma}})$, showing that

$\exists y\phi(x,y,p,S(\kappa_2))$ does not define the complement of $S(\kappa_1)$.

If $\beta \in S(\kappa_2) \cap H$ then there is a cofinal $f:\kappa_2 \to \beta$, $f \in H$ since H is a Σ_1-elementary substructure of L_λ. Then $\pi \circ f$ is cofinal in $\pi(\beta)$ so $\pi(\beta) \in S(\kappa_2)$ Conversely, suppose $\pi(\beta) \in S(\kappa_2)$. Choose an α-finite, increasing function $f:\kappa_2 \to H$ so that $\pi \circ f$ is cofinal in $\pi(\beta)$. Then for some $\delta < \kappa_1$ and unbounded α-finite $X \subseteq \kappa_2$, $g = f \restriction X$ has range contained in $H_{\delta+1}$. By the lemma $g \in H_{\delta+1}$ But g is cofinal in β as otherwise for some $\overline{\beta} \in H$, $\cup\text{Range}(g) < \overline{\beta} < \beta$ and $\cup\text{Range}(\pi \circ f) < \pi(\overline{\beta}) < \pi(\beta)$, contradicting the fact that $\pi \circ f$ is cofinal in $\pi(\beta)$. —|

Some Remarks and Questions

1) The proof of Theorem 2 is easily extended to show: If there is no largest α-cardinal and $\kappa, \kappa_1, \ldots, \kappa_n$ are distinct α-cardinals then $S(\kappa)$ is not α-recursive in $S(\kappa_1) \vee \ldots \vee S(\kappa_n)$. Moreover $S(\kappa_1) \vee \ldots \vee S(\kappa_n)$ is low and hyperregular.

2) Suppose there is no largest α-cardinal and κ_1, κ_2 are distinct α-cardinals. Then is $S(\kappa_1)$, $S(\kappa_2)$ a minimal pair (i.e., does $A \leq_\alpha S(\kappa_1)$, $A \leq_\alpha S(\kappa_2)$ imply A is α-recursive)?

3) Are there any incomplete α-RE degrees greater than all of the α-degrees of $S(\kappa)$ for regular α-cardinals κ (assuming there is no largest α-cardinal)?

4) Problem: Find natural, intermediate α-RE degrees when there is a largest α-cardinal. If $\alpha^* = \alpha$ then there is no largest α-stable; is there a way of making use of the α-stables similar to the above use of α-cardinals?

References

Gödel, K [1939] Consistency Proof for the Generalized Continuum Hypothesis, Proc. Nat. Acad. Sci. U.S.A.

Sacks, G. and Simpson. S. [1972] The α-Finite Injury Method, Ann. Math. Logic 4.

Simpson S. [1974] Degree Theory on Admissible Ordinals, in "Generalized Recursion Theory" (Fenstad, Hinman, Eds.), North Holland.

ELEMENTARY THEORY OF AUTOMORPHISM GROUPS OF DOUBLY HOMOGENEOUS CHAINS

A. M. W. Glass[1], Yuri Gurevich[2], W. Charles Holland and Michèle Jambu-Giraudet[2]

1. INTRODUCTION.

At the Logic Meeting in Storrs, Connecticut (November 1979), the first author presented a survey of the research being done classifying linearly ordered sets by the elementary theory of their automorphism groups. In this paper we wish to present some aspects of our more recent research which may be of interest to logicians. A more up-to-date survey will appear in the Proceedings of the Algebra Conference, Carbondale (1980)--to be published by Springer-Verlag in their lecture note series.

In the early 1970's, (unordered) sets were classified by the elementary (first order) group-theoretic properties of their symmetric groups--i.e., automorphism groups (see [10], [11] and [13]). A natural extension of this work is to try to classify structures of a given signature by the first order properties of their automorphism groups. One such problem is to take the models as $\langle \Omega, \leq \rangle$, linearly ordered sets (or chains, for short). Besides its obvious naturalness, there are two further reasons to study the automorphisms of chains. The first is that if Σ is any set of sentences (of a first order language) having an infinite model and $\langle \Omega, \leq \rangle$ is a chain, there is a model $\mathfrak{A}_\Omega$ of Σ containing Ω as a subset such that each automorphism of $\langle \Omega, \leq \rangle$ extends to an automorphism of the model $\mathfrak{A}_\Omega$; i.e., $\mathrm{Aut}(\langle \Omega, \leq \rangle)$ is a subgroup of $\mathrm{Aut}(\mathfrak{A}_\Omega)$ (See [12] for this and further motivating reasons for model theorists). The second reason is provided by the theorem [3; Appendix I] that every lattice-ordered group can be embedded in the automorphism group of a chain. We will write $\mathcal{A}(\Omega)$ for $\mathrm{Aut}(\langle \Omega, \leq \rangle)$; i.e., $\mathcal{A}(\Omega)$ is the group of all order-preserving permutations of the chain Ω. The classification in this case was begun in [7], [8] and [5]. This paper provides further results.

---

[1]A. M. W. Glass wishes to thank N.S.F. for providing his expenses at Storrs, and the University of Connecticut for its hospitality--especially Manny Lerman and Jim Schmerl.

[2]Yuri Gurevich and Michèle Jambu-Giraudet wish to thank Bowling Green State University for its hospitality in the Spring and Fall Quarters (respectively) of 1980.

The presence of linear ordering complicates matters in two ways. Whereas the symmetric group on a set is always transitive, the same is not necessarily true of $A(\Omega)$--e.g., if $\Omega = \omega$, then $A(\Omega) = \{e\}$ (e is the identity element of the group $A(\Omega)$). So $A(T) \cong A(\omega \overleftarrow{\cup} T)$ for any chain T, where $\omega \overleftarrow{\cup} T$ is $\omega \cup T$ ordered by: $n < \tau$ for all $\tau \in T$, $n \in \omega$. In order to obtain any nice classification of chains Ω by the elementary properties of the group $A(\Omega)$, we will assume that Ω is <u>homogeneous</u> (i.e., for each $\alpha,\beta \in \Omega$, there exists $f \in A(\Omega)$ such that $f(\alpha) = \beta$; so homogeneous in our sense means 1-homogeneous in the usual model-theoretic sense). If Ω is homogeneous, we will say that $A(\Omega)$ is <u>transitive</u>. The second complication is that the symmetric group on a set is always primitive (i.e., there is no non-trivial equivalence relation on the set which is respected by the symmetric group). However, even when Ω is homogeneous, there may exist non-trivial equivalence relations on Ω (having convex classes) which are respected by $A(\Omega)$. For example, let $\Omega = \mathbb{R} \vec{\times} \mathbb{Z}$, the lexicographic product of the real line, $\mathbb{R}$, and the integers, $\mathbb{Z}$ (i.e., $\mathbb{R} \otimes \mathbb{Z}$ ordered by: $(r,m) > (s,n)$ if $r > s$ or $(r = s$ & $m > n)$). Then $(r,m) \sim (s,n)$ if $r = s$ is an equivalence relation of the desired kind. (Two points of Ω are equivalent only if there are only finitely many points of Ω between them.) Such chains are said to be <u>non-primitive</u>. Fortunately, there is a group-theoretic sentence which is satisfied in a transitive $A(\Omega)$ if and only if Ω is primitive (see [3; Theorem 4D] or [7; Lemma 4]); so we will confine ourselves to primitive chains in this article. The non-primitive case will be investigated in a later paper.

If Ω is primitive, then [3; Theorem 4.B] either

(i) $A(\Omega)$ is abelian, or

(ii) Ω is <u>doubly homogeneous</u> (for each $\alpha_i,\beta_i \in \Omega$ $(i = 1,2)$ with $\alpha_1 < \alpha_2$ and $\beta_1 < \beta_2$, there exists $f \in A(\Omega)$ such that $f(\alpha_i) = \beta_i$ $(i = 1,2)$).

Moreover, (i) and (ii) are disjoint ([3, Lemma 1.6.8]) and so can be distinguished by a group-theoretic sentence about $A(\Omega)$. Hence we may deal with them separately in attempting to classify homogeneous chains by the elementary properties of their automorphism groups. Case (i), the rigidly homogeneous case, was completely studied in [5]. So we will confine ourselves to doubly homogeneous chains in this article. Our main thrust will be to establish that

<u>the first order language of automorphism groups of doubly homogeneous chains is extremely rich.</u>

Although we are primarily interested in the group $\mathcal{A}(\Omega)$, an auxiliary relation on it will simplify matters. This auxiliary relation is the pointwise ordering on $\mathcal{A}(\Omega)$, with respect to which $\mathcal{A}(\Omega)$ becomes a lattice-ordered group: $f \leq g$ if $f(\alpha) \leq g(\alpha)$ for all $\alpha \in \Omega$; so $g \geq e$ if g moves no points down. In [8] (or [9]) it was shown that there is a formula $\psi(x,y)$ of the group language such that $\mathcal{A}(\Omega) \models \psi(f,g)$ if and only if $e \leq f,g$ or $e \geq f,g$. Hence if $\mathcal{A}(\Omega) \equiv \mathcal{A}(\Lambda)$ as groups, $\langle\mathcal{A}(\Omega),\leq\rangle \equiv \langle\mathcal{A}(\Lambda),\leq\rangle$ or $\langle\mathcal{A}(\Omega),\leq\rangle \equiv \langle\mathcal{A}(\Lambda),\leq^*\rangle$ as lattice-ordered groups where $\leq^*$ is the reverse of the pointwise ordering. Since in most of our results, $\langle\mathcal{A}(\Lambda),\leq\rangle$ satisfies the desired properties if and only if $\langle\mathcal{A}(\Lambda),\leq^*\rangle$ does, we will assume that the pointwise ordering and the inherited lattice operations $\vee$ and $\wedge$ are explicitly in the language. (We use "&" and "or" for the conjunction and disjunction of the language.)

For $g \in \mathcal{A}(\Omega)$, let <u>supp(g)</u> $= \{\alpha \in \Omega: g(\alpha) \neq \alpha\}$, the <u>support</u> of g. If $f,g \in \mathcal{A}(\Omega)$ and $\text{supp}(f) < \text{supp}(g)$ (i.e., $\alpha < \beta$ for all $\alpha \in \text{supp}(f)$ and $\beta \in \text{supp}(g)$), we say that f is to the <u>left</u> of g. Since $\text{supp}(hgh^{-1}) = h(\text{supp}(g))$, it follows that if $f,g > e$, then f is to the left of g if and only if $\mathcal{A}(\Omega) \models (\forall h)(h \geq e \rightarrow f \wedge hgh^{-1} = e)$. We abbreviate this formula to <u>L(f,g)</u>. Hence $g > e$ has bounded support can be expressed in our language by the formula $(\exists f_1 > e)(\exists f_2 > e)(L(f_1,g) \,\&\, L(g,f_2))$.

Let $\bar{\Omega}$ be the Dedekind completion of Ω. Each $g \in \mathcal{A}(\Omega)$ has a unique extension $\bar{g}$ to an element of $\mathcal{A}(\bar{\Omega})$ given by: $\bar{g}(\bar{\alpha}) = \sup\{g(\alpha): \alpha \in \Omega \,\&\, \alpha \leq \bar{\alpha}\}$ $(\bar{\alpha} \in \bar{\Omega})$. We will identify g with $\bar{g}$. If Ω is doubly homogeneous and $\bar{\alpha} \in \bar{\Omega}$, there is $e < g \in \mathcal{A}(\Omega)$ of bounded support such that $\bar{\alpha} = \sup(\text{supp}(g))$. Moreover, if $e < g' \in \mathcal{A}(\Omega)$ and $\bar{\alpha}' = \sup(\text{supp}(g'))$, $\bar{\alpha} = \bar{\alpha}'$ if and only if $\mathcal{A}(\Omega) \models (\forall h > e)(L(g,h) \leftrightarrow L(g',h))$; and $\bar{\alpha} \leq \bar{\alpha}'$ precisely when $\mathcal{A}(\Omega) \models (\forall h > e)(L(g',h) \rightarrow L(g,h))$. We can therefore interpret $\bar{\Omega}$ in $\mathcal{A}(\Omega)$ in this uniform way (the formulae are independent of the particular doubly homogeneous chain Ω). That is, we can interpret (uniformly) $\langle\mathcal{A}(\Omega),\bar{\Omega},\cdot,^{-1},e,\leq_{\mathcal{A}(\Omega)},\leq_{\bar{\Omega}}\rangle$ in $\langle\mathcal{A}(\Omega),\cdot,^{-1},e,\leq_{\mathcal{A}(\Omega)}\rangle$. Moreover, $f(\bar{\alpha}) = \bar{\alpha}'$ if and only if $\mathcal{A}(\Omega) \models (\forall h > e)(L(fgf^{-1},h) \leftrightarrow L(g',h))$.

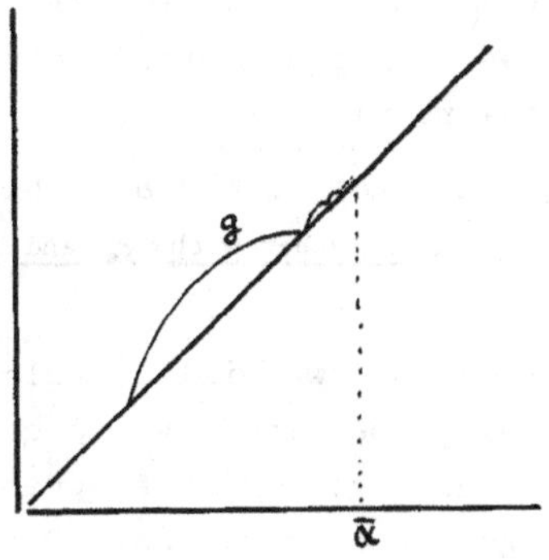

Consequently, we will assume that our language is equipped explicitly with variables for the points of $\bar{\Omega}$ together with relations for the order on $\bar{\Omega}$ and the action of $\mathcal{A}(\Omega)$ on $\bar{\Omega}$. If $\mathcal{A}(\Omega) \cong \mathcal{A}(\Lambda)$ as groups implies only $\langle\mathcal{A}(\Omega),\leq\rangle \equiv \langle\mathcal{A}(\Lambda),\leq^*\rangle$, we would obtain $\langle\mathcal{A}(\Omega),\bar{\Omega},\leq_{\mathcal{A}(\Omega)},\leq_{\bar{\Omega}}\rangle \equiv \langle\mathcal{A}(\Lambda),\bar{\Lambda},\leq^*_{\mathcal{A}(\Lambda)},\leq^*_{\bar{\Lambda}}\rangle$. This means that any results of the form: $\mathcal{A}(\Omega) \equiv \mathcal{A}(\Lambda)$ implies $\Omega \cong \Lambda$, really should have the weaker conclusion that the homogeneous chains Ω and Λ ordermorphic or anti-ordermorphic. This makes a difference in Theorem 12 (Cases (a) and (b) become one since $\mathcal{A}(\vec{\mathbb{R}}) \cong \mathcal{A}(\overleftarrow{\mathbb{R}})$ as groups; an isomorphism is furnished by conjugating by an anti-ordermorphism between $\vec{\mathbb{R}}$ and $\overleftarrow{\mathbb{R}}$).

Points $\bar{\alpha},\bar{\beta} \in \bar{\Omega}$ lie in the same <u>orbit</u> of $\mathcal{A}(\Omega)$ if $f(\bar{\alpha}) = \bar{\beta}$ for some $f \in \mathcal{A}(\Omega)$. As we saw above, this is recognizable in our language. Hence the orbits of $\mathcal{A}(\Omega)$ in $\bar{\Omega}$ are interpretable in our language. Now Ω is an orbit of $\mathcal{A}(\Omega)$. We may not always be able to distinguish it in our language from another orbit T of $\mathcal{A}(\Omega)$ in $\bar{\Omega}$ since it is possible that $\mathcal{A}(\Omega) \cong \mathcal{A}(T)$ as lattice-ordered groups, with the isomorphism being furnished by extending an element of $\mathcal{A}(\Omega)$ to its unique extension in $\mathcal{A}(\bar{\Omega})$ and then restricting the domain to T. However, we will assume that variables for points of Ω are included in our language; thus we can distinguish Ω from any other orbit of $\mathcal{A}(\Omega)$. This means that any results of the form: $\mathcal{A}(\Omega) \equiv \mathcal{A}(\Lambda)$ implies $\Omega \cong \Lambda$, really should have the weaker conclusion that the homogeneous chain Λ is ordermorphic (or anti-ordermorphic) to an orbit of $\mathcal{A}(\Omega)$ in $\bar{\Omega}$; i.e., $\Lambda \cong (\mathcal{A}(\Omega))(\bar{\alpha})$ for some $\bar{\alpha} \in \bar{\Omega}$. For example, in Theorem 5, the conclusion should be: Λ is ordermorphic to the rationals or irrationals.

Throughout this paper, then, our language will be the first order language of lattice-ordered groups, together with variables for points of Ω and for points of $\bar{\Omega}$, and a symbol $f(\bar{\alpha})$ for the action of $f \in \mathcal{A}(\Omega)$ on $\bar{\alpha} \in \bar{\Omega}$. (So if $\bar{\alpha} \in \Omega$, $f(\bar{\alpha}) \in \Omega$.) However, we will use $\mathcal{A}(\Omega)$ as a shorthand for the structures of this language. Our most powerful result is:

THEOREM A: <u>Let Ω be a doubly homogeneous chain. Then countable subsets of Ω</u> **<u>together with $\subseteq$ and</u>** <u>membership in them are interpretable in $\mathcal{A}(\Omega)$</u>.

From this we will be able to characterize many chains whose defining properties involve countability; e.g., Suslin, Luzin, Specker. We will also give simpler proofs than those in [7] to show that the language can express that Ω is (isomorphic to) the real line $\mathbb{R}$, the rational line $\mathbb{Q}$, the real long lines, and certain long rational lines.

Also, we will be able to tell (in the language) if Ω can be embedded in $\mathbb{R}$, and whether it can bear the arithmetic structure of an additive subgroup or a subfield of $\mathbb{R}$.

If this background is inadequate, see [3], [4], [7], [8] or [9].

2. RESULTS.

The language $\mathcal{L}$ is the first order language with the usual logical symbols, $=$, variables $f,g,h,\ldots$ for members of $\mathcal{A}(\Omega)$, $\alpha,\beta,\gamma,\ldots$ for members of Ω and $\bar\alpha,\bar\beta,\bar\gamma,\ldots$ for members of $\bar\Omega(\supseteq \Omega)$; a constant e for the identity element of $\mathcal{A}(\Omega)$, symbols for multiplication, inverse, least upper bound ($\vee$) and greatest lower bound ($\wedge$) for $\mathcal{A}(\Omega)$ as well as the pointwise order $\leq$ on $\mathcal{A}(\Omega)$ (as a shorthand: $f \leq g$ stands for $f \wedge g = g$); the total order relation ($\leq$) on $\bar\Omega$ and the inherited order on Ω; and the action of $\mathcal{A}(\Omega)$ on $\bar\Omega$.

An element $e < g \in \mathcal{A}(\Omega)$ is said to have <u>one bump</u> if whenever $\alpha < \bar\beta < \gamma$ with $\alpha,\gamma \in \operatorname{supp}(g)$, $g(\bar\beta) \neq \bar\beta$. This is equivalent to $\mathcal{A}(\Omega) \models (\forall u)(\forall v)(u \wedge v = e \;\&\; u \vee v = g \rightarrow u = e$ or $v = e)$. (See [3], [7], [8] or [9].) We will write <u>Bump(g)</u> for this formula of $\mathcal{L}$ (in one free variable g).

Let $e < f \in \mathcal{A}(\Omega)$ and $e < b \in \mathcal{A}(\Omega)$. b is a <u>bump of f</u> if b has just one bump and $f|\operatorname{supp}(b) = b|\operatorname{supp}(b)$. Note that if b_1,b_2 are distinct bumps of f, then $b_1 \wedge b_2 = e$.

LEMMA 0: <u>Let Ω be a homogeneous chain and</u> $e < b,f \in \mathcal{A}(\Omega)$. <u>Then</u> b <u>is a bump of f if and only if</u> $\mathcal{A}(\Omega) \models (b \wedge b^{-1}f = e) \;\&\; \mathrm{Bump}(b)$.

<u>Proof</u>: Let b be a bump of f and $\alpha \in \operatorname{supp}(b)$. Since $f(\alpha) = b(\alpha)$, $b^{-1}f(\alpha) = \alpha$. If $\beta \notin \operatorname{supp}(b)$, $b(\beta) = \beta$. Thus $b \wedge b^{-1}f = e$. Conversely, if $b \wedge b^{-1}f = e$ and $\alpha \in \operatorname{supp}(b)$, $b^{-1}f(\alpha) = \alpha$ so $f(\alpha) = b(\alpha)$. Hence $f|\operatorname{supp}(b) = b|\operatorname{supp}(b)$ and b is a bump of f.

LEMMA 1: <u>Let</u> $\theta_T \equiv (\exists f > e)[\mathrm{Bump}(f) \;\&\; (\forall g)(\neg L(g,f) \;\&\; \neg L(f,g))]$, $\theta_I \equiv (\exists f > e)[\mathrm{Bump}(f) \;\&\; (\forall g)\neg L(g,f)]$ <u>and</u> $\theta_F \equiv (\exists f > e)[\mathrm{Bump}(f) \;\&\; (\forall g)\neg L(f,g)]$. <u>If Ω is a doubly homogeneous, then</u> (i) $\mathcal{A}(\Omega) \models \theta_T$ <u>if and only if Ω has countable coterminality</u>, (ii) $\mathcal{A}(\Omega) \models \theta_I$ <u>if and only if Ω has countable coinitiality</u>, (iii) $\mathcal{A}(\Omega) \models \theta_F$ <u>if and only if Ω has countable cofinality</u>.

<u>Proof</u>: (i) Since $\{f^n(\alpha): n \in \mathbb{Z}\}$ is coterminal in $\operatorname{supp}(f)$ if has one bump, it remains to prove that $\mathcal{A}(\Omega) \models \theta_T$ if Ω is doubly homogeneous and has countable coterminality. Let $\{\alpha_n: n \in \mathbb{Z}\}$ be coterminal in Ω. Since Ω is doubly homogeneous, there is an order-morphism $f_n: [\alpha_n,\alpha_{n+1}] \cong [\alpha_{n+1},\alpha_{n+2}]$. Let $f = \bigcup\{f_n: n \in \mathbb{Z}\}$. Then

$e < f \in \mathcal{A}(\Omega)$ and fixes no point of $\bar{\Omega}$. Moreover, $\mathrm{supp}(f) = \Omega$ so $\mathcal{A}(\Omega) \models \theta_T$.

If $X \subseteq \mathcal{A}(\Omega)$, let $\underline{C(X)} = \{f \in \mathcal{A}(\Omega)\colon (\forall g \in X)(fg = gf)\}$, the centralizer of X in $\mathcal{A}(\Omega)$. If $f \in \mathcal{A}(\Omega)$, write $\underline{\langle f\rangle}$ for the subgroup generated by f; i.e., $\langle f\rangle = \{f^n\colon n \in \mathbb{Z}\}$.

The following lemma is very similar to Lemma 16 of [6] where the condition of having one bump is removed and $\mathcal{A}(\Omega)$ is replaced by an existentially complete lattice-ordered group.

LEMMA 2: *If* $f \in \mathcal{A}(\Omega)$ *and has one bump, then* $g \in \langle f\rangle$ *if and only if* $g \in C(C(f))$.

Proof: Since f has one bump, $f(\alpha) > \alpha$ for all $\alpha \in \mathrm{supp}(f)$. We must show that if $g \in C(C(f))$, then $g \in \langle f\rangle$, the other direction being obvious. If $g \in C(C(f))\setminus\langle f\rangle$, let $\bar{B}$ be the single open interval of support of f in $\bar{\Omega}$; i.e., $\bar{B}$ is the convexification (in $\bar{\Omega}$) of $\mathrm{supp}(f)$. For each $n \in \mathbb{Z}$, the sets $\{\bar{\beta} \in \bar{B}\colon g(\bar{\beta}) \le f^n(\bar{\beta})\}$ and $\{\bar{\beta} \in \bar{B}\colon g(\bar{\beta}) \ge f^{n+1}(\bar{\beta})\}$ are closed (in $\bar{B}$) and disjoint (since $f^n(\bar{\beta}) = f^{n+1}(\bar{\beta})$ implies $\bar{\beta} = f(\bar{\beta})$, contradicting $\bar{\beta} \in \bar{B}$). Since $\bar{B}$ is connected, there exists $\bar{\alpha} \in \bar{B}$ such that $g(\bar{\alpha}) \notin \{f^n(\bar{\alpha})\colon n \in \mathbb{Z}\}$. Either $g(\bar{\alpha}) \notin \bar{B}$ or, for some $n \in \mathbb{Z}$, $f^n(\bar{\alpha}) < g(\bar{\alpha}) < f^{n+1}(\bar{\alpha})$. In the first case, there exists $h \in \mathcal{A}(\Omega)$ such that $hg(\bar{\alpha}) \neq g(\bar{\alpha})$ and $\mathrm{supp}(h) \cap \bar{B} = \emptyset$. Then $h \in C(f)$ but $g \notin C(h)$ $(hg(\bar{\alpha}) \neq g(\bar{\alpha}) = gh(\bar{\alpha})$, a contradiction. In the second case there is $h \in \mathcal{A}(\Omega)$ such that $g(\bar{\alpha}) \in \mathrm{supp}(h) \subseteq (f^n(\bar{\alpha}), f^{n+1}(\bar{\alpha}))$. Let $h^* \in \mathcal{A}(\Omega)$ be the identity of $\bar{B}$ and agree with $f^m h f^{-m}$ on $(f^{n+m}(\bar{\alpha}), f^{n+m+1}(\bar{\alpha}))$ (for all $m \in \mathbb{Z}$). Since $\alpha = f^0(\alpha)$, $h^*(\alpha) = \alpha$. Then $h^* \in C(f)$ but $g \notin C(h^*)$, the desired contradiction.

We have shown that $g \in \langle f\rangle$ is expressible in our language if f has one bump. We can therefore assume "$g \in \langle f\rangle$" is in $\mathcal{L}$ if f has one bump.

LEMMA 3: *The statement "B is a countable bounded subset of* Ω*" is interpretable in* $\mathcal{A}(\Omega)$*, as is the formula "*$\alpha \in B$*". Hence* $\subseteq$ *between countable bounded subsets of* Ω *is interpretable in* $\mathcal{A}(\Omega)$.

Proof: Let B be a countable bounded non-empty subset of Ω. Let $\beta \in B$ and $\gamma < B$. Since $\mathcal{A}(\Omega)$ is transitive, there exists $e < g \in \mathcal{A}(\Omega)$ such that $B < g(\gamma)$ and g has one bump. Let B be enumerated $\{\beta_n\colon n \in \omega\}$ with $\beta_0 = \beta$. By double homogeneity, there exists $e < f \in \mathcal{A}(\Omega)$ such that f has one bump and for all $n \in \omega$, $f^n(\beta) = g^n(\beta_n)$ and $f^n(\gamma) = g^n(\gamma)$.

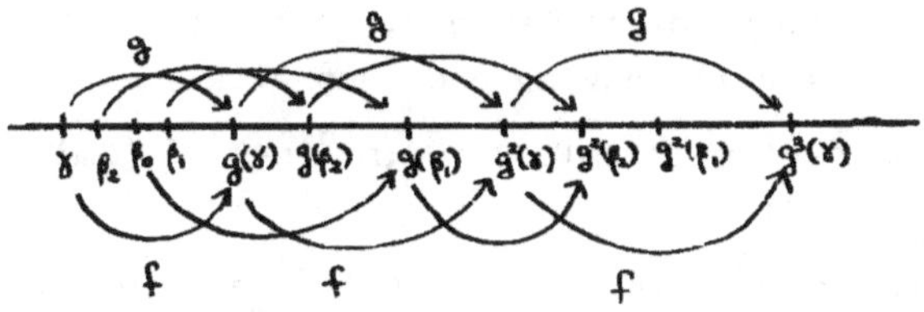

It follows that $B = \{g^{-n}f^{n}(\beta) : n \in \omega\} = \{k^{-1}h(\beta) : e < h \in \langle f\rangle$ & $e < k \in \langle g\rangle\} \cap (\gamma, g(\gamma))$. By Lemma 2, this last expression is an interpretation of B in $\mathcal{A}(\Omega)$ via the quadruple (β,γ,f,g), where $\beta,\gamma \in \Omega$ with $\gamma < \beta$, and $e < f,g$ have one bump with $\{h(\gamma) : h \in \langle f\rangle\} = \{k(\gamma) : k \in \langle g\rangle\}$. This last condition can be expressed in our language by $(\forall h \in \langle f\rangle)(\exists k \in \langle g\rangle)(h(\gamma) = k(\gamma))$ & $(\forall k \in \langle g\rangle)(\exists h \in \langle f\rangle)(h(\gamma) = k(\gamma))$. Thus we can determine in $\mathcal{L}$ which quadruples determine the same bounded countable set, etc. Since any quadruple satisfying the above conditions yields a countable bounded set, the lemma is proved.

Note that the interpretation is uniform; i.e., the formulae of $\mathcal{L}$ for interpretation are independent of the particular doubly homogeneous chain Ω. Also, we could choose f,g of the proof of the lemma so that their supports are contained in any open interval (σ,τ) with $\sigma < \inf(B) \leq \sup(B) < \tau$.

We now prove Theorem A.

Let Ω be a doubly homogeneous chain and $\Delta \subseteq \Omega$ be countable. If Λ is bounded, $\Lambda = \{b^{-1}a(\beta) : e < a \in \langle f\rangle$ & $e < b \in \langle g\rangle\} \cap (\gamma, g(\gamma))$ for some β,γ,f,g where f and g have one bump. If Δ is unbounded in Ω, let $\{\beta_m : m \in \mathbb{Z}\}$, $\{\gamma_m : m \in \mathbb{Z}\}$, $\{\sigma_m : m \in \mathbb{Z}\}$ and $\{\tau_m : m \in \mathbb{Z}\}$ be coterminal in Δ (and hence in Ω) with $\beta_m < \sigma_m < \gamma_m < \tau_m < \beta_{m+1}$ $(m \in \mathbb{Z})$. Let $\Delta_{0,m} = \Delta \cap [\gamma_{2m-1}, \gamma_{2m}]$ and $\Delta_{1,m} = \Delta \cap (\gamma_{2m-1}, \gamma_{2m})$.

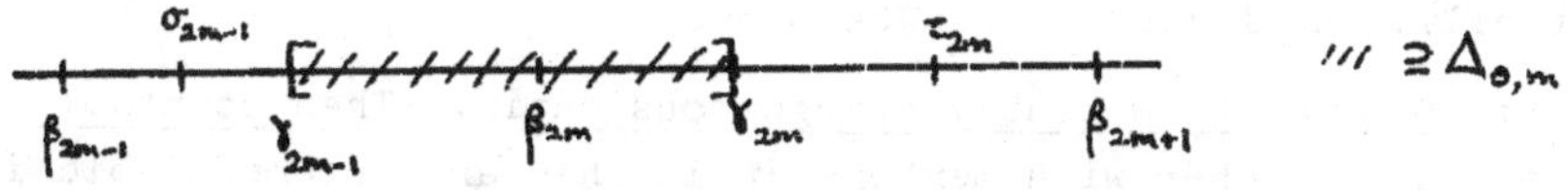

By the remark following the proof of Lemma 3, there are $e < f_{i,m}, g_{i,m} \in \mathcal{A}(\Omega)$ $(i = 0,1)$ having one bump with $\operatorname{supp}(f_{0,m}) \cup \operatorname{supp}(g_{0,m}) \subseteq (\beta_{2m-1}, \beta_{2m+1})$, $g_{0,m}(\sigma_{2m-1}) = \tau_{2m}$, $f^n_{0,m}(\sigma_{2m-1}) = g^n_{0,m}(\sigma_{2m-1})$ $(n \in \mathbb{Z})$,
$\Delta_{0,m} = \{g^{-n}_{0,m} f^n_{0,m}(\beta_{2m}) : n \in \omega\} = (\sigma_{2m-1}, \tau_{2m}) \cap \{b^{-1}_{0,m} a_{0,m}(\beta_{2m}) :$
$e < a_{0,m} \in \langle f_{0,m} \rangle \ \& \ e < b_{0,m} \in \langle g_{0,m} \rangle\}$, and
$\operatorname{supp}(f_{1,m}) \cup \operatorname{supp}(g_{1,m}) \subseteq (\beta_{2m}, \beta_{2m+2})$, $g_{1,m}(\sigma_{2m}) = \tau_{2m+1}$,
$f^n_{1,m}(\sigma_{2m}) = g^n_{1,m}(\sigma_{2m})$ $(n \in \mathbb{Z})$,
$\Delta_{1,m} = \{g^{-n}_{1,m} f^n_{1,m}(\beta_{2m+1}) : n \in \omega\} = (\sigma_{2m}, \tau_{2m+1}) \cap \{b^{-1}_{1,m} a_{1,m}(\beta_{2m+1}) :$
$e < a_{1,m} \in \langle f_{1,m} \rangle \ \& \ e < b_{1,m} \in \langle g_{1,m} \rangle\}$ $(m \in \mathbb{Z})$. Let f_i and g_i be the supremum of the pairwise disjoint set of elements $\{f_{i,m} : m \in \mathbb{Z}\}$ and $\{g_{i,m} : m \in \mathbb{Z}\}$ respectively $(i = 0,1)$. Then
$\Delta = \bigcup_{m \in \mathbb{Z}} \{g^{-n}_0 f^n_0(\beta_{2m}) : n \in \omega\} \cup \bigcup_{m \in \mathbb{Z}} \{g^{-n}_1 f^n_1(\beta_{2m+1}) : n \in \omega\}$. As in the proof of Lemma 1 (i), there is $e < h \in \mathcal{A}(\Omega)$ having one bump such that $h(\beta_m) = \beta_{m+1}$, $h(\sigma_m) = \sigma_{m+1}$ and $h(\tau_m) = \tau_{m+1}$ $(m \in \mathbb{Z})$. So
$\Delta = \{g^{-n}_0 f^n_0 h^{2m}(\beta_0) : m \in \mathbb{Z}, n \in \omega\} \cup \{g^{-n}_1 f^n_1 h^{2m} h(\beta_0) : m \in \mathbb{Z}, n \in \omega\}$. We have coded Δ by $(g_0, g_1, f_0, f_1, h, \beta_0, \sigma_0, \tau_0)$ in the following sense: $\delta \in \Delta$ if and only if there are $k \in \langle h^2 \rangle$, $e < a_i \in \langle c_i \rangle$ and $e < b_i \in \langle d_i \rangle$ with $\delta = b^{-1}_i a_i k(\beta_i)$ for $i = 0$ or 1, where $\beta_1 = h(\beta_0)$, $\sigma_1 = h(\sigma_0)$, $\tau_1 = h(\tau_0)$, and c_i and d_i are the unique bumps of f_i and g_i respectively with $c_i k(\sigma_i) = \tau_i = d_i k(\sigma_i)$ and $h^{-1}k(\sigma_i) < b^{-1}_i a_i k(\beta_i) < k(\tau_i)$. By Lemmas 0 and 2, this is expressible in $\mathcal{L}$. Moreover, any octuple $(g_0, g_1, f_0, f_1, h, \beta_0, \sigma_0, \tau_0)$ with $\beta_0 < \sigma_0 < \tau_0 < h(\beta_0)$, $e < h$ has one bump, and $e < g_0, g_1, f_0, f_1$ with $g_i h^{2n}(\sigma_i) = h^{2n+1}(\tau_i) = f_i h^{2n}(\sigma_i)$ $(i = 0,1)$ where $\sigma_1 = h(\sigma_0)$ and $\tau_1 = h(\tau_0)$ gives a countable subset of Ω via
$\{g^{-n}_0 f^n_0 h^{2m}(\beta_0) : m \in \mathbb{Z}, n \in \omega\} \cup \{g^{-n}_1 f^n_1 h^{2m+1}(\beta_0) : m \in \omega, n \in \mathbb{Z}\}$. Hence Theorem A is proved.

From now on, we will assume that $\mathcal{L}$ includes variables for countable subsets of $\mathcal{L}$ as well as $=$ and $\subseteq$ between them, and membership of elements of Ω in them.

Actually, the following can be proved:

THEOREM B: Let Ω be a doubly homogeneous chain. Then countable subsets of $\bar{\Omega}$ together with membership in them are interpretable in $\mathcal{A}(\Omega)$.

We sketch the proof of Theorem B:

Let $\Delta \subseteq \bar{\Omega}$ be countable. Let $\Delta_1 = \{\delta \in \Delta: (\exists\alpha \in \Omega)[\alpha,\delta) \cap \Delta = \emptyset\}$, $\Delta_2 = \{\delta \in \Delta\setminus\Delta_1: (\exists\beta \in \Omega)(\delta,\beta] \cap \Delta = \emptyset\}$ and $\Delta_3 = \Delta\setminus(\Delta_1 \cup \Delta_2)$. For each $\delta_1 \in \Delta_1$, choose $\alpha = \alpha(\delta_1) \in \Omega$ with $[\alpha_1,\delta_1) \cap \Delta_1 = \emptyset$. Let $e < g_{\delta_1} \in \mathcal{A}(\Omega)$ with the closure of $\mathrm{supp}(g_{\delta_1}) = [\alpha_1,\delta_1]$; so $(\forall f > e)(f \wedge g_{\delta_1} = e \rightarrow L(f,g_{\delta_1})$ or $L(g_{\delta_1},f))$. Let $g_1 \in \mathcal{A}(\Omega)$ be the pointwise supremum of the pairwise disjoint set $\{g_{\delta_1}: \delta_1 \in \Delta_1\}$. Now $A_1 = \{\alpha = \alpha(\delta_1): \delta_1 \in \Delta_1\}$ is a countable subset of Ω. Thus A_1 can be recognized by Theorem A. Hence, using A_1 and g_1 we can recognize Δ_1 and that it is countable. Dually for Δ_2. Now each point of Δ_3 is the supremum and infimum of a countable subset of Δ (and hence of Ω). But, by double--and hence m for all $m \in \omega$--transitivity, there is at most one $\mathcal{A}(\Omega)$ orbit of such points of $\bar{\Omega}\setminus\Omega$. Therefore, Δ_3 comprises at most two countable orbits of $\mathcal{A}(\Omega)$. Hence Δ_3 can be captured in $\mathcal{L}$ by Theorem A. Consequently, so can Δ and Theorem B is proved.

We can now express that Ω is separable (i.e., has a countable subset whose topological closure is Ω) by:
$(\exists \text{ countable } \Delta)(\forall\alpha)(\forall\beta)(\alpha < \beta \rightarrow (\exists\delta_1,\delta_2,\delta_3 \in \Delta)(\delta_1 < \alpha < \delta_2 < \beta < \delta_3))$.

But Ω can be embedded in $\mathbb{R}$ if and only if it is separable. Hence

COROLLARY 4. _There is a sentence_ σ _such that, for homogeneous_ Ω, $\mathcal{A}(\Omega) \models \sigma$ _if and only if_ Ω _can be embedded in_ $\mathbb{R}$.

We next give easy proofs of the main theorems of [7].

THEOREM 5. _There is a sentence_ ψ _of_ $\mathcal{L}$ _such that, for any homogeneous chain_ Ω, $\mathcal{A}(\Omega) \models \psi$ _if and only if_ $\Omega \cong \mathbb{Q}$.

Proof: Apply Theorem A. (The only doubly homogeneous countable chain is $\mathbb{Q}$.)

THEOREM 6. _There is a sentence_ ρ _of_ $\mathcal{L}$ _such that, for any homogeneous chain_ Ω, $\mathcal{A}(\Omega) \models \rho$ _if and only if_ $\Omega \cong \mathbb{R}$.

Proof: It was shown in [7] that a homogeneous chain Ω is Dedekind complete if and only if $\mathcal{A}(\Omega) \models (\forall f > e)(\mathrm{supp}(f)\text{bounded} \rightarrow (\exists h)[L(f,h^{-1}fh) \,\&\, (\forall g > e)\neg(L(f,g) \,\&\, L(g,h^{-1}fh))])$. The theorem now follows from Corollary 4.

A chain Ω is said to enjoy the _Suslin property_ if every pairwise disjoint collection of open intervals of Ω is countable. A homogeneous Dedekind complete chain other than $\mathbb{R}$ that satisfies the Suslin property is called a _Suslin line_. If they exist at all, they exist in profusion (see [1]).

Recall that the bumps of $e < f \in \mathcal{A}(\Omega)$ are pairwise disjoint and recognizable in $\mathcal{L}$--with parameter f (Lemma 0).

THEOREM 7. <u>A doubly homogeneous Ω has the Suslin property if and only if for all $e < f \in \mathcal{A}(\Omega)$, there is a countable $\Delta \subseteq \mathrm{supp}(f)$ such that each bump of f moves exactly one point of Δ. Consequently, there are sentences θ_1, θ_2 of $\mathcal{L}$ such that (1) $\mathcal{A}(\Omega) \models \hat{\sigma}_1$ if and only if Ω enjoys the Suslin property, (2) $\mathcal{A}(\Omega) \models \hat{\sigma}_2$ if and only if Ω is a Suslin line.</u>

<u>Proof</u>: Since the bumps of f have disjoint open intervals of support, the condition is clearly necessary. For sufficiency, by double homogeneity, each open interval A contains the support of some one bump $e < f_A \in \mathcal{A}(\Omega)$. Let $f \in \mathcal{A}(\Omega)$ have $\{f_A\colon A \in \mathcal{F}\}$ as its set of bumps. Hence $|\mathcal{F}| \leq \aleph_0$, and Ω enjoys the Suslin property.

As noted in the introduction, we may assume that $f \in \mathcal{A}(\Omega)$ is identified with its unique extension to $\mathcal{A}(\bar{\Omega})$. The set of fixed points of f in $\bar{\Omega}$ is always a closed set; the complementary set is a disjoint union of open intervals each being the convexification (in $\bar{\Omega}$) of the support of a bump of f. Hence each connected component of this complementary set has countable coterminality. Conversely, if Ω is doubly homogeneous and $\bar{\Delta}$ is a closed subset of $\bar{\Omega}$ with each connected component of its complement having countable coterminality, then $\bar{\Delta}$ is the fixed point set (in $\bar{\Omega}$) of some $f \in \mathcal{A}(\Omega)$. For, as in the proof of Lemma 1, we may construct a one bump $f_A \in \mathcal{A}(\Omega)$ on each component A of the complement with $\mathrm{supp}(f_A) = A$. The desired function is $f \in \mathcal{A}(\Omega)$ whose set of bumps is just the set of f_A's. If Ω is separable, every interval has countable coterminality. Hence we have proved:

LEMMA 8: <u>If Ω is a separable doubly homogeneous chain, the closed subsets of $\bar{\Omega}$ are precisely the fixed point sets of functions $f \in \mathcal{A}(\Omega)$. Hence the closed subsets of $\bar{\Omega}$ are interpretable in $\mathcal{A}(\Omega)$ for such chains Ω.</u>

A subset $\bar{C}$ of a chain $\bar{\Omega}$ is said to be a <u>Cantor set</u> if $\bar{C}$ is closed, nowhere dense, and has no isolated points.

LEMMA 9: <u>If Ω is a separable doubly homogeneous chain, the Cantor sets of $\bar{\Omega}$ are interpretable in $\mathcal{A}(\Omega)$.</u>

<u>Proof</u>: If Ω is doubly homogeneous and $\bar{C}$ is the fixed point set of $e < f \in \mathcal{A}(\Omega)$, then $\bar{C}$ is nowhere dense if and only if $\bar{C}$ contains no non-empty open interval, which is equivalent to $(\forall g)(f \wedge g = e \rightarrow g = e)$ since every open interval contains the support

of some $e < h \in \mathcal{A}(\Omega)$. $\bar{C}$ has no isolated points precisely when the following formula of $\mathcal{L}$ holds in $\mathcal{A}(\Omega)$:
$(\forall b_1,b_2$ bumps of $f)(L(b_1,b_2) \rightarrow (\exists b_3$ bump of $f)(L(b_1,b_3)\ \&\ L(b_3,b_2)))$.
Lemma 9 now follows from Lemma 8.

Again the interpretation given is uniform (for such Ω) in $\mathcal{L}$.

A chain Ω is said to have the Luzin property if every Cantor set of $\bar{\Omega}$ meets Ω in a countable set.

By Theorem A and Lemma 9 we have:

THEOREM 10: There is a sentence θ of $\mathcal{L}$ such that if Ω is a separable doubly homogeneous chain, then $\mathcal{A}(\Omega) \models \theta$ if and only if Ω has the Luzin property.

$\mathbb{Q}$ clearly is separable, doubly homogeneous and has the Luzin property. We now show that there are uncountable chains Ω enjoying these properties, by modifying the standard Luzin construction.

We assume the Continuum Hypothesis. Enumerate the Cantor subsets of $\mathbb{R}$, $\{C_\mu : \mu < \omega_1\}$. For each $\mu < \omega_1$, let $M_\mu = \cup\{C_\lambda : \lambda \leq \mu\}$. Then each M_μ is meager; that is, M_μ is a countable union of nowhere dense sets. Let $\langle\langle S\rangle\rangle$ denote the rational subspace generated by $S \subseteq \mathbb{R}$. Now choose, inductively, $x_\mu \in \mathbb{R}$ $(\mu < \omega_1)$ so that $x_0 \notin \mathbb{Q}M_0$, the rational multiples of members of M_0, and for each $\mu < \omega_1$, $x_\mu \notin \mathbb{Q}M_\mu + \langle\langle\{x_\lambda : \lambda < \mu\}\rangle\rangle = N_\mu$. This is possible since each N_μ is meager. We now claim that $\Lambda = \langle\langle\{x_\mu : \mu < \omega_1\}\rangle\rangle$ has the Luzin property. Note that for each $\nu < \omega_1$, if $0 \neq x \in \Lambda \cap M_\nu$, then $x = q_1x_{\nu_1} + \ldots + q_nx_{\nu_n}$ for $0 \neq q_i \in \mathbb{Q}$ and $\nu_1 < \ldots < \nu_n < \omega_1$. This implies that $x_{\nu_n} \in \mathbb{Q}M_\nu + \langle\langle\{x_\lambda : \lambda < \nu_n\}\rangle\rangle$ and hence $\nu_n < \nu$. Therefore $M_\nu \cap \Lambda \subseteq \langle\langle\{x_\lambda : \lambda < \nu\}\rangle\rangle$, a countable set. Since $C_\nu \subseteq M_\nu$, $\Lambda \cap C_\nu$ is countable, as required. Moreover, Λ is uncountable--it contains $\{x_\mu : \mu < \omega_1\}$. Since Λ is a rational vector space, it is homogeneous ($\mathcal{A}(\Lambda)$ contains translations by members of Λ). Multiplication by 2 is also an automorphism of $(\Lambda,\leq)$, so $\mathcal{A}(\Lambda)$ is not abelian. Because of the small translations (say by arbitrary qx_0 $(q \in \mathbb{Q})$), for every interval I of Λ, there is $f \in \mathcal{A}(\Lambda)$ such that $\emptyset \neq I \cap f(I) \neq I$. This is enough to ensure that Λ is primitive and hence doubly homogeneous (see the introduction). Λ is obviously separable as required.

A doubly homogeneous chain Ω is short if it has countable coterminality and for each $\bar{\alpha} \in \bar{\Omega}$, there exist countable bounded $\Gamma,\Delta \subseteq \Omega$ such that $\sup \Gamma = \bar{\alpha} = \inf \Delta$. By Lemma 1 (i) and Lemma 3, shortness is definable in $\mathcal{L}$ and we will assume that it is explicitly in $\mathcal{L}$. An uncountable short chain Ω that contains no uncountable separable subset is said to enjoy the *Specker property*. "Separable" here means separable in the interval topology of the subset (not necessarily the interval topology of Ω). Thus Ω has the Specker property if and only if it is uncountable, short and doubly homogeneous, and for all countable subsets Γ of Ω, there exists a countable subset Δ of Ω such that for all $\alpha \in \Omega \setminus \Gamma$ there is $\delta \in \Delta$ with no member of Γ between α and δ. Thus

THEOREM 11: *There is a sentence χ of $\mathcal{L}$ such that if Ω is homogeneous, $\mathcal{A}(\Omega) \models \chi$ if and only if Ω has the Specker property.*

The *long* (to the right) *real line* $\vec{\mathbb{R}}$ is constructed by removing the smallest point from the antilexicographically ordered chain $I \overleftarrow{\times} \omega_1$, where I is the half open real interval $[0,1)$. We can define $\overleftarrow{\mathbb{R}}$ and $\overleftrightarrow{\mathbb{R}}$ similarly.

THEOREM 12: *Let Ω be a homogeneous set. There are sentences ϕ_i $(i = 1,2,3)$ such that*

(a) $\mathcal{A}(\Omega) \models \phi_1$ *if and only if* $\Omega \cong \vec{\mathbb{R}}$.

(b) $\mathcal{A}(\Omega) \models \phi_2$ *if and only if* $\Omega \cong \overleftarrow{\mathbb{R}}$.

(c) $\mathcal{A}(\Omega) \models \phi_3$ *if and only if* $\Omega \cong \overleftrightarrow{\mathbb{R}}$.

Proof: (a) $\vec{\mathbb{R}}$ is completely characterized by the statement that it is Dedekind complete and not separable, yet every subset $\{\alpha: \alpha < \beta\}$ is separable. All of these clauses, together with double homogeneity, are describable in $\mathcal{L}$. ϕ_1 is their conjunction.

(b) and (c) are similar.

The long (to the right) rational lines Ω are constructed as follows. Let I_0 be the set of all rational numbers in the real open interval $(0,1)$. Choose $M \subseteq \omega_1$ with $0 \notin M$. Let $\Omega = \{(q,\nu): q \in I_0, \nu \in \omega_1\} \cup \{(0,\mu): \mu \in M\} \subseteq \vec{\mathbb{R}}$, with the induced order. All the constructions in which M contains a closed unbounded subset of ω_1 (or club for short)--*the long rationals with internal club*--give rise to ordermorphic chains. Likewise, all constructions in which the complement of M contains a club (*the long rationals with external club*). The two cases are not ordermorphic and are distinct from all other cases (in which neither M nor its complement contains a club).

THEOREM 13: *There are sentences ξ_1, ξ_2, ξ_3 of $\mathcal{L}$ such that if Ω is homogeneous, then*

(a) $\mathcal{A}(\Omega) \models \xi_1$ *if and only if Ω is ordermorphic to the long rational line with internal club,*

(b) $\mathcal{A}(\Omega) \models \xi_2$ *if and only if Ω is ordermorphic to the long rational line with external club,*

(c) $\mathcal{A}(\Omega) \models \xi_3$ <u>if and only if</u> Ω <u>is ordermorphic to a long rational line with neither internal nor external club</u>.

<u>Proof</u>: Consider Ω a long rational line with internal club. We may assume that $M = \omega_1 \setminus \{0\}$. Then Ω is uncountable but every subset $\{\alpha \in \Omega: \alpha < \beta\}$ is countable. Moreover, $\Gamma = \{(0,\mu): \mu \in M\}$ is a closed unbounded above subset of Ω that is well-ordered; and if $\gamma \in \Gamma$ and $\gamma = \sup_\Gamma\{\delta \in \Gamma: \delta < \gamma\}$, then $\gamma = \sup_\Omega\{\delta \in \Gamma: \delta < \gamma\}$. Now let $e < f \in \mathcal{A}(\Omega)$ have one bump on each interval $\{\alpha: (0,\mu) < \alpha < (0,\mu + 1)\}$ $(\mu \in \omega_1)$. Then the set of bumps of f is well-ordered and $(\forall h)(f \wedge h = e \rightarrow h = e)$. Furthermore, for each bump b of f (except the left-most), there exist $\alpha,\beta \in \Omega$ such that $b(\overline{\delta}) \neq \overline{\delta}$ if and only if $\alpha < \overline{\delta} < \beta$. All of these facts are expressible in $\mathcal{L}$.

Conversely, suppose that Ω is a doubly homogeneous uncountable chain with $\{\alpha \in \Omega: \alpha < \beta\}$ countable for each $\beta \in \Omega$. Assume there exists $e < f \in \mathcal{A}(\Omega)$ such that $(\forall h)(f \wedge h = e \rightarrow h = e)$, the set of bumps of f is well-ordered, and for each bump b of f (except the left-most), there exist $\alpha,\beta \in \Omega$ such that $b(\overline{\delta}) \neq \overline{\delta}$ if and only if $\alpha < \overline{\delta} < \beta$. Let Γ be the set of left endpoints of supports of bumps of f (other than the left-most). Then Γ is a well-ordered uncountable subset of Ω and hence $\Gamma \cong \omega_1$. Also if $\gamma \in \Gamma$ and $\gamma = \sup_\Gamma\{\delta \in \Gamma: \delta < \gamma\}$, then $\gamma = \sup_\Omega\{\delta \in \Gamma: \delta < \gamma\}$. For a bump b of f, we may call the end points of the support of b, μ and $\mu + 1$ $(\mu \in \omega_1)$. Then Ω is the disjoint union of the intervals $[\mu,\mu + 1)$ together with $(-\infty,0)$, each of which is countable and so ordermorphic to the rational interval $[0,1)$ or $(0,1)$. Hence Ω is ordermorphic to $[0,1) \overleftarrow{\times} \omega_1$ with least point $(0,0)$ removed.

(b) is proved similarly, except that we need an f such that all bumps of f have supporting intervals with no endpoints in Ω.

(c) now follows from (a) and (b).

We may also obtain long rational lines inside $\overleftarrow{\mathbb{R}}$ and $\overleftrightarrow{\mathbb{R}}$. Analogous results then hold in these cases.

The characterization of $\mathbb{R}$ and $\mathbb{Q}$ in [7] was achieved with heavy reliance on the arithmetic structure. We have avoided that here by using Theorem A. Still, it is of interest to know whether, for a given chain Ω, it is possible to define an arithmetic on Ω so that Ω becomes an ordered group or an ordered field. We call a chain Ω <u>Archimedean groupable</u> if it is possible to define an operation + on Ω so that $(\Omega,+)$ becomes an Archimedean ordered group. Such groups are abelian and are isomorphic to subgroups of $\mathbb{R}$ (see [2, p. 45]). If, in addition, it is possible to define an operation $\times$ on Ω so that $(\Omega,+,\times)$ becomes an Archimedean ordered field, we say that Ω is Archimedean fieldable. The "field" part of the next theorem is due to Greg Cherlin.

THEOREM 14: There are sentences τ,τ' of $\mathcal{L}$ such that if Ω is a homogeneous chain, $\mathcal{A}(\Omega) \models \tau$ ($\mathcal{A}(\Omega) \models \tau'$) if and only if Ω is Archimedean groupable (Archimedean fieldable).

Proof: Let Ω be Archimedean groupable; so Ω is a subgroup of $\mathbb{R}$ without loss of generality. If $\mathcal{A}(\Omega)$ is abelian, the result follows from [5], and if Ω is countable the result follows from Theorem 4 (or [5] if Ω is discrete). Since Ω is primitive, we are reduced to the doubly homogeneous uncountable case. We may assume $1,\zeta \in \Omega$ for some irrational $\zeta > 0$. Let $t_1,t_\zeta \in \mathcal{A}(\Omega)$ be translations by 1 and ζ, respectively; i.e., $t_1\colon \alpha \mapsto \alpha + 1$, $t_\zeta\colon \alpha \mapsto \alpha + \zeta$. As in [7, proof of Theorem], the centralizer $C = C(t_1,t_\zeta)$ in $\mathcal{A}(\Omega)$ consists exactly of the translations $f\colon \alpha \mapsto \alpha + \beta$ $(\beta \in \Omega)$. In particular, C is transitive on Ω, is an abelian totally ordered subgroup of $\mathcal{A}(\Omega)$, and no element of C except e fixes any point of $\bar{\Omega}$ $(= \mathbb{R})$. Now let Λ be any uncountable doubly homogeneous chain such that there exist $e < r,s \in \mathcal{A}(\Lambda)$ with $C(r,s)$, the centralizer of r and s in $\mathcal{A}(\Lambda)$, transitive on Λ, abelian and totally ordered, and no element of $C(r,s)$ other than e fixes any point of $\bar{\Lambda}$. Choose any $\alpha \in \Lambda$. For each $\beta \in \Lambda$, there is a unique $f_\beta \in C(r,s)$ such that $f_\beta(\alpha) = \beta$. The correspondence $\beta \leftrightarrow f_\beta$ provides an ordermorphism between Λ and $C(r,s)$, so it is enough to show that $C(r,s)$ is Archimedean groupable. But $C(r,s)$ is an Archimedean ordered group, for if $e < f < g$ with $f,g \in C(r,s)$, then for any $\beta \in \Lambda$, $\{f^n(\beta)\colon n \in \omega\}$ can have no upper bound (otherwise f would fix the least upper bound in $\bar{\Lambda}$). So for some $n \in \omega$, $f^n(\beta) > g(\beta)$. Since $C(r,s)$ is totally ordered, $f^n > g$.

Now assume that Ω is an Archimedean ordered field. Then Ω is doubly homogeneous and we may assume that it is a subfield of $\mathbb{R}$. If Ω is countable, the result follows from Theorem 4, so assume Ω is uncountable. For any $0 < \beta \in \Omega$, the function $h_\beta\colon \alpha \mapsto \alpha\beta$ belongs to $\mathcal{A}(\Omega)$. Indeed, $h_\beta \in N(C(t_1,t_\zeta))$, the normalizer of $C(t_1,t_\zeta)$, since if $g \in C$ (say $g\colon \alpha \mapsto \alpha + \gamma$), then $(h_\beta g h_\beta^{-1})(\alpha) = (\alpha/\beta + \gamma)\beta = \alpha + \gamma\beta$. So for any $h \in N = N(C)$, the map $g \mapsto hgh^{-1}$ is an order-preserving automorphism of the Archimedean ordered group C, and so must correspond to multiplication by a positive real number by Hion's Lemma [2, p. 46]. In particular, if $e < f \in C$, there is $h \in N$ such that $f = ht_1h^{-1}$ (if

$f: \alpha \mapsto \alpha + \gamma$, $f = h_\gamma t_1 h_\gamma^{-1}$); and if $h_1, h_2 \in N$ and $h_1 t_1 h_1^{-1} = h_2 t_1 h_2^{-1}$, then $h_1 g h_1^{-1} = h_2 g h_2^{-1}$ for all $g \in C$. Now suppose that in $\mathcal{A}(\Lambda)$, for every $e < f \in C(r,s)$ there exists $h \in N'$, the normalizer of $C(r,s)$, such that $hrh^{-1} = f$, and that if $h_1, h_2 \in N'$ with $h_1 r h_1^{-1} = h_2 r h_2^{-1}$, then $h_1 g h_1^{-1} = h_2 g h_2^{-1}$ for all $g \in C(r,s)$. We can define a product $\otimes$ on $C(r,s)$ as follows: Let $f, g \in C(r,s)$ with $e < f$. There exists $h \in N'$ such that $hrh^{-1} = f$. Define $g \otimes f = hgh^{-1}$. This is well-defined and the extension of $\otimes$ to all products (when $f \leq e$) is done in the obvious way. It is straightforward to check that this makes $C(r,s)$ an Archimedean ordered field. Since Λ is ordermorphic to $C(r,s)$, this completes the proof.

So far we have been able to capture every property we want. However, since there are only $2^{\aleph_0}$ complete theories, there exist non-ordermorphic doubly homogeneous chains Ω and Λ with $\mathcal{A}(\Omega) \equiv \mathcal{A}(\Lambda)$; indeed, such Ω and Λ exist with $|\Omega| \neq |\Lambda|$. As yet, we have been unable to explicitly obtain such Ω and Λ. The problem is that the Ehrenfeucht game to be played between $\mathcal{A}(\Omega)$ and $\mathcal{A}(\Lambda)$ (to prove $\mathcal{A}(\Omega) \equiv \mathcal{A}(\Lambda)$) is rather complicated. This is a big gap in our work to date.

REFERENCES

1. K. J. Devlin and H. Johnsbråten, The Souslin Problem, Lecture Notes Series No. 405, Springer-Verlag, 1974.

2. L. Fuchs, Partially Ordered Algebraic Systems, Academic Press, 1963.

3. A. M. W. Glass, Ordered Permutation Groups, London Mathematical Society Lecture Notes Series, Cambridge University Press (to appear in 1981).

4. A. M. W. Glass, Elementary types of automorphisms of linearly ordered sets--a survey, in Proc. Carbondale Algebra Conference, 1980· Springer-Verlag (to appear).

5. A. M. W. Glass, Yuri Gurevich, W. Charles Holland and Saharon Shelah, Rigid homogeneous chains, Math. Proc. of Cambridge Philosophical Soc. (to appear).

6. A. M. W. Glass and Keith R. Pierce, Existentially complete lattice-ordered groups, Israel J. Math. (to appear).

7. Yuri Gurevich and W. Charles Holland, Recognizing the real line, Trans. Amer. Math. Soc. (to appear).

8. Michèle Jambu-Giraudet, Théories des Modèles des Groupes d'automorphismes d'ensembles Totalement Ordonnés, Thèse 3ème Cycle, Univ. Paris VII, 1979.

9. Michèle Jambu-Giraudet, Bi-interpretable groups and lattices, Trans. Amer. Math. Soc. (to appear).

10. Ralph N. McKenzie, On elementary types of symmetric groups, Alg. Universalis, 1(1971), 13-20.

11. A. G. Pinus, On elementary definability of symmetric groups, Alg. Universalis, 3(1973), 59-66.

12. M. Rabin, Universal groups of automorphisms of models, in Symp. on the Theory of Models, (ed. Addison, Henkin & Tarski), North-Holland, 1965, pp. 274-284.

13. Saharon Shelah, First order theory of permutation groups, Israel J. Math., 14(1973), 149-162; 15(1973), 437-441.

Glass and Holland:
Bowling Green State University
Bowling Green, Ohio 43403

Jambu-Giraudet
Université Paris X,
200 Ave. de la République
92001 Nanterre; Cedex, France

Gurevich:
Ben-Gurion University of the Negev
Beer-Sheva
Israel

THREE EASY CONSTRUCTIONS OF RECURSIVELY ENUMERABLE SETS

Carl G. Jockusch, Jr.[1]

University of Illinois
Urbana, Illinois 61801

The study of recursively enumerable (r.e.) degrees is regarded (with some justification) as one of the most difficult areas of mathematical logic. On the other hand, our purpose here is to present three results in this area which are quite easy. The first is due to Lawrence Welch [11] and asserts the existence of low r.e. degrees $\underset{\sim}{a}$, $\underset{\sim}{b}$ such that every r.e. degree is of the form $\underset{\sim}{a}_0 \underset{\sim}{\cup} \underset{\sim}{b}_0$ with $\underset{\sim}{a}_0 \leq_T \underset{\sim}{a}$ and $\underset{\sim}{b}_0 \leq_T \underset{\sim}{b}$. This, as Welch has observed, gives an easy proof of the existence of a low r.e. degree which can be nontrivially cupped up to every r.e. degree above itself. The second is a simple direct proof of the result of A. H. Lachlan [4, p.569] and C. E. M. Yates [13] that the r.e. degrees do not form a lattice. The final result is that not every r.e. nonrecursive truth-table degree contains a simple set

Our notation is quite standard. In particular we use symbols such as e, x for natural numbers, A, B for sets of natural numbers, $\underset{\sim}{a}$, $\underset{\sim}{b}$ for degrees, and Φ, Ψ for operators (i.e. functions from subsets of 2^ω into 2^ω). We write $\Phi(A;x)$ for $\Phi(A)(x)$. If Φ is a partial recursive operator (i.e. Turing reduction procedure) and $\Phi(A;x)$ is defined, then $use(\Phi(A;x))$ denotes the smallest number exceeding all numbers whose membership or nonmembership in A is used to compute $\Phi(A;x)$. Let $<.,.>$ be a recursive pairing function. We identify sets and their characteristic functions so that $A(x) = 1$ iff $x \in A$. A degree $\underset{\sim}{a}$ is called <u>low</u> if $\underset{\sim}{a}' = \underset{\sim}{0}'$, where $\underset{\sim}{a}'$ denotes the jump of $\underset{\sim}{a}$.

The following result shows that the recursively enumerable degrees are generated under the l.u.b. operation $\underset{\sim}{\cup}$ by the union of two proper principal ideals.

<u>Theorem</u> 1. (L. Welch [11]). There are low r.e. degrees $\underset{\sim}{a}$, $\underset{\sim}{b}$ such that every r.e. degree $\underset{\sim}{c}$ is of the form $\underset{\sim}{a}_0 \underset{\sim}{\cup} \underset{\sim}{b}_0$ for some $\underset{\sim}{a}_0 \leq_T \underset{\sim}{a}$, $\underset{\sim}{b}_0 \leq_T \underset{\sim}{b}$.

---

[1]This research was supported by a grant from the National Science Foundation.

Proof. For any set C, let $C^{[e]} = \{n : <e,n> \in C\}$. Let $K_0 = \{<e,n> : n \in W_e\}$, so that $K_0^{[e]} = W_e$ for all e. (Here W_e is the eth r.e. set.) Apply Sacks' splitting theorem ([9, §6] or [10, Theorem 1.2 and Remark 4.5]) to obtain disjoint r.e. sets A, B of low degree with $A \cup B = K_0$ (and hence $A^{[e]} \cap B^{[e]} = \phi$ and $A^{[e]} \cup B^{[e]} = W_e$ for all e). Let $\underset{\sim}{a}$, $\underset{\sim}{b}$ be the degrees of A, B respectively, and let $\underset{\sim}{c}$ be any given r.e. degree. Choose e so that W_e has degree $\underset{\sim}{c}$, and let $\underset{\sim}{a}_0$, $\underset{\sim}{b}_0$ be the degrees of $A^{[e]}$, $B^{[e]}$ respectively. Then $\underset{\sim}{a}_0 \cup \underset{\sim}{b}_0 = \underset{\sim}{c}$ since $A^{[e]} \oplus B^{[e]} \equiv_T A^{[e]} \cup B^{[e]} = W_e$.

It is not known whether there are r.e. degrees $\underset{\sim}{a}, \underset{\sim}{b} < \underset{\sim}{0}'$ such that every nonzero r.e. degree $\underset{\sim}{c}$ is of the form $\underset{\sim}{a}_0 \cup \underset{\sim}{b}_0$ with $\underset{\sim}{a}_0 \lneq \underset{\sim}{a}$, $\underset{\sim}{b}_0 \lneq \underset{\sim}{b}$, and $\underset{\sim}{a}_0$, $\underset{\sim}{b}_0$ incomparable.

L. Harrington [3] proved that there is a nonzero r.e. degree $\underset{\sim}{a}$ such that every nonzero r.e. degree $\underset{\sim}{c} \lneq \underset{\sim}{a}$ can be non-trivially cupped up to each r.e. degree $\underset{\sim}{d} \gneq \underset{\sim}{a}$. The following Corollary gives a very simple proof of the case $\underset{\sim}{c} = \underset{\sim}{a}$ of this result. A proof of the special case $\underset{\sim}{d} = \underset{\sim}{0}'$ of Harrington's unpublished result may be found in the paper of Fejer and Soare in this volume.

Corollary 1. (L. Harrington). There is a nonzero low r.e. degree $\underset{\sim}{a}$ such that for every r.e. degree $\underset{\sim}{c} \gneq \underset{\sim}{a}$ there exists an r.e. degree $\underset{\sim}{e} < \underset{\sim}{c}$ with $\underset{\sim}{a} \cup \underset{\sim}{e} = \underset{\sim}{c}$.

Proof. (L. Welch). Let $\underset{\sim}{a}$ be as in Theorem 1. Given an r.e. degree $\underset{\sim}{c} \gneq \underset{\sim}{a}$, choose $\underset{\sim}{a}_0 \lneq \underset{\sim}{a}$, $\underset{\sim}{b}_0 \lneq \underset{\sim}{b}$ with $\underset{\sim}{a}_0 \cup \underset{\sim}{b}_0 = \underset{\sim}{c}$. Then clearly also $\underset{\sim}{a} \cup \underset{\sim}{b}_0 = \underset{\sim}{c}$ since $\underset{\sim}{a} \lneq \underset{\sim}{c}$. Suppose for a contradiction that $\underset{\sim}{b}_0 = \underset{\sim}{c}$. Then $\underset{\sim}{b} \gneq \underset{\sim}{b}_0 = \underset{\sim}{c} \gneq \underset{\sim}{a}$, yet the theorem (applied with $\underset{\sim}{c} = \underset{\sim}{0}'$) implies that $\underset{\sim}{a} \cup \underset{\sim}{b} = \underset{\sim}{0}'$. This contradiction shows that $\underset{\sim}{b}_0 < \underset{\sim}{c}$, and thus that the Corollary holds with $\underset{\sim}{e} = \underset{\sim}{b}_0$.

Lachlan [4] proved that the r.e. degrees are not a lattice by combining his "non-diamond" theorem [4, Theorem 5] with Sacks' splitting theorem and a lemma on the absoluteness of the partial inf operation $\underset{\sim}{\cap}$ on r.e. degrees. Yates [13] also outlined a proof that the r.e. degrees are not a lattice. Yates' argument involved relativizing the r.e. minimal pair construction to a certain uniformly ascending sequence of r.e. degrees. The proof below will employ some of the basic ideas of the non-diamond and minimal pair constructions, but will be easier and more direct than either.

Theorem 2. (Lachlan, Yates). The r.e. degrees do not form a lattice.

Proof. We must construct r.e. sets A, B whose degrees have no infimum in the r.e. degrees. Let $\{(W_e,\Phi_e,\Psi_e)\}_{e\in\omega}$ be an effective enumeration of all triples (W,Φ,Ψ) with W an r.e. set and Φ, Ψ Turing reduction procedures. For each e we shall define an r.e. set V_e which satisfies for all i the following requirement:

$$R_{e,i} : \Phi_e(A) = \Psi_e(B) = W_e \Rightarrow V_e \neq \{i\}^{W_e}.$$

We will arrange in addition under the hypotheses of $R_{e,i}$ that V_e is recursive in each of A, B. This and the satisfaction of $R_{e,i}$ for all i imply that the degree of W_e is not the infimum of those of A and B.

The method for insuring $V_e \neq \{i\}^{W_e}$ is the basic Friedberg-Muchnik technique of choosing a witness x, awaiting a computation $\{i\}^{W_e^s}(x) = 0$, putting x into V_e, and "restraining" W_c so as to preserve the computation. Of course we actually have no control over W_e, but by restraining A or B we can either prevent the relevant numbers from entering W_e or else satisfy $R_{e,i}$ vacuously by insuring $\Phi_e(A) \neq W_e$ or $\Psi_e(B) \neq W_e$ respectively. To arrange that $V_e \leq_T A,B$ we require that x be enumerated in both A and B if x is enumerated in V_e. This creates the obstacle that putting x into V_e (and thus into A and B) ruins the computations by which A, B each control W_e and thus allows the computation $\{i\}^{W_e^s}(x) = 0$ to be destroyed (while the agreement $\Phi_e(A) = \Psi_e(A) = W_e$ is preserved). This obstacle is avoided by first enumerating x in A (but not in B or V_e) at some stage s when B controls the computation $\{i\}^{W_e^s}(x) = 0$. We then wait for a later stage t at which sufficient agreement of $\Phi_e^t(A^t)$ and W_e^t occurs so that A controls the computation $\{i\}^{W_e^t}(x) = 0$. Then at stage t, x is enumerated in V_e and B, and restraints on A are used to insure (modulo requirements of higher priority) that $\{i\}^{W_e}(x) = 0$ if $\Phi_e(A) = W_e$. Thus the alternating control of A and B over the computation $\{i\}^{W_e}(x) = 0$ is reminiscent of the minimal pair construction, and the "delayed permitting" method for arranging that $V_e \leq_T A$ is one ingredient of the proof of the non-diamond theorem.

The construction and proof are straightforward. Although we supply some of the details of these here, we recommend that the reader work these out for himself.

Suppose the various requirements are assigned a priority ranking in standard fashion. We say that a requirement $R_{e,i}$ is satisfied at stage s if there is a stage $s' < s$ such that $R_{e,i}$ receives attention at s' and such that no requirement of higher priority than $R_{e,i}$ receives attention at any stage s'', $s' \leq s'' < s$. We write A^s for the finite subset of A enumerated before stage s and use analogous notation for other sets and for operators. For any set X, let $X^{(e)} = \{<e,n> : <e,n> \in X\}$.

A number x is called an eligible witness for $R_{e,i}$ at stage s if

(i) $x \in N^{(<e,i>)} - A^s$

(ii) x is not restrained from A or B at s by any requirement of higher priority than $R_{e,i}$

(iii) $\{i\}_s^{W_e^s}(x) = 0$,

and

(iv) $\Psi_e^s(B^s;u) = W_e^s(u)$ for each $u <$ use $(\{i\}^{W_e^s}(x))$.

If $s \in N^{(<e,i>)}$ and $R_{e,i}$ is not satisfied at s but has an eligible witness at s, let x be the least such witness. At stage s, enumerate x in A, restrain B for $R_{e,i}$ so as to preserve all the computations $\Psi_e^s(B^s;u)$ mentioned in (iv), and say that $R_{e,i}$ receives attention.

The requirement $R_{e,i}$ may receive attention again through the same witness x as follows. If $t \in N^{(<e,i>)}$ and $R_{e,i}$ is satisfied at t via attention at stage $s < t$ and a witness $x \notin B^t$ such that $\Phi_e^t(A^t;u) = W_e^t(u) = W_e^s(u)$ for all $u <$ use$(\{i\}^{W_e^s}(x))$, then at stage t enumerate x in V_e and in B and restrain A for $R_{e,i}$ so as to preserve the computations $\Phi_e^t(A^t;u)$ for all $u <$ use$(\{i\}^{W_e^s}(x))$. (The requirement $R_{e,i}$ then receives no further attention unless some requirement of higher priority receives attention after s, in which case $R_{e,i}$ starts over with a new witness.)

Each requirement receives attention at most twice after requirements of higher priority stop receiving attention. Thus each requirement receives attention only finitely often. Hence each requirement restrains only finitely many numbers and $A^{(<e,i>)}$, $V^{(<e,i>)}$ are finite for each e,i. Suppose now for a contradiction that $R_{e,i}$ is not satisfied, so that $\Phi_e(A) = \Psi_e(B) = W_e$ and $V_e = \{i\}^{W_e}$. Then each sufficiently large x in $N^{(<e,i>)^e}$ is an eligible witness for $R_{e,i}$ at all sufficiently large stages. Let s_0 be the last stage at which some requirement of higher priority receives attention. Then $R_{e,i}$ receives attention at some stage $s > s_0$ through some witness x, where we take s as small as possible. If there is a stage $t > s$ such that $R_{e,i}$ receives attention at t, the restraints on A imposed at t insure that $\Phi_e(A) \neq W_e$ or $\{i\}^{W_e}(x) \neq V_e(x)$. Otherwise some number $u < use(\{i\}^{W_e^s}(x))$ is enumerated in W_e at some stage $t' > s$, and then the restraints on B imposed at s insure that $\Psi_e(B) \neq W_e$. Thus the requirement $R_{e,i}$ is satisfied.

It remains to check that $V_e \leq_T A,B$ if $\Phi_e(A) = \Psi_e(B) = W_e$. Clearly $V_e \leq_T B$ since $V_e = \cup_i B^{(<e,i>)}$. To show how to compute V_e recursively from A, let a number x be given. If $x \notin \cup_i A^{(<e,i>)}$, then $x \notin V_e$. Suppose now that $x \in A^{(<e,i>)}$ and x is enumerated in A at s_x. Let t_x be the first stage $t > s_x$ such that either some requirement of higher priority than $R_{e,i}$ receives attention at t or $x \in V_e^{t+1}$. Such a stage t exists because $\Phi_e(A) = W_e$. By construction $x \in V_e$ iff $x \in V_e^{t_x+1}$. Since t_x is found recursively in A, it follows that $V_e \leq_T A$. (The argument in effect shows the existence of an r.e. set C_e such that $\cup_i A^{(<e,i>)}$ is the disjoint union of V_e and C_e. We may not claim that C_e is finite, though, because as i varies there may be infinitely many requirements $R_{e,i}$ which cause some $x \in N^{(<e,i>)}$ to enter A but do not later cause x to enter V_e because of interference from some requirement of higher priority.)

The following question was included in the lecture on which this paper was based: is every r.e. degree except $\underset{\sim}{0}$ and $\underset{\sim}{0}'$ part of a pair of r.e. degrees which has no infimum in the r.e. degrees? The question has recently been answered in the affirmative by K. Ambos and L. Harrington independently (private communication). They in fact showed the existence of an r.e. degree $\underset{\sim}{a} \neq \underset{\sim}{0}, \underset{\sim}{0}'$ such that $\underset{\sim}{a} \cap \underset{\sim}{b}$ fails to exist for every r.e. degree $\underset{\sim}{b}$ incomparable with $\underset{\sim}{a}$. Furthermore they proved that for each r.e. degree $\underset{\sim}{c}$ except $\underset{\sim}{0}$ and $\underset{\sim}{0}'$, there exists such a degree $\underset{\sim}{a}$ which is incomparable with $\underset{\sim}{c}$.

We now consider the question raised by P. G. Odifreddi of whether every r.e. nonrecursive truth-table (tt) degree contains a simple set. Several known results go in the direction of a positive answer. For instance J. C. E. Dekker ([1], [3, p. 140]) showed that for every r.e. nonrecursive set A there is a hypersimple set W such that $W \leq_{tt} A$ and $A \leq_T W$. Also a direct combination of E. L. Post's construction of a tt-complete simple set ([7], [8, p. 112]) and C. E. M. Yates' construction [12] of a simple, non hypersimple set of given nonzero r.e. Turing degree shows that for every r.e. nonrecursive set A there is a simple set W such that $W \leq_T A$ and $A \leq_{tt} W$. Finally J. Myhill and S. Tennenbaum [2, p.365] showed in effect that every nonrecursive tt-degree contains an immune set (although the result is stated only for T-degrees). Nonetheless we now obtain a negative answer to the question raised above. A proof of this theorem is also sketched in Odifreddi's expository paper [6].

Theorem 3. There is an r.e. nonrecursive set A whose tt-degree contains no simple set.

Proof. Let $\{(\Phi_e,\Psi_e,W_e)\}_{e\in\omega}$ be a recursive enumeration of all triples (Φ,Ψ,W) with Φ, Ψ tt-reduction procedures and W and r.e. set. We use notation such as A^s, Ψ_e^s, $A^{(e)}$ as in Theorem 2. Since Φ_e is a tt-reduction procedure, if $\Phi_e^s(X;x)$ is defined for any set X, then $\Phi_e^s(Y;x)$ is defined with the same use function for all sets Y. The Ψ_e^s have the corresponding property. For each e we will define an r.e. set V_e which satisfies the requirement

$$R_e\colon\ \Phi_e(A) = W_e \text{ and } \Psi_e(W_e) = A \Rightarrow V_e \text{ infinite } \& \ V_e \cap W_e \text{ finite.}$$

Clearly it suffices to make A r.e. and to satisfy all these requirements R_e. (In particular A is nonrecursive because otherwise we could choose W_e to be cofinite.) Assign priorities to the R_e's as usual. Each R_e will receive attention only finitely often and will cause only elements of $N^{(\geq e)}$ to enter A. (Here, for any set X, we write $X^{(\geq e)}$ for $\cup_{i\geq e}X^{(i)}$ and define $X^{(>e)}$, $X^{(<e)}$, $X^{(\leq e)}$ analogously.) Thus $A^{(\leq e)}$ will be finite for each e. In giving attention to R_e at stage s we assume that $A^{(\leq e)} = (A^{s+1})^{(\leq e)}$ since this will be true if s is sufficiently large. We also pretend that $A \supseteq N^{(>e)}$ since any element of $N^{(>e)}$ can be put into A for R_e if necessary. This pretense pre-empts any requirements of lower priority than R_e from injuring R_e by enumerating elements of $N^{(>e)}$ in A, so it is not necessary for R_e to restrain any numbers from entering A. In line with this pretense, let $A_e^s = A^s \cup N^{(>e)}$.

In attempting to satisfy R_e we pursue two strategies in parallel. The first strategy is aimed at insuring $\Phi_e(A) \neq W_e$ and the second at insuring $\Psi_e(W_e) \neq A$. The first strategy is very simple. Let $V_e^s = \{u : \Phi_e^s(A_e^s;u) = 0\}$. If $u \in V_e^s \cap W_e^s$, enumerate in A^{s+1} all numbers $z \in N^{(>e)}$ with $z < \mathrm{use}(\Phi_e^s(A_e^s;u))$. Give no further attention to R_e (under either strategy) unless some higher priority requirement subsequently receives attention. If the latter does not occur, then A and A_e^s coincide below $use(\Phi_e^s(A_e^s;u))$ so $\Phi_e(A;u) = \Phi_e^s(A_e^s;u) = 0 \neq W_e^s(u) = W_e(u)$.

We suppose now that R_e never receives attention under the first strategy (after the last stage s_0 at which any requirement of higher priority receives attention) and consider the second strategy. R_e will receive attention at most twice after s_0 under this strategy so there is a stage s_1 such that $(A^{s_1})^{(\leq e)} = A^{(\leq e)}$. For $s \geq s_1$, $V_e^s \cap W_e^s = \phi$ and $V_e^s \subseteq V_e^{s+1}$ since $A_e^s = A_e^{s+1}$. It follows that $V_e \cap W_e$ is finite, where $V_e = \cup_s V_e^s$. Thus R_e is satisfied by this V_e unless V_e is finite. If V_e is finite and $\Phi_e(A) = W_e$, the second strategy will insure that $A \neq \Psi_e(W_e)$ in Friedberg-Muchnik fashion, using A to "control" W_e via Φ_e, as in Theorem 2. Again there is an obstacle, namely that the enumeration of a witness x for $A \neq \Psi_e(W_e)$ in A can cause A to lose its control of W_e. The obstacle may be overcome by using a witness $x \in N^{(e)}$ such that $x > use(\Phi_e(A;u))$ for all u in the finite set V_e. (Any $u \notin V_e$ with $u < use(\Psi_e(W_e;x))$ will be "forced" to be enumerated in W_e, so that it will not be necessary to keep x from entering A to control $W_e(u)$.) To carry this out, let $P_e^s = \{u : \Phi_e^s(A_e^s;u) = 1\}$. We say that x is an eligible witness for R_e at stage s if

(i) $x \in N^{(e)} - A^s$

(ii) $\Psi_e^s(P_e^s;x)$ is defined, say with use $u_{e,x}^s$

and

(iii) for each $u < u_{e,x}^s$, either $u \in P_e^s$ or $\Phi_e^s(A_e^s;u)$ is defined with use less than x.

At stage s, if x is an eligible witness for R_e, we enumerate in A all elements of $N^{(>e)}$ which are less than $use(\Phi_e^s(A_e^s;u))$ for any $u < u_{e,x}^s$. We then do nothing further for R_e until we come to a stage $t > s$ such that P_e^s and W_e^t agree on all arguments below $u_{e,x}^s$ and $\Psi_e^t(W_e^t;x) = 0$. At the first such stage t (assuming no higher priority requirement intervenes) enumerate x in A. Give no further attention to R_e unless a higher priority requirement requires attention. It is clear, as previously claimed, that each requirement receives attention only finitely often and therefore that $A^{(\leq e)}$ is finite for each e.

Suppose now for a contradiction that R_e is not satisfied, so that $\Phi_e(A) = W_e$, $\Psi_e(W_e) = A$, and V_e is finite. If $u \notin V_e$, then $u \in P_e^s$ for all sufficiently large s because Φ_e, Ψ_e are total and 0-1 valued. Thus each sufficiently large x in $N^{(e)}$ is an eligible witness for R_e at all sufficiently large stages. Therefore R_e will receive attention at some stage s (say under the second strategy) after all higher priority requirements have stopped receiving attention. We will then come to a stage $t > s$ as described since $\Phi_e(A) = W_e$. To conclude that $\Psi_e(W_e) \neq A$ at x, it suffices to show that W_e and P_e^s agree below $u_{e,x}^s$. Let $u < u_{e,x}^s$ be given. If $u \in P_e^s$, then $u \in W_e^t$ so $u \in W_e$. If $u \notin P_e^s$, then use $(\Phi_e^s(A_e^s;u)) < x$ by (iii) and so A and A_e^s agree below $\mathrm{use}(\Phi_e^s(A_e^s;u))$. It follows that $\Phi_e(A;u) = \Phi_e^s(A_e^s;u) = 0$ (since $u \notin P_e^s$ and $\Phi_e^s(A_e^s;u)$ is defined). Since $\Phi_e(A) = W_e$ we conclude that $u \notin W_e$ as required. Further details of the construction and proof are completely straightforward and are omitted.

For the reader familiar with weak truth-table reducibility, denoted $\leq_w$, (see [5]) we remark that in the preceding proof we may assume that the Ψ_e's are merely w reduction procedures. (In clause (ii) of the definition of "eligible witness" we no longer require that $\Psi_e^s(P_e^s;x)$ be defined but only that a bound $u_{e,x}^s$ for its use shall have been computed by stage s.) Thus there is an r.e. nonrecursive set A such that no simple set W satisfies $W \leq_{tt} A$ and $A \leq_w W$. On the other hand, the constructions of Post and Yates mentioned before the statement of the theorem show that for every r.e. nonrecursive set A there is a simple set W such that $W \leq_w A$ and $A \leq_{tt} W$.

REFERENCES

1. J. C. E. Dekker, A theorem on hypersimple sets, Proc. Amer. Math. Soc. 5(1954), 791-796.

2. J. C. E. Dekker and J. Myhill, Retraceable sets, Canadian J. Math. 10(1958), 357-373.

3. L. Harrington, Plus-cupping in the r.e. degrees, unpublished manuscript.

4. A. H. Lachlan, Lower bounds for pairs of recursively enumerable degrees, Proc. London Math. Soc. 16(1966), 537-569.

5. R. Ladner and L. Sasso, The weak truth table degrees of recursively enumerable sets, Annals of Math. Logic 8(1975), 429-448.

6. P. G. Odifreddi, Strong reducibilities, to appear in Bull. Amer. Math. Soc.

7. E. L. Post, Recursively enumerable sets of positive integers and their decision problems, Bull. Amer. Math. Soc. 50(1944), 284-316.

8. H. Rogers, Jr., Theory of Recursive Functions and Effective Computability, McGraw-Hill, New York, 1967.

9. G. E. Sacks, Degrees of unsolvability, Annals of Math. Studies 55, 1966.

10. R. I. Soare, The infinite injury priority method, J. Symbolic Logic 41(1976), 513-530.

11. L. Welch, A hierarchy of families of recursively enumerable degrees and a theorem on bounding minimal pairs, Doctoral Dissertation, University of Illinois, 1980.

12. C. E. M. Yates, Three theorems on the degrees of recursively enumerable sets, Duke Math. J. 32(1965), 461-468.

13. C. E. M. Yates, A minimal pair of recursively enumerable degrees, J. Symbolic Logic 31(1966), 159-168.

ON EXISTENCE OF Σ_n END EXTENSIONS

Matt Kaufmann[1]
Purdue University
West Lafayette, IN 47907

Recall the Keisler-Morley Theorem from [2], which implies that every countable model of ZF has an elementary end extension. In Theorem 1 a refinement of that result is presented. For countable structures $\mathfrak{A}$: possessing an end extension which is elementary for Σ_n formulas is equivalent to Σ_n-collection holding in $\mathfrak{A}$ (all $n \geq 2$). A similar result has been obtained independently by Paris/Kirby [5] for models of arithmetic. Theorem 1 also relates the above criteria to the existence of a certain filter on the definable sets. This idea goes back to Skolem [7], and related work appears in Keisler/Silver [3].

Theorem 2 uses the idea of a normal filter to give a criterion for the existence of "blunt" Σ_n end extension ($n\geq 2$). (This construction is well known from e.g. the theory of measurable cardinals.) A related construction follows in a remark which ties these ultrapowers together with Σ_1 well-founded compactness, which has been studied in Cutland/Kaufmann [0].

We also present some observations on "largeness" properties of those α for which L_α has a Σ_2 end extension. The concluding remark considers the exceptional case of Σ_1 end extensions.

Most of the work for this paper was done while a graduate student at the University of Wisconsin (Madison). I would especially like to thank my advisor, Professor Jon Barwise, for his encouragement. Some of these results have since been extended by E. Kranakis [4].

We review some standard definitions. We consider structures $\mathfrak{A} = (A,E)$ and $\mathfrak{B} = (B,F)$ for the language $\{\in\}$. $\mathfrak{B}$ is a Σ_n <u>end extension of</u> $\mathfrak{A}$, $\mathfrak{A} \prec_n \mathfrak{B}$, if $\mathfrak{B}$ is a proper end extension of $\mathfrak{A}$ and $\mathfrak{A}$ and $\mathfrak{B}$ satisfy the same Σ_n sentences with parameters in $\mathfrak{A}$.

A function or relation on A is said to be Σ_n-<u>definable</u> (<u>over</u> $\mathfrak{A}$) if it is defined in $\mathfrak{A}$ by a Σ_n formula which may contain parameters in A. (We adopt a similar convention for Π_n and Δ_n.)

$\mathfrak{A}$ is <u>resolvable</u> if $\mathfrak{A} \models \forall x\, \exists \alpha$ ("α is an ordinal" $\wedge$ "$x \in f(\alpha)$"), where f is some Δ_1-definable function over $\mathfrak{A}$; we say f <u>resolves</u> $\mathfrak{A}$. $\mathfrak{A}$ has Δ_n <u>Skolem functions</u> if the following criterion is met,

[1] Partially supported by NSF grant 043-50-13955.

for every $X \subseteq A^{k+1}$ $(k \in \omega)$. Suppose that X is Δ_n-definable with parameters over $\mathfrak{A}$, such that for all $\vec{a} \in A^k$ there exists $b \in A$ such that $(\vec{a},b) \in X$. Then for some function $f: A^k \longrightarrow A$ which is Σ_n-definable over $\mathfrak{A}$, $(\vec{a},f(\vec{a})) \in X$ for all $\vec{a} \in A^k$. A well-known result of Jensen and Karp (see e.g. Devlin [1], p.39, Lemma 24) says that for $\alpha \geq \omega$, L_α has Δ_n Skolem functions, all $n \geq 1$. (Simply let $f(\vec{a})$ be the least b, in the canonical Δ_1 well-ordering of L, such that $(\vec{a},b) \in X$, if α is Σ_n-admissible.)

Let $\mathcal{U}$ be a filter on the Π_n-definable subsets of $\mathfrak{A}$, that is, a collection closed under finite intersections. (Recall our convention that parameters are allowed in the definitions.) $\mathcal{U}$ is an <u>$\mathfrak{A}$-complete ultrafilter on the Δ_n subsets of $\mathfrak{A}$</u> iff the following four conditions are met.

(a) For all X which are Δ_n over $\mathfrak{A}$, $X \in \mathcal{U}$ or $A \backslash X \in \mathcal{U}$.

(b) For all $a \in A$, $\{a\} \notin \mathcal{U}$.

(c) For all $X \subseteq Y \subseteq A$, $X \in \mathcal{U}$ implies $Y \in \mathcal{U}$,

(d) For every $X \subseteq A^2$ which is Δ_n over $\mathfrak{A}$, and for all $a \in A$, set $X_a = \{b: (a,b) \in X\}$. Then for all $d \in A$, if $X_a \in \mathcal{U}$ for all $a \mathrel{E} d$ then $\bigcap_{a E d} X_a \in \mathcal{U}$.

If in addition the following condition is met, we say that $\mathcal{U}$ <u>is closed under Δ_n diagonal intersections</u>:

(e) Choose X, X_a as in (d), and suppose $X \in \mathcal{U}$ for all $a \in A$. Then $\{b \in A: (\forall a \mathrel{E} b)(b \in X_a)\} \in \mathcal{U}$.

<u>Σ_n-collection</u> is the axiom schema

$$\forall x \in u \, \exists y_1 \exists y_2 \cdots \exists y_k \phi \longrightarrow \exists w \, \forall x \in u \, \exists y_1 \in w \cdots \exists y_k \in w \phi,$$

abbreviated $\forall x \in u \, \exists \vec{y} \phi \longrightarrow \exists w \, \forall x \in u \, \exists \vec{y} \in w$, for all $\Sigma_n \phi$ (with parameters); similarly for Π_n-collection. We use implicitly the well-known observation that for each $n \geq 1$, Σ_n-collection is equivalent to Π_{n-1}-collection.

<u>Theorem</u> 1. Suppose $\mathfrak{A} = (A,E)$ is a structure for $\{\in\}$, such that all axioms of KP (admissible set theory) hold in $\mathfrak{A}$ excepting possibly Foundation and Δ_0-collection. Consider the following properties,

(i) $\mathfrak{A}$ has a Σ_n end extension.

(ii) $\mathfrak{A}$ satisfies every instance of Σ_n-collection.

(iii) There is an $\mathfrak{A}$-complete ultrafilter on the Δ_{n-1} subsets of $\mathfrak{A}$.

Then for every $n \geq 2$:

If $\mathfrak{A}$ is resolvable, (i) implies (ii).

If $\mathfrak{A}$ is countable, (ii) implies (i).

If $\mathfrak{A}$ has Δ_{n-1} Skolem functions, (iii) implies (i).

Without additional hypotheses, (i) implies (iii).

(And if $\mathfrak{A}$ is resolvable, we could require that the set of ordinals of $\mathfrak{A}$ belongs to the filters.)

In particular, if $\mathfrak{A} = (L_\alpha, \in)$ and $\alpha \geq \omega$ is countable, then (i), (ii), and (iii) are equivalent.

To start with (i) $\Rightarrow$ (ii) we prove three lemmas.

Lemma 1. For every Π_m formula ϕ there is a Π_m formula ψ such that: Π_{m-1}-collection $\vdash \exists x \in u\phi \leftrightarrow \psi$.

Proof. By induction on m. For $m = 0$ we take Π_{m-1}-collection to be empty and let ψ equal $\exists x \in u\phi$. Now suppose the result holds for all $m' < m$. Then assuming Π_{m-1}-collection, the following are equivalent if ϕ is $\forall\vec{y}\theta$ and θ is Σ_{m-1}: $\exists x \in u\phi \leftrightarrow \exists x \in u\, \forall\vec{y}\, \theta \leftrightarrow \neg\, \forall x \in u\, \exists y\, \neg\theta \leftrightarrow \neg\, \exists w \forall x \in u \exists\vec{y} \in w \neg\theta \leftrightarrow \exists w \forall x \in u\eta$ (for some $\Pi_{m-1}\eta$, by Π_{m-2}-collection and the inductive hypothesis) $\leftrightarrow \forall w\, \exists x \in u\, \neg\eta$, which is Π_m. □

Lemma 2. If $\mathfrak{A} \prec_2 \mathfrak{B}$ and $\mathfrak{A}$ is resolvable, then for some $b \in |\mathfrak{B}|\backslash A$, $\mathfrak{B} \models$ "b is an ordinal".

Proof. Let f be a Σ_1 function witnessing the resolvability of $\mathfrak{A}$, and let $\phi(x,y)$ be a Σ_1 definition of "$x \in f(y)$". By definition of $\mathfrak{A} \prec_2 \mathfrak{B}$, there exists $c \in |\mathfrak{B}|\backslash A$. Now the following Π_2 sentence holds in $\mathfrak{A}$ and therefore holds in $\mathfrak{B}$: $\forall x\, \exists y[\phi(x,y) \wedge$ "y is an ordinal"]. So for some $b \in |\mathfrak{B}|$, $\mathfrak{B} \models \phi(c,b) \wedge$ "b is an ordinal". It remains to show $b \notin A$. Suppose otherwise, and let $a = f(b)$. Now the following Π_1 sentence holds in $\mathfrak{A}$ and hence in $\mathfrak{B}$: $\forall x(\phi(x,b) \rightarrow x \in a)$. Therefore $\mathfrak{B} \models c \in a$, so since $\mathfrak{B}$ is an end extension of $\mathfrak{A}$, $c \in A$. This contradicts the choice of c. □

Lemma 3. If $\mathfrak{A} \prec_2 \mathfrak{B}$ and $\mathfrak{A}$ is resolvable, then for some $c \in |\mathfrak{B}|$, $\mathfrak{B} \models a \in c$ for all $a \in A$.

Proof. Choose b as in Lemma 2, and let f resolve $\mathfrak{A}$ as before. Let $\phi(x,y)$ be a Σ_1 definition of "$x \in f(y)$", and let $\psi(z,u)$ be a Σ_1 definition of "$u = f(z) \wedge$ 'z is an ordinal'." Then the following Π_2 sentence holds in $\mathfrak{A}$ and therefore holds in $\mathfrak{B}$: $\forall x\, \exists y$ ("x is an ordinal" $\rightarrow \psi(x,y)$). So we may choose c so that $\mathfrak{B} \models \psi(b,c)$. Fix $a \in A$; we must show that $\mathfrak{B} \models a \in c$. Choose $\beta \in A$ such that $\mathfrak{A} \models \phi(a,\beta)$. Now the following sentence holds in $\mathfrak{A}$ and hence in $\mathfrak{B}$: $\forall x \forall y \forall z \forall u\, [\phi(x,y) \wedge y \in z \wedge \psi(z,u) \rightarrow x \in u]$. Plugging in $x = a$, $y = \beta$, $z = b$, and $u = c$, we obtain $\mathfrak{B} \models a \in c$.□

Proof of (i) ⇒ (ii): by induction on $n \geq 2$. Let ϕ be Π_{n-1} over $\mathfrak{A}$, and suppose $\mathfrak{A} \models \forall x \in a\, \exists \vec{y}\, \phi(x,\vec{y})$ and $\mathfrak{A} <_n \mathfrak{B}$, where $\mathfrak{A}$ is resolvable. By Lemma 3, pick $c \in |\mathfrak{B}|$ such that $\mathfrak{B} \models a \in c$ for all $a \in A$. Since $\mathfrak{A} <_n \mathfrak{B}$, for all $x_0 E a$ there exists $\vec{y}_0$ from A such that $\mathfrak{B} \models \phi(x_0,\vec{y}_0)$; so since $\mathfrak{B}$ is an end extension of $\mathfrak{A}$,

(*) $$\mathfrak{B} \models \forall x \in a\, \exists \vec{y} \in c\, \phi(x,\vec{y}).$$

Let ψ be Π_{n-1} such that (Π_{n-2}-collection) $\vdash \exists \vec{y} \in z \phi(x,y) \leftrightarrow \psi(x,z)$, by Lemma 1. By the inductive hypothesis (or verify directly if $n = 2$), $\mathfrak{A} \models \Pi_{n-2}$-collection. So by choice of ψ, the following sentence θ holds in $\mathfrak{A}$:

$$\theta \equiv \forall z\left[\exists \vec{y} \in z\ \phi(x,\vec{y}) \rightarrow \psi(x,z)\right].$$

Since θ is Π_n (or more precisely, provably equivalent to its Π_n prenex form), $\mathfrak{B} \models \theta$. By (*) and choice of ψ, $\mathfrak{B} \models \forall x \in a\, \psi(x,c)$. Thus the following Σ_n sentence holds in $\mathfrak{B}$, and hence in $\mathfrak{A}$: $\exists z\, \forall x \in a\, \psi(x,z)$. By choice of ψ and because $\mathfrak{A} \models \Pi_{n-2}$-collection, $\mathfrak{A} \models \exists z\, \forall x \in a\, \exists \vec{y} \in z\, \phi$. □

To prove (ii) ⇒ (i), we assume $\mathfrak{A}$ is countable and $\mathfrak{A} \models \Sigma_n$-collection, where $n \geq 2$. For convenience, set $p = n - 2$. This is the direction which is similar to the Keisler/Morley Theorem in [2], referred to before. Hence, we need an appropriate lemma for omitting types, which is an easy modification of the standard Omitting Types Theorem.

Lemma 4. Let T be a consistent theory consisting of Π_{p+2} sentences in a countable language L. Also let $\{\Sigma_i : i \in I\}$ be a countable family of sets of Σ_{p+1} L-formulas with only x free. Suppose that for every Σ_{p+1} L-formula $\phi(x)$ and $i \in I$, if ϕ is consistent with T then so is $\exists x(\phi(x) \wedge \neg\sigma(x))$, for some $\sigma \in \Sigma_i$. Then T has a model omitting each Σ_i.

Proof. Let $C = \{c_m : m \in \omega\}$ be a countable set of new constant symbols. Enumerate the Σ_p sentences of $L \cup C$ as $\{\phi_m : m \in \omega\}$; the Σ_{p+1} sentences of $L \cup C$ as $\{\psi_m : m \in \omega\}$; and $I \times \omega$ as $\{\langle i_m, j_m\rangle : m \in \omega\}$. Set $T_0 = \emptyset$, and do the following at stages $m \geq 1$ to get $T_m \supseteq T_{m-1}$, where each T_m is a finite set of Σ_{p+1} sentences of $L \cup C$. The reader can check that each stage of the construction preserves consistency with T.

1) If ϕ_{m-1} is consistent with $T \cup T_{m-1}$, set $T_m^1 = T_m \cup \{\phi_{m-1}\}$; otherwise set $T_m^1 = T_{m-1} \cup \{\neg\phi_{m-1}\}$.

2) If ϕ_{m-1} is $\exists x\psi(x)$ and $\phi_{m-1} \in T_m^1$, choose $c \in C$ not occuring in T_m^1 and set $T_m^2 = T_m^1 \cup \{\psi(c)$. Otherwise, set $T_m^2 = T_m^1$.

3) Let $i = i_{m-1}$ and $j = j_{m-1}$. We now ensure that c_j doesn't realize Σ_i. Choose $\theta(x,\vec{y}) \in \Sigma_{p+1}$ such that $\models \wedge T_m^2 \leftrightarrow \theta(c_j,\vec{c})$. Since T_m^2 is consistent with T, we may use the hypothesis of the lemma to choose $\sigma \in \Sigma_i$ such that $\exists x(\exists\vec{y}\theta(x,\vec{y}) \wedge \neg\sigma(x))$ is consistent with T. Set $T_m^3 = T_m^2 \cup \{\neg\sigma(c_j)\}$.

4) If $T \vdash \psi_{m-1}$ and $\psi_{m-1} = \exists\vec{y}\theta(\vec{y})$, $\theta \in \Sigma_{m-2}$, set $T_m^4 = T_m^3 \cup \{\theta(\vec{c})\}$ for any $\vec{c} \in C^{<\omega}$ disjoint from the constants in T_m^3. Otherwise, $T_m^4 = T_m^3$.

Let $T_\omega = \bigcup_m T_m$. As usual, we can form a Henkin model $\mathfrak{B}$ such that for all Σ_p ϕ of $L \cup C$, $\phi \in T_\omega$ iff $\mathfrak{B} \models \phi$, by 1) and 2). This, together with 4), shows that $\mathfrak{B} \models T$. By 3), $\mathfrak{B}$ omits each Σ_i. □

To apply Lemma 4, we let $T = \{\phi(\vec{a}) : \vec{a} \in A^{<\omega}$, ϕ is Π_n, and $\mathfrak{A} \models \phi\} \cup \{$"c is an ordinal"$\} \cup \{\beta \in c : \mathfrak{A} \models$ "β is an ordinal"$\}$. We want a model of T which omits, for each $a \in A$, $\Sigma_a = \{x \in a \wedge x \neq b : bEa\}$.

<u>Lemma 5</u>. If $\phi(x)$ is Σ_n over $\mathfrak{A}$, then $\phi(c)$ is consistent with T iff $\mathfrak{A} \models (\exists$ arbitrarily large ordinals $\beta)$ $\phi(\beta)$.

<u>Proof</u>. ($\Leftarrow$) is clear by compactness. For ($\Rightarrow$), suppose $\mathfrak{B} \models T$ and $\mathfrak{B} \models \phi(c)$. For each "ordinal" β of $\mathfrak{A}$, $\mathfrak{B} \models \exists\ \delta(\beta \in \delta \wedge$ "δ is an ordinal" $\wedge\ \phi(\delta))$, so this sentence is true in $\mathfrak{A}$ since $\mathfrak{B} \models T$ and ϕ is Σ_n over $\mathfrak{A}$. □

Proof of (ii) $\Rightarrow$ (i): Form T as above and assume $\mathfrak{A} \models \Sigma_n$-collection. It suffices to show that T and $\{\Sigma_a : a \in A\}$ satisfy the "local omitting" hypothesis of Lemma 4. Suppose $\phi(x,c)$ is Σ_{n-1} and for each bEa, $T \vdash \forall x(\phi(x,c) \rightarrow (x \in a \wedge x \neq b))$; we show $T \vdash \neg\exists x\phi(x,c)$. By Lemma 5 (or, its contrapositive),

$$\mathfrak{A} \models \forall y \in a\ \exists\beta\ \forall\delta \geq \beta\forall x(\phi(x,\delta) \rightarrow (x \in a \wedge x \neq y)),$$

where here and henceforth Greek letters $\alpha,\beta,\gamma,\delta$ refer to ordinals of the model. By Π_{n-1}-collection in $\mathfrak{A}$ (and the basic closure properties of $\mathfrak{A}$), there exists an "ordinal" α of $\mathfrak{A}$ such that

$$\mathfrak{A} \models \forall y \in a\ \exists\beta \in \alpha\forall\delta \geq \beta\forall x(\phi(x,\delta) \rightarrow (x \in a \wedge x \neq y)).$$

So for all δ, if $\mathfrak{A} \models \delta \geq \alpha$ then $\mathfrak{A} \models \neg\exists x\phi(x,\delta)$. By Lemma 5, since $\exists x\phi$ is Σ_{n-1} , $\phi(x,c)$ is not consistent with T. □

Proof of (i) ⇒ (iii). Suppose $\mathfrak{A} \prec_n \mathfrak{B}$; pick $c \in |\mathfrak{B}|\backslash A$. If $\phi(x)$ is Π_{n-1} over $\mathfrak{A}$ and $X = \{a: \mathfrak{A} \models \phi(a)\}$, set $X \in \mathcal{U}$ iff $\mathfrak{B} \models \phi(c)$. We check that this (admittedly standard) definition of $\mathcal{U}$ is well-defined, and satisfies properties (a) through (d) of the definition of "$\mathfrak{A}$-complete ultrafilter on the Δ_{n-1} subsets of $\mathfrak{A}$." $\mathcal{U}$ is well-defined and satisfies properties (a) and (c), because if $\phi(x)$ and $\psi(x)$ are $\Sigma_{n-1} \cup \Pi_{n-1}$ over $\mathfrak{A}$ and both define X, then: $\mathfrak{A} \models \forall x(\phi \leftrightarrow \psi)$, so $\mathfrak{B} \models \phi(c) \Leftrightarrow \mathfrak{B} \models \psi(c)$. Property (b) is clear since $\mathfrak{B} \models a \neq c$ for all $a \in A$. For (d), suppose $\phi(x,y)$ is a Π_{n-1} definition of $X \subseteq A^2$ over $\mathfrak{A}$, and that for all $a \in d$, $X_a = \{b: (a,b) \in X\} \in \mathcal{U}$. Then for all $a \in d$, $\mathfrak{B} \models \phi(a,c)$. So $\mathfrak{B} \models \forall y \in d\ \phi(y,c)$. Since $\forall y \in d\ \phi(y,x)$ is a Π_{n-1} definition of $\bigcap_{a \in d} X_a$, that set belongs to $\mathcal{U}$, as desired.

(If in fact $\mathfrak{A}$ is resolvable, then by Lemma 2 we may assume $\mathfrak{B} \models$ "c is an ordinal", and then the set of ordinals of $\mathfrak{A}$ belongs to $\mathcal{U}$.)□

Proof of (iii) ⇒ (i). Suppose $\mathfrak{A}$ has Δ_{n-1} Skolem functions and assume that $\mathcal{U}$ is an $\mathfrak{A}$-complete ultrafilter on the Δ_{n-1} subsets of $\mathfrak{A}$. We will take a definable ultrapower of $\mathfrak{A}$, mod $\mathcal{U}$. That is, for Δ_{n-1} f and g on $\mathfrak{A}$, $f \sim g$ iff $\{a: f(a) = g(a)\} \in \mathcal{U}$. Define $\mathfrak{B}$ as follows: $|\mathfrak{B}| = \{[f]: f \text{ is } \Delta_{n-1} \text{ on } \mathfrak{A}\}$, where [f] is the equivalence class of f under ~ ; and $[f] \in^{\mathfrak{B}} [g]$ iff $\{a: f(a) \in g(a)\} \in \mathcal{U}$.

Lemma 6. For any Σ_{n-2} or Π_{n-2} $\phi(x_1,\dots,x_k)$, and Δ_{n-1} $f_1,\dots,f_k$, $\mathfrak{B} \models \phi([f_1],\dots,[f_k])$ iff $\{a: \mathfrak{A} \models \phi(f_1(a),\dots,f_k(a))\} \in \mathcal{U}$. (Recall that $\mathfrak{A}$ has Δ_{n-1} Skolem functions.)

Proof. By induction on complexity of ϕ. It's clear for atomic ϕ, by definition of $\mathfrak{B}$. The negation step is clear by property (a) for $\mathcal{U}$, and the conjunction step is all right because $\mathcal{U}$ is a filter. Now suppose ϕ is $\exists x\ \psi(x,\vec{y})$, where ψ is Σ_{n-2}.

Assume $\mathfrak{B} \models \phi([\vec{g}])$, say $\mathfrak{B} \models \psi([f],[\vec{g}])$. Then for some $X \in \mathcal{U}$, $\mathfrak{A} \models \psi(f(a), \vec{g}(a))$ for all $a \in X$, by the inductive hypothesis. So by closure of $\mathcal{U}$ under supersets (property (c)), $\{a: \mathfrak{A} \models \exists x\ \psi(x,\vec{g}(a))\} \in \mathcal{U}$.

Conversely, suppose that $\{a: \mathfrak{A} \models \exists x\ \psi(x,\vec{g}(a)\} \in \mathcal{U}$. Let f be a Δ_{n-1} Skolem function for $\{(x,a): \mathfrak{A} \models \psi(x,\vec{g}(a)) \vee [\neg\exists x\,\psi(x,\vec{g}(a)) \wedge x = 0]\}$ Then $\{a: \mathfrak{A} \models \psi(f(a),g(a))\} \in \mathcal{U}$, so by the inductive hypothesis, $\mathfrak{B} \models \psi([f],[\vec{g}])$ and hence $\mathfrak{B} \models \phi([\vec{g}])$. This concludes the proof of Lemma 6.

Returning to the proof of (iii) ⇒ (i), suppose $\mathfrak{A} \models \forall \vec{x}\, \exists \vec{y}\, \phi(\vec{x},\vec{y})$, where ϕ is Π_{n-2}. We may assume $\vec{x}$ and $\vec{y}$ are single variables, by pairing and that $\mathfrak{A} \subseteq \mathfrak{B}$ by treating $a \in A$ as a constant function (so now we can ignore the parameters in ϕ). Let $f \in |\mathfrak{B}|$; we show $\mathfrak{B} \models \exists y\, \phi([f],y)$. Let g be a Δ_{n-1} Skolem function for $\{(x,y): \mathfrak{A} \models \phi(f(x),y)\}$. Then $\mathfrak{A} \models \phi(f(a),g(a))$ for all $a \in A$, so since $A \in \mathcal{U}$ (and by Lemma 6), $\mathfrak{B} \models \phi([f],[g])$.□

We wish to present next a version of the equivalence (i) $\Leftrightarrow$ (iii) of Theorem 1, but where we require the end extension of L_α to have a least new ordinal. The following definition generalizes this notion, as we check in Proposition A below.

Definition. For $\mathfrak{B} \supseteq_{end} \mathfrak{A}$, $\mathfrak{B}$ is a weakly blunt end extension of $\mathfrak{A}$ if for some $b \in B \setminus A$, $\{a \in B: \mathfrak{B} \models a \in b\} \subseteq A$. $\mathfrak{B}$ is a blunt end extension of $\mathfrak{A}$ if we may also require: $\mathfrak{B} \models b \not\subseteq c$ for all $c \in A$. Notice that if $\mathfrak{A} = (L_\alpha, \in)$, then a Σ_2 end extension $\mathfrak{B}$ of $\mathfrak{A}$ is blunt iff there is a least new ordinal in $\mathfrak{B}$; this isn't hard and follows from:

Proposition A. Let $\mathfrak{A}$ be resolvable and suppose $\mathfrak{B}$ is a blunt Σ_2 end extension of $\mathfrak{A}$. Also assume that $\mathfrak{A} \models$ KP (though in light of Theorem 1, we need not actually make Σ_1-collection a hypothesis). Then for some "ordinal" c of $\mathfrak{B}$ and some $d \in B$, the following conditions hold.

(i) For all $x \in B$, $\mathfrak{B} \models x \in c$ iff $x \in A$ and $\mathfrak{A} \models$ "x is an ordinal", and

(ii) for all $x \in B$, $\mathfrak{B} \models x \in d$ iff $x \in A$.

Proof. (ii) follows from (i) as Lemma 3 follows from Lemma 2. For (i), first choose $b \in B$ such that $\{a \in B: \mathfrak{B} \models a \in b\} \subseteq A$, but $\mathfrak{B} \models b \not\subseteq c$ for all $c \in A$. Also let f be a function which resolves $\mathfrak{A}$. We choose the least ordinal α of $\mathfrak{B}$ such that $b \subseteq f^{\mathfrak{B}}(\alpha)$. More precisely, let $\phi(x,y)$ be the following Σ formula:

"y is an ordinal" $\wedge\ x \subseteq f(y) \wedge (\forall z \in y)\, x \not\subseteq f(z)$.

We may set $\phi_0(x,y) \equiv (\exists$ transitive $u)\, \phi^u(x,y)$. Notice $\mathfrak{A} \models \forall x \forall y(\phi \rightarrow \phi_0)$ since $\mathfrak{A} \models \Sigma_1$-collection, and $\mathfrak{B} \models \forall x \forall y(\phi_0 \rightarrow \phi)$.
Now $\mathfrak{A} \models \forall x \exists y \phi(x,y)$, so $\mathfrak{A} \models \forall x \exists y \phi_0$; hence $\mathfrak{B} \models \forall x \exists y \phi_0$, hence $\mathfrak{B} \models \forall x \exists y \phi$. Choose $c \in B$ such that $\mathfrak{B} \models \phi(b,c)$. Clearly $c \notin A$, since otherwise, as $\mathfrak{B} \models b \subseteq f(c)$ where $f^{\mathfrak{B}}(c) = f^{\mathfrak{A}}(c) \in A$, this would contradict our choice of b. Finally, suppose $x \in B$ and $\mathfrak{B} \models x \in c$. Clearly then $\mathfrak{B} \models$ "x is an ordinal"; so we're done if we can show $x \in A$. But this follows from the minimality of c.□

Definition. (i) For $\mathfrak{A} = (A,E)$, we call $X \subseteq A$ unbounded if $(\forall a \in A)(\exists b \in X)(\neg bEa)$.

(ii) Let $\mathcal{U}$ be a Π_n ultrafilter on $\mathfrak{A} = (A,E)$. $\mathcal{U}$ is large if $\{a \in A: \{b \in A: aEb\} \in \mathcal{U}\}$ is unbounded.

Theorem 2. Suppose $\mathfrak{A}$ is a resolvable model of KP which has Δ_{n-1}

Skolem functions, where $n \geq 2$.

(i) $\mathfrak{A}$ has a weakly blunt Σ_n end extension iff there is an $\mathfrak{A}$-complete ultrafilter on the Π_{n-1} subsets of $\mathfrak{A}$, which is closed under Δ_{n-1} diagonal intersections.

(ii) $\mathfrak{A}$ has a blunt Σ_n end extension iff for some $\mathcal{U}$ as above in (i), $\mathcal{U}$ is large.

<u>Proof of</u> (i). (⇒): Choose $c \in B\setminus A$ such that $\{a \in B: \mathfrak{B} \models a \in c\} \subseteq A$. Then define $\mathcal{U}$ as in the proof of Theorem 1, (i) ⇒ (iii). We check that $\mathcal{U}$ is closed under Δ_{n-1} diagonal intersections. Suppose $X \subseteq A^2$ is Δ_{n-1} over $\mathfrak{A}$, and let $X_a = \{b: (a,b) \in X\}$. Let $\phi(x,y)$ be a Π_{n-1} definition of X over $\mathfrak{A}$. Then we must show that $\{b \in A: \forall a\ Eb(b \in X_a)\} \in \mathcal{U}$, i.e. that $\mathfrak{B} \models \forall x \in c\phi(x,c)$, if $X_a \in \mathcal{U}$ (i.e. $\mathfrak{B} \models \phi(a,c)$) for all $a \in A$. But this is clear.

(⇐) Given such $\mathfrak{A}$ and $\mathcal{U}$, form $\mathfrak{B}$ as in the proof of Theorem 1, (iii) ⇒ (i). Let $i: A \to A$ be the identity function. Suppose $\mathfrak{B} \models [f] \in [i]$; it suffices to show that f is equivalent to a constant function. Otherwise set $X = \{(x,y): \mathfrak{A} \models f(y) \neq x\}$; then $X_a \in \mathcal{U}$ for all $a \in A$. So by hypothesis, $\{y: (\forall aEy)\ (f(y) \neq a)\} \in \mathcal{U}$; that is $\{y: \mathfrak{A} \models f(y) \notin y\} \in \mathcal{U}$. Therefore $\langle [f],[i]\rangle \notin \in^{\mathfrak{B}}$, contradicting $\mathfrak{B} \models [f] \in [i]$.□

<u>Proof of</u> (ii). (⇒): Suppose $\mathfrak{A}$ has a blunt Σ_n end extension $\mathfrak{B}$. Define $\mathcal{U}$ as in the proof of (i), where now we also assume that for all $a \in A$, $\mathfrak{B} \models c \not\subseteq a$. We check that $\mathcal{U}$ is large. Suppose not; say $\{a \in A: \{b \in A: aEb\} \in \mathcal{U}\} \subseteq d_E$, where $d_E = \{x \in A: \mathfrak{A} \models x \in d\}$. That is, $\{a \in A: \mathfrak{B} \models a \in c\} \subseteq d_E$; this contradicts $\mathfrak{B} \models c \not\subseteq d$.

(⇐): Given such $\mathcal{U}$, form $\mathfrak{B}$ as before; we check that $\mathfrak{B}$ is a <u>blunt</u> extension of $\mathfrak{A}$ by showing that for all $d \in A$, $\mathfrak{B} \models [i] \not\subseteq d$. But if $d \in A$ and $\mathfrak{B} \models [i] \subseteq d$, then by the "Łos Lemma" (Lemma 6), $\{x \in A: \mathfrak{A} \models x \subseteq d\} \in \mathcal{U}$. Now if $a \in A$ and $\{b \in A: aEb\} \in \mathcal{U}$, then $\{x \in A: \mathfrak{A} \models x \subseteq d\} \cap \{x \in A: aEx\} \in \mathcal{U}$, so since $\emptyset \notin \mathcal{U}$, $\mathfrak{A} \models (a \in x \wedge x \subseteq d)$ for some x, and therefore $\mathfrak{A} \models a \in d$. Hence $\{a \in A: \{b \in A: a \in b\} \in \mathcal{U}\}$ is contained in d_E, contradicting the hypothesis that $\mathcal{U}$ is large.□

<u>Remark</u> 1. Of course, to get a well-founded end extension from a filter $\mathcal{U}$ (as above), it suffices that no countable intersection from $\mathcal{U}$ be empty. For countable $\mathfrak{A}$, no such $\mathcal{U}$ exists, since $\bigcap \{A\setminus\{a\}: a \in A\} = \emptyset$. However, we can still get a well-founded Σ_2 end extension via a definable ultrapower, in certain special cases.

Let $\mathcal{L}_\alpha^{wf}$ be infinitary logic over L_α, except that for some fixed binary relation symbol E we only consider models in which E is well-founded. α is Σ_1-well-founded compact ($\Sigma_1$wfc) if $\mathcal{L}_\alpha^{wf}$ is Σ_1-compact over L_α. This notion has been studied in Cutland/Kaufmann [0], where the following results were proved (along with a number of others):

(1) α is $\Sigma_1$wfc iff for every Π_1 formula $\phi(x)$ over L_α, $V \models \phi(\alpha)$ implies $V \models \exists\beta < \alpha\phi(\beta)$.

(2) If α is $\Sigma_1$wfc, then $L_\alpha <_1 L$ and L_α has a well-founded Σ_2 end extension.

We can now check the second part of (2) in a somewhat more "constructive" manner, and in the process show how to "construct" well-founded models using our definable ultrapowers. If $\phi(x)$ is a formula with parameters in L_α, let $\phi^{L_\alpha} = \{a: L_\alpha \models \phi(a)\}$. Let $\mathcal{U}_0 = \{\phi^{L_\alpha}: \phi(x)$ is Π_1 over L_α and $V \models \phi(\alpha)\}$, and let $\mathcal{U} = \{X \subseteq L_\alpha: X \supseteq Y$ for some $Y \in \mathcal{U}_0\}$. Construct the definable ultrapower $\mathfrak{B}$ of L_α mod $\mathcal{U}$ using Δ_1 functions, exactly as in the proof of Theorem 1, (iii) $\Rightarrow$ (i). The proof of Lemma 6 is easily adapted to show that for any Δ_0 formula ϕ, $\mathfrak{B} \models \phi([\vec{f}])$ iff $\{a: L_\alpha \models \phi(\vec{f}(a))\} \in \mathcal{U}$. Then as before, it's not hard to show that L_α and $\mathfrak{B}$ satisfy the same Σ_2 and Π_2 sentences over L_α.

Now we can show that $\mathfrak{B}$ is a blunt end extension of $(L_\alpha, \in)$. Let $i: L_\alpha \to L_\alpha$ be the identity function: note that $\mathfrak{B} \models$ "$[i]$ is an ordinal". So, it suffices to show that if $[f] \in [i]$, then $[f] = \beta$ for some $\beta < \alpha$. Given $[f] \in [i]$, let $\theta(x,y)$ be a Σ_1 definition of f over L_α. By replacing θ with θ' defined below, we may assume that

$$(*) \qquad V \models \forall x\forall y\forall z(\theta(x,y) \wedge \theta(x,z) \to y = z).$$

θ' is a Σ_1 formula equivalent to $\exists\beta(L_\beta \models \theta(x,y) \wedge \forall z \in L_\beta(L_\beta \models$ "$z <_L y$" $\to \neg\theta(x,z))$, where "$z <_L y$" is a canonical Δ_1 well-ordering of L. Now since $[f] \in [i]$, there is a Π_1 formula $\phi(x)$ over L_α such that $V \models \phi(\alpha)$ and $\phi^{L_\alpha} \subseteq \{x \in L : f(x) \in x\}$. Then

$$L_\alpha \models \forall x(\phi(x) \to \exists y(\theta(x,y) \wedge y \in x)).$$

So $V \models \phi(\alpha) \to \exists y(\theta(\alpha,y) \wedge y \in \alpha)$ (by the contrapositive of Π_1-indescribability (1)).

Hence $V \models \theta(\alpha,\beta) \wedge \beta \in \alpha$ for some β, since $V \models \phi(\alpha)$.

By (*), $V \models \forall y(\theta(\alpha,y) \to y = \beta)$ (some $\beta \in \alpha$).

Then $\{x: L_\alpha \models \forall y(\theta(x,y) \to y = \beta)\} \in \mathcal{U}$.

Therefore $\{x: f(x) = \beta\} \in \mathcal{U}$.

That is, $[f] = \beta$.

A similar argument shows that $\mathfrak{B}$ is well-founded. For suppose (to the contrary) that $[f_{i+1}] \in [f_i]$ for all $i \in \omega$. Let $\theta_i(x,y)$ be a good Σ_1 definition of f_i over L_α (all $i \in \omega$); in other words, assume (*) holds for each θ_i. Then each θ_i defines a function $\overline{f_i}$ in V. Now given $i \in \omega$, choose a Π_1 formula $\phi(x)$ over L_α such that $V \models \phi(\alpha)$ and $\phi^{L_\alpha} \subset \{x \in L_\alpha: f_{i+1}(x) \in f_i(x)\}$. Then $L_\alpha \models \forall x(\phi(x) \to \exists y \exists z(\theta_{i+1}(x,y) \wedge \theta_i(x,z) \wedge y \in z))$. So (by Π_1-indescrib.), $V \models \phi(\alpha) \to \exists y \exists z(\theta_{i+1}(\alpha,y) \wedge \theta_i(\alpha,z) \wedge y \in z)$. That is, $\bar{f}_{i+1}(\alpha) \in \bar{f}_i(\alpha)$. Since this holds for all $i \in \omega$, we have a contradiction; so $\mathfrak{B}$ is well-founded.

What does the above construction have to do with Σ_1-compactness of $\mathcal{L}_\alpha^{WF}$? Let α be $\Sigma_1$wfc, and suppose T is a Σ_1 theory of $\mathcal{L}_\alpha^{WF}$ which is α-finitely satisfiable in constructible models. (For example, T is α-finitely satisfiable, and $\alpha < \omega_1^L$ or $V = L$.) As mentioned before, $L_\alpha \prec_1 L$; so T is α-finitely satisfiable inside L_α. Let $\phi(x)$ be a Σ_1 definition (over L_α) of T.

We show how to "construct" a model of T using a definable ultrapower. Choose $\mathfrak{B} \succ_2 (L_\alpha, \in)$ as constructed above; since $\mathfrak{B}$ is a well-founded model of extensionality, we identify $\mathfrak{B}$ with a transitive set. Define a Σ_1 function f over L_α as follows. If x is an ordinal, then f(x) is the $<_L$-least (well-founded) model of ϕ^{L_x}; otherwise, set f(x) = 0. We'll show $[f] \models T$. Fix $\beta < \alpha$. Define a Σ_1 function g as follows. If x is an ordinal, g(x) is the $<_L$-least transitive set y such that $L_\alpha \models [a \models \phi^{L_\beta}]^y$, where a = f(x). (If x is not an ordinal, set g(x) = 0.) Since $\alpha \in \mathcal{U}$, and since $L_\alpha \models (f(\gamma) \models \phi^{L_\beta})^{g(\gamma)}$ for all $\gamma < \alpha$, then $\mathfrak{B} \models ([f] \models \phi^{L_\beta})^{[g]}$ by our Łos theorem. Also, $\mathfrak{B} \models$ "[g] is transitive". Since $\mathfrak{B}$ is transitive, $[f] \models \phi^{L_\beta}$. Since $\beta < \alpha$ is arbitrary, $[f] \models T$. □

Here are some results which show that if L_α has sufficiently nice blunt end extensions, then α must be fairly large. For related results see §4 of [0].

Lemma 7 (essentially [0,4.13]). If $\mathfrak{B}$ is a blunt Σ_2 end extension of L_α which satisfies KP, then $\mathfrak{B} \models$ "$L_\alpha \prec_2 L_\delta$" for some $\delta \in B$. □

Theorem 3. Suppose $\mathfrak{B}$ is a blunt Σ_2 end extension of $(L_\alpha, \in)$ which satisfies KP. Then there exist arbitrarily large $\beta < \alpha$ such that for some $\gamma < \alpha$, $(L_\beta, \in) \prec_2 (L_\gamma, \in)$.

Proof. By Lemma 7, $\mathfrak{B} \models$ "$L_\alpha \prec_2 L_\delta$" for some $\delta \in B$. So if $\eta < \alpha$, $\mathfrak{B} \models \exists x \exists y($"$L_x \prec_2 L_y$"$ \wedge \eta < x)$ and hence $L_\alpha \models$ "$L_\beta \prec_2 L_\gamma$" for some $\beta, \gamma > \eta$. □

Proposition B. If $\mathfrak{B}$ is any blunt Σ_2 end extension of $(L_\alpha, \in)$, then α is a limit of Σ_2 admissibles.

Proof. By Theorem 1, α is Σ_2-admissible. So for all $\gamma < \alpha$ $\mathfrak{B} \models \exists\beta(\gamma < \beta \wedge$ "β is Σ_2-admissible"). Hence this holds in L_α.□

Further details of this proof are left to the reader. A similar argument shows that if L_α is the minimal model of ZF, then L_α has no blunt Σ_2 end extension.

Remark 2. The case $n = 1$, which is left out of Theorem 1, is of some interest. For arithmetic, Paris and Kirby showed in [5] that if $\mathfrak{A}$ is countable and satisfies very minimal hypotheses and $\mathfrak{A} \prec_1 \mathfrak{B}$, then $\mathfrak{A} \models \Sigma_2$-collection and hence $\mathfrak{A}$ has a Σ_2 end extension. This result does not carry over completely for models of set theory, since if α is the least stable ordinal then $L_\alpha \prec_1 L_{\omega_1}$, but $L_\alpha \not\models \Sigma_2$-collection since α is projectible. However, Simpson[1] [6] has shown that for countable limit α, $L_\alpha \models \Sigma_2$-collection iff L_α has a Σ_1 end extension $\mathfrak{B} \models V = L$ with a new ordinal, which is not blunt, and satisfies KP. So putting Theorem 1 and (a slight improvement of) Simpson's result together, we get the following somewhat surprising result, analogous to the one above of Paris and Kirby: if α is countable and L_α has an admissible Σ_1 end extension with a new ordinal but no least new ordinal, then L_α has a Σ_2 end extension.

Question. If L_α has a Σ_2 end extension, does it necessarily have one satisfying Σ_1-collection? Or would that imply that α is a limit of Σ_2-admissibles?

REFERENCES

[0] N. Cutland and M. Kaufmann, Σ_1-well-founded compactness, Annals of Math. Logic (to appear).

[1] K. Devlin, Aspects of Constructibility, Lecture Notes in Mathematics, vol. 354 (Springer-Verlag, 1973).

[2] H. J. Keisler and M. Morley, Elementary extensions of models of set theory, Israel J. Math. 5(1968), 49-65.

[3] H. J. Keisler and J. Silver, End extensions of models of set theory, Axiomatic Set Theory (Part 1), Proceedings of Symposia in Pure Mathematics 13(1970), 177-187.

[4] E. Kranakis, notes (personal communication).

[5] J. Paris and L. Kirby, Σ_n-Collection schemas in arithmetic, preprint.

[6] S. Simpson, notes (personal communication).

[7] T. Skolem, Über die Nicht-Charakterisierbarkeit der Zahlenreihe mittels endlich oder abzählbar unendlicht vieler Aussagen mit ausschliesslich Zahlervariablen, Fund. Math. 23, 150-161.

MODEL THEORETIC CHARACTERIZATIONS IN GENERALIZED RECURSION THEORY

Phokion G. Kolaitis[1]

Department of Mathematics
The University of Chicago
Chicago, Illinois 60637

Our main purpose in this paper is to establish model theoretic and invariant definability characterizations of recursion in $\mathbf{E}$ and positive elementary induction (recursion in $\mathbf{E}^{\#}$) on abstract structures. These theories are two different natural generalizations of the hyperarithmetic theory on the integers, while recursion in $\mathbf{E}$ is in addition an extension of the theory of recursion in normal higher type objects.

Moschovakis [1969c], [EIAS] characterized the "boldface" hyperelementary (recursive in $\mathbf{E}^{\#}$) relations using the Schema of Δ^1_1-Comprehension. Here we study finer "lightface" analogs of this schema in terms of which we characterize both recursion in $\mathbf{E}$ and positive elementary induction. Moreover we obtain hierarchies for these theories, which provide constructions of the recursive in $\mathbf{E}$ and the recursive in $\mathbf{E}^{\#}$ relations with the respective relations of lower levels as basis.

Due to the limitations of space, in this paper we restrict ourselves to the statements of the results and to a few comments about the tools used in the proofs. However in the first two sections we have included some of the basic definitions, as well as a survey of the results in the hyperarithmetic theory and the theory of positive elementary induction. The characterizations of recursion in $\mathbf{E}$ are presented in §3, while the last section contains the results about recursion in $\mathbf{E}^{\#}$ and a comparison between the two theories. The proofs of the main theorems and extensions of this work to recursion in generalized quantifiers will appear elsewhere.

§1. Recursion in $\mathbf{E}$ and Recursion in $\mathbf{E}^{\#}$ on a Structure

1.1. Let $\mathfrak{A} = \langle A, R_1, \ldots, R_n, f_1, \ldots, f_m, c_1, \ldots, c_\ell \rangle$ be a structure such that $\omega \subseteq A$ and for $k \in \omega$ let $\mathcal{PF}_k$ be the set of all k-ary partial functions from A into ω. A functional (on A with values in ω) is a partial mapping

$$\Phi: A^s \times \mathcal{PF}_{k_1} \times \ldots \times \mathcal{PF}_{k_t} \to \omega$$

which is monotone, i.e., if $g_1 \subseteq h_1, g_2 \subseteq h_2, \ldots, g_t \subseteq h_t$ and $\Phi(\bar{x}, g_1, \ldots, g_t) = w$, then $\Phi(\bar{x}, h_1, \ldots, h_t) = w$.

[1] During the preparation of this paper, the author was partially supported by NSF Grant #MCS-8002763.

If $\overline{\Phi} = (\Phi_1, \ldots, \Phi_s)$ is a sequence of functionals on A, then we can define the notion of a recursive in $\overline{\Phi}$ k-ary partial function from A into ω. Inductive definability provides a conceptually simple way to do this. We associate first with the sequence $\overline{\Phi}$ the class $\mathcal{F}[\overline{\Phi}]$, which is the smallest collection of functionals on A containing $\Phi_1, \ldots, \Phi_s$ and satisfying certain minimal closure properties (closed under composition, definition by cases, functional substitution, etc.). Then the recursive in $\overline{\Phi}$ partial functions are obtained by iterating to the transfinite the operative functionals in $\mathcal{F}[\overline{\Phi}]$ (i.e., the functionals in $\mathcal{F}[\overline{\Phi}]$ of the form $\Psi: A^s \times \mathcal{PF}_s \to \omega$). The first few sections of Kechris-Moschovakis [1977] and Kolaitis [1978], [1979] contain a detailed account of these definitions.

A relation is semirecursive in $\overline{\Phi}$ if it is the domain of a recursive in $\overline{\Phi}$ partial function; a relation is recursive in $\overline{\Phi}$ if its characteristic function is recursive in $\overline{\Phi}$. The class $\mathrm{ENV}[\overline{\Phi}]$ (the Envelope of $\overline{\Phi}$) is the collection of all semirecursive in $\overline{\Phi}$ relations, while the class $\mathrm{SEC}[\overline{\Phi}]$ (the Section of $\overline{\Phi}$) is the collection of all recursive in $\overline{\Phi}$ relations. These notions relativize directly to a finite sequence $\bar{x} = (x_1, \ldots, x_k)$ from A, so that we put

$$\mathrm{ENV}[\overline{\Phi}, \bar{x}] = \text{all semirecursive relations in } \overline{\Phi} \text{ from } \bar{x},$$
$$\mathrm{SEC}[\overline{\Phi}, \bar{x}] = \text{all recursive relations in } \overline{\Phi} \text{ from } \bar{x}.$$

If we take the union of the above classes as $\bar{x}$ varies over all finite sequences from A, then we obtain respectively the collections of the "boldface" semirecursive in $\overline{\Phi}$ and "boldface" recursive in $\overline{\Phi}$ relations, namely

$$\underset{\sim}{\mathbf{ENV}}[\overline{\Phi}] = \bigcup\{\mathrm{ENV}[\overline{\Phi}, \bar{x}]: \bar{x} \in A^{<\omega}\},$$
$$\underset{\sim}{\mathbf{SEC}}[\overline{\Phi}] = \bigcup\{\mathrm{SEC}[\overline{\Phi}, \bar{x}]: \bar{x} \in A^{<\omega}\},$$

where $A^{<\omega}$ is the set of all finite sequences from A.

1.2. In this paper we are mainly concerned with the functionals $E: \mathcal{PF}_1 \to \omega$ and $E^{\#}: \mathcal{PF}_1 \to \omega$ defined as follows:

$$E(f) = \begin{cases} 0 & \text{, if } f \text{ is total and } (\exists x)(f(x) = 0), \\ 1 & \text{, if } f \text{ is total and } (\forall x)(f(x) \neq 0), \\ \text{undefined} & \text{, if } f \text{ is not total;} \end{cases}$$

$$E^{\#}(f) = \begin{cases} 0 & \text{, if } (\exists x)(f(x) = 0), \\ 1 & \text{, if } (\forall x)(f(x)\downarrow \neq 0), \\ \text{undefined} & \text{, otherwise.} \end{cases}$$

Both these functionals embody existential quantification over A. It is very easy to show that E is recursive in $E^{\#}$, and therefore it is always true that

$$\mathrm{ENV}[\mathbf{E}] \subseteq \mathrm{ENV}[\mathbf{E}^{\#}] .$$

The functional $\mathbf{E}$ appeared first in the theory of recursion in higher types. In fact, if we take $A = \omega = \mathrm{Tp}(0)$, then $\mathbf{E}$ is nothing else but the Kleene type 2 object ${}^2\mathbf{E}$. Moreover, if $A = \omega^\omega = \mathrm{Tp}(1)$, then $\mathbf{E}$ becomes the type 3 object ${}^3\mathbf{E}$ which expresses equality of sets of reals or equivalently existential quantification over the reals.

1.3. It is well known that recursion in $\mathbf{E}^{\#}$ on an arbitrary structure $\mathfrak{A}$ coincides with <u>positive elementary induction</u> on $\mathfrak{A}$. In order to make this statement more precise we review briefly the main definitions involved.

Let $\mathfrak{A} = \langle A, R_1, \ldots, R_n, f_1, \ldots, f_m, c_1, \ldots, c_\ell \rangle$ be a structure such that $\omega \subseteq A$. The <u>"lightface" first order language $\mathcal{L}^{\mathfrak{A}}$</u> of the structure $\mathfrak{A}$ has an infinite list $x, y, z, \ldots$ of individual variables, an infinite list $S, T, V, \ldots$ of k-ary relation variables for each $k \geq 1$, constant symbols $\mathbf{c}_1, \ldots, \mathbf{c}_\ell$, a constant symbol $\mathbf{k}$ for each $k \in \omega$, relation symbols $\mathbf{R}_1, \ldots, \mathbf{R}_n$, function symbols $\mathbf{f}_1, \ldots, \mathbf{f}_m$, the equality symbol $=$, and the logical symbols $\neg$, &, $\vee, \rightarrow, \forall, \exists$. The <u>"boldface" first order language $\boldsymbol{\mathcal{L}}^{\mathfrak{A}}$</u> of the structure $\mathfrak{A}$ has in addition a constant symbol $\mathbf{a}$ for each $a \in A$. The <u>formulas</u> of both $\mathcal{L}^{\mathfrak{A}}$ and $\boldsymbol{\mathcal{L}}^{\mathfrak{A}}$ are defined in the standard way <u>with the quantifiers $\forall$ and $\exists$ ranging over the individual variables only</u>.

If $\varphi(x_1, \ldots, x_n, S)$ is a formula of the language $\mathcal{L}^{\mathfrak{A}}$ in which S is an n-ary relation symbol occurring positively, then we can iterate φ to the transfinite and define <u>the fixed point</u> $\varphi^\infty = \bigcup_\xi \varphi^\xi$ of φ, where

$$\varphi^\xi = \{\bar{x} : \varphi(\bar{x}, \bigcup_{\eta<\xi} \varphi^\eta)\} .$$

A relation $R \subseteq A^m$ is <u>inductive on $\mathfrak{A}$</u> if there is a positive formula φ as above and a sequence $\bar{k}$ from ω such that $(R(\bar{y}) \Longleftrightarrow (\bar{k}, \bar{y}) \in \varphi^\infty)$. A relation $R \subseteq A^m$ is <u>hyperelementary on $\mathfrak{A}$</u> if both R and $A^m - R$ are inductive on $\mathfrak{A}$.

It is easy to show that the inductive relations on $\mathfrak{A}$ coincide with the semirecursive in $\mathbf{E}^{\#}$ relations, so that

$$\mathrm{ENV}[\mathbf{E}^{\#}] = \mathrm{IND} = \text{all inductive relations on } \mathfrak{A}$$
$$\mathrm{SEC}[\mathbf{E}^{\#}] = \mathrm{HYP} = \text{all hyperelementary relations on } \mathfrak{A}.$$

1.4. Kleene [1959a] and Spector [1961] established that on <u>the structure of arithmetic</u> $\mathbb{N} = \langle \omega, +, \cdot \rangle$

$$\mathrm{ENV}[\mathbf{E}] = \mathrm{ENV}[\mathbf{E}^{\#}] = \Pi^1_1 .$$

We see therefore that over the integers recursion in $\mathbf{E}$ coincides with positive elementary induction. The picture changes completely on the structure of analysis $\mathbb{R} = \langle\omega \cup \omega^{\omega}, +, \cdot, Ap\rangle$ (here $Ap(\alpha, n) = \alpha(n)$), where as Moschovakis [1967] showed

$$\mathrm{ENV}[\mathbf{E}] \subsetneq \mathrm{ENV}[\mathbf{E}^{\#}] \subsetneq \Pi^1_1 .$$

The preceding results led to the study of recursion in $\mathbf{E}$ and positive elementary induction (i.e., recursion in $\mathbf{E}^{\#}$) on abstract structures, as natural extensions of the hyperarithmetic theory on the integers. At the same time they also gave rise to the program whose goal is to understand the similarities and the differences between recursion in $\mathbf{E}$ and positive elementary induction.

1.5. We conclude this section by mentioning that both recursion in $\mathbf{E}$ and recursion in $\mathbf{E}^{\#}$ on a structure $\mathfrak{A}$ can be characterized in terms of the theory of set recursion or E-recursion, which is a recursion theory on the universe of sets introduced by Normann [1978]. The exact connection between these theories is exhibited by the following results:

i) a partial function $\varphi: A^k \to \omega$ is recursive in $\mathbf{E}$ if and only if it is E-recursive using the structure $\mathfrak{A}$ as a constant;

ii) a partial function $\varphi: A^k \to \omega$ is recursive in $\mathbf{E}^{\#}$ if and only if it is E-recursive relative to $\mathbf{E}^{\#}$ using the structure $\mathfrak{A}$ as a constant.

These characterizations can also serve as alternative definitions of recursion in $\mathbf{E}$ and recursion in $\mathbf{E}^{\#}$ on a structure $\mathfrak{A}$.

§2. Model Theoretic and Invariant Definability Characterizations of Recursion in $\mathbf{E}^{\#}$

A. The Classical Theory

2.1. Grzegorczyk, Mostowski and Ryll-Nardzewski [1958] proved that the hyperarithmetic relations on the integers are exactly the invariantly definable ones over all ω-models of analysis by arbitrary formulas of second order number theory. This means that a relation $R \subseteq \omega^k$ is hyperarithmetic if and only if there is a formula of the second order language of the structure $\mathbb{N} = \langle\omega, +, \cdot\rangle$ which defines R on every ω-model of analysis.

Kleene [1959b] characterized the collection HYP of the hyperarithmetic relations on $\mathbb{N} = \langle\omega, +, \cdot\rangle$ as the smallest model of the Schema of Σ^1_1-Comprehension with basis, i.e., HYP is the smallest class Δ of relations on the integers such that for any relation $R \subseteq \omega^k$ and any formula $\varphi(Y, \bar{y})$ of the first order language $\mathcal{L}^{\mathbb{N}}$ of the structure of arithmetic

if $$(R(\bar{y}) \iff (\exists Y \in \Delta)\varphi(Y, \bar{y}) \iff (\exists Y)\varphi(Y, \bar{y})) ,$$
then $$R \in \Delta .$$

However, in that paper Kleene established much more. He related the hyperarithmetic sets to the ramified analytic ones by showing that the ramified analytic hierarchy up to level ω_1^{ck} consists precisely of the hyperarithmetic sets. Moreover, all hyperarithmetic sets can be obtained by using a single existential or a single universal second order quantifier applied to the hyperarithmetic sets of lower levels as basis. From these results it follows also that the class HYP of the hyperarithmetic relations is the smallest model of the Schema of Δ_1^1-Comprehension, i.e., HYP is the smallest class Δ of relations on the integers such that for any relation $R \subseteq \omega^k$ and any formulas $\varphi(Y, \bar{y})$, $\psi(Y, \bar{y})$ of the first order language $\mathcal{L}^{\mathbb{N}}$ of the structure of arithmetic

if $$(R(\bar{y}) \iff (\exists Y \in \Delta)\varphi(Y, \bar{y}) \iff (\forall Y \in \Delta)\psi(Y, \bar{y}))\ ,$$
then $$R \in \Delta\ .$$

It should be pointed out that the preceding characterization of HYP in terms of the Δ_1^1-Comprehension Schema is only implicit in Kleene [1959b]. It is however explicitly stated and proved in Kreisel [1961], where in addition he characterized the hyperarithmetic relations as the ones which are Δ_1^1 invariantly definable over all models of the Δ_1^1-Comprehension Schema.

B. The Theory of Recursion in $E^\#$ on Abstract Structures

2.2. Moschovakis [1969a], [1969b], [1969c] studied the theory of recursion in $E^\#$ on almost arbitrary structures and showed that it is a successful generalization of the hyperarithmetic theory on the integers. A comprehensive account of this work is contained in the monograph Moschovakis [EIAS], where the theory is developed in the context of inductive definability.

The problem of finding model theoretic and invariant definability characterizations of recursion in $E^\#$ is attacked in Moschovakis [1969c]. It turns out that both the invariant definability characterization of Grzegorczyk, Mostowski, Ryll-Nardzewski and Kleene's characterization in terms of the Σ_1^1-Comprehension Schema fail to generalize to abstract structures. In particular, these results do not hold on the structure of analysis, as well as on any structure on which the notion of wellfoundedness is first-order. On the other hand, Moschovakis [1969c], [EIAS] established that the "boldface" recursive in $E^\#$ relations on an acceptable structure can be characterized in terms of the Δ_1^1-Comprehension Schema.

A structure $\mathfrak{U}$ is acceptable if, roughly speaking, it possesses a first-order coding machinery, so that the notions of finite sequence, length of finite sequence etc. can be coded in a first-order way. Most structures of recursion or set theoretic interest are acceptable, for example the structure of arithmetic, the structure of analysis, and for each ordinal λ the structure $\mathbf{V}_\lambda = \langle V_\lambda, \in \rangle$. For such structures we have then the following model theoretic characterization of the class $\underset{\sim}{\mathbf{SEC}}[\mathbf{E}^{\#}] = \bigcup\{SEC[\mathbf{E}^{\#}, \bar{x}]: \bar{x} \in A^{<\omega}\}$ of the "boldface" recursive in $\mathbf{E}^{\#}$ relations (i.e., the "boldface" hyperelementary relations on $\mathfrak{A}$).

2.3. Theorem (Moschovakis). If $\mathfrak{A}$ is an acceptable structure, then the class $\underset{\sim}{\mathbf{SEC}}[\mathbf{E}^{\#}]$ of the "boldface" recursive in $\mathbf{E}^{\#}$ relations is the smallest model of the Δ^1_1-Comprehension Schema with parameters, i.e., it is the smallest collection Δ of relations on A such that for any relation R on A, any formulas $\varphi(Y, \bar{Z}, \bar{y})$, $\psi(Y, \bar{Z}, \bar{y})$ of the first order language $\underset{\sim}{\mathcal{L}}^{\mathfrak{A}}$ of the structure $\mathfrak{A}$ and any relations $\bar{S}$ in Δ

if $\qquad (R(\bar{y}) \iff (\exists Y \in \Delta)\varphi(Y, \bar{S}, \bar{y}) \iff (\forall Y \in \Delta)\psi(Y, \bar{S}, \bar{y}))$,

then $\qquad R \in \Delta$. ⊣

In addition to the above Theorem 2.3 Moschovakis [1969c], [EIAS] obtained the following invariant definability result about recursion in $\mathbf{E}^{\#}$.

2.4. Theorem. If $\mathfrak{A}$ is an acceptable structure, then a relation R is "boldface" recursive in $\mathbf{E}^{\#}$ if and only if it is Δ^1_1 invariantly definable over all models of the Δ^1_1-Comprehension Schema with parameters, i.e., if and only if there are formulas $\varphi(Y, \bar{y})$, $\psi(Y, \bar{y})$ of the first order language $\underset{\sim}{\mathcal{L}}^{\mathfrak{A}}$ of the structure $\mathfrak{A}$ such that for every model Δ of the Δ^1_1-Comprehension Schema with parameters

$$R(\bar{y}) \iff (\exists Y \in \Delta)\varphi(Y, \bar{y}) \iff (\forall Y \in \Delta)\psi(Y, \bar{y}) .$$ ⊣

2.5. The preceding Theorems 2.3 and 2.4 are proved with substantially different methods from the ones used for the classical results in the hyperarithmetic theory. This is due to the fact that the classical proofs used heavily the hierarchy $\langle H_\alpha, \alpha < \omega_1^{ck}\rangle$ of the hyperarithmetic sets on the integers, while there is no such analog for the recursive in $\mathbf{E}^{\#}$ relations on an acceptable structure in general. In order to overcome this difficulty Moschovakis [1969c], [EIAS] had to bring out and analyze the second order properties of positive elementary induction. He developed in particular certain technical tools, such as the second stage comparison theorem, which link recursion in $\mathbf{E}^{\#}$ with second order definability. As an outcome of this investigation he discovered also a second order hierarchy for recursion in $\mathbf{E}^{\#}$ on an acceptable structure $\mathfrak{A}$, which he related to the ramified second order hierarchy over $\mathfrak{A}$.

Let $\mathfrak{A} = \langle A, R_1, \ldots, R_n, f_1, \ldots, f_m, c_1, \ldots, c_\ell\rangle$ be a structure with $\omega \subseteq A$, let $\underset{\sim}{\mathcal{L}}^{\mathfrak{A}}$ be the "boldface" first order language of $\mathfrak{A}$ (i.e., there is a constant symbol $\mathbf{a}$ for each $a \in A$) and let Δ be a collection of relations on A. We say that a relation $R \subseteq A^s$ is Σ^1_1-definable with basis Δ and parameters from Δ if there is a formula $\varphi(Y, Z_1, \ldots, Z_k, \bar{y})$ of the language $\underset{\sim}{\mathcal{L}}^{\mathfrak{A}}$ and relations $S_1, \ldots, S_k$ in Δ such that

$$R(\bar{y}) \iff (\exists Y \in \Delta)\varphi(Y, S_1, \ldots, S_k, \bar{y}) \iff (\exists Y)\varphi(Y, S_1, \ldots, S_k, \bar{y}) .$$

We say that $R \subseteq A^s$ is <u>Δ^1_1-definable with basis Δ and parameters from Δ</u> if both R and $A^s - R$ are Σ^1_1-definable with basis Δ and parameter from Δ. In terms of these notions Moschovakis [1969c], [EIAS] established the following hierarchy result about the recursive in $\mathbf{E}^\#$ relations on an acceptable structure.

2.6. Theorem. Let $\mathfrak{A} = \langle A, R_1, \ldots, R_n, f_1, \ldots, f_m, c_1, \ldots, c_\ell \rangle$ be an acceptable structure. For each ordinal ξ define the class of relations $\mathcal{B}^\xi$ by the induction

$\mathcal{B}^0$ = the class of first order definable relations on $\mathfrak{A}$,

$\mathcal{B}^\xi$ = the class of the Δ^1_1- definable relations with basis $\bigcup_{\eta<\xi} \mathcal{B}^\eta$ and parameters from $\bigcup_{\eta<\xi} \mathcal{B}^\eta$.

Then a relation R is "boldface" recursive in $\mathbf{E}^\#$ if and only if there is an ordinal ξ such that $R \in \mathcal{B}^\xi$, i.e.,

$$\underset{\sim}{\text{SEC}}[\mathbf{E}^\#] = \bigcup_\xi \mathcal{B}^\xi. \qquad \dashv$$

The above Theorem 2.6 is the key result for proving the preceding model theoretic and invariant definability characterizations of recursion in $\mathbf{E}^\#$. In addition it gives a level by level construction "from below" of the "boldface" recursive in $\mathbf{E}^\#$ relations on an acceptable structure, which is of particular significance especially in the absence of a hierarchy analogous to the $\langle H_\alpha, \alpha < \omega_1^{ck} \rangle$ one for the hyperarithmetic sets.

§3. Model Theoretic and Invariant Definability Characterizations of Recursion in E

3.1. The study of recursion in $\mathbf{E}$ on abstract structures was pursued mainly for two reasons. First it provides a natural extension of the hyperarithmetic theory, which in general is different from the one given by recursion in $\mathbf{E}^\#$. Second it is intimately connected with the theory of recursion in higher types, since recursion in $\mathbf{E}$ on the $Tp(n)$ structure coincides with Kleene recursion in the type $(n + 2)$ object ${}^{n+2}\mathbf{E}$.

In view of the model theoretic and invariant definability results about recursion in $\mathbf{E}^\#$, the question was raised if there are similar characterizations of recursion in $\mathbf{E}$ on abstract structures. Moreover, are there results in this direction which contribute to comparison of these two theories and explain their differences?

As the study of recursion in $\mathbf{E}$ progressed it was realized that in many cases the correct analogs for this theory are "lightface" versions of results about recursion in $\mathbf{E}^{\#}$. This is due to the fact that in general the semirecursive in $\mathbf{E}$ relations are not closed under existential quantification, so that one has to keep track of the first-order parameters involved and introduce "lightface" notions. In connection to our problem this observation suggests that one should try to characterize the collection

$$\mathcal{R} = \{\mathrm{SEC}[\mathbf{E}, \bar{x}]: \bar{x} \in A^{<\omega}\}$$

of the classes of the recursive in $\mathbf{E}$ relations from $\bar{x}$ (as $\bar{x}$ varies over all finite sequences from A), rather than the class

$$\underset{\sim}{\mathbf{SEC}}[\mathbf{E}] = \bigcup\{\mathrm{SEC}[\mathbf{E}, \bar{x}]: \bar{x} \in A^{<\omega}\}$$

of the "boldface" recursive in $\mathbf{E}$ relations.

3.2. The collection $\mathcal{R} = \{\mathrm{SEC}[\mathbf{E}, \bar{x}]: \bar{x} \in A^{<\omega}\}$ possesses certain combinatorial properties which now we turn into a definition. This definition should be thought of as capturing the combinatorial properties of relativization to a finite sequence.

Let $\mathfrak{A} = \langle A, R_1, \ldots, R_n, f_1, \ldots, f_m, c_1, \ldots, c_\ell \rangle$ be a structure such that $\omega \subseteq A$. An <u>indexed family on $\mathfrak{A}$</u> is a collection $\mathcal{L} = \{\Lambda(\bar{x}): \bar{x} \in A^{<\omega}\}$ of nonempty classes of relations on A with the following three properties:

i) If $\bar{x} \in A^{<\omega}$ and $\bar{k} \in \omega^{<\omega}$, then $\Lambda(\bar{x}, \bar{k}) = \Lambda(\bar{x})$.

ii) If $\bar{x} = (x_1, \ldots, x_k)$, $\bar{y} = (y_1, \ldots, y_\ell)$ are finite sequences from A and $\{y_j: j = 1, 2, \ldots, \ell\} \subseteq \{x_i: i = 1, 2, \ldots, k\}$, then $\Lambda(\bar{y}) \subseteq \Lambda(\bar{x})$.

iii) If $\bar{x} = (x_1, \ldots, x_k)$, $\bar{y} = (y_1, \ldots, y_\ell)$ are finite sequences from A and $R \subseteq A^{\ell+s}$ is an element of $\Lambda(\bar{x})$, then the relation $R_{\bar{y}} = \{\bar{z}: R(\bar{y}, \bar{z})\}$ is an element of $\Lambda(\bar{x}, \bar{y})$.

The typical examples of indexed families we have in mind are of course the collections

$$\mathcal{R} = \{\mathrm{SEC}[\mathbf{E}, \bar{x}]: \bar{x} \in A^{<\omega}\} \text{ and } \mathcal{H} = \{\mathrm{SEC}[\mathbf{E}^{\#}, \bar{x}]: \bar{x} \in A^{<\omega}\} .$$

Let $\mathcal{L} = \{\Lambda(\bar{x}): \bar{x} \in A^{<\omega}\}$ be an indexed family on $\mathfrak{A}$. We say that $\mathcal{L}$ is a model of the <u>Δ^1_1-Comprehension Schema without parameters</u> if for any relation R on A, any $\bar{x} \in A^{<\omega}$ and any formulas $\varphi(Y, \bar{u}, \bar{v})$, $\psi(Y, \bar{u}, \bar{v})$ of the "lightface" first order language $\mathcal{L}^{\mathfrak{A}}$ of the structure $\mathfrak{A}$

if $\quad (R(\bar{y}) \iff (\exists Y \in \Lambda(\bar{x}, \bar{y}))\varphi(Y, \bar{x}, \bar{y}) \iff (\forall Y \in \Lambda(\bar{x}, \bar{y}))\psi(Y, \bar{x}, \bar{y}))$,

then $\quad R \in \Lambda(\bar{x})$.

We say that the indexed family $\mathcal{L} = \{\Lambda(\bar{x}): \bar{x} \in A^{<\omega}\}$ is a model of the <u>Δ^1_1-Comprehension Schema with parameters</u> if for any relation R on A, any $\bar{x} \in A^{<\omega}$, any formulas $\varphi(Y, \bar{Z}, \bar{u}, \bar{v})$, $\psi(Y, \bar{Z}, \bar{u}, \bar{v})$ of the language $\mathcal{L}^{\mathfrak{A}}$ and any relations $\bar{S}$ in $\Lambda(\bar{x})$

if $\quad (R(\bar{y}) \iff (\exists Y \in \Lambda(\bar{x}, \bar{y}))\varphi(Y, \bar{S}, \bar{x}, \bar{y}) \iff (\forall Y \in \Lambda(\bar{x}, \bar{y}))\psi(Y, \bar{S}, \bar{x}, \bar{y}))$,

then $\quad R \in \Lambda(\bar{x})$.

If $\mathfrak{A}$ is an acceptable structure, then one can show that both the indexed families

$$\mathcal{R} = \{\mathrm{SEC}[\mathbf{E}, \bar{x}]: \bar{x} \in A^{<\omega}\} \quad \text{and} \quad \mathcal{H} = \{\mathrm{SEC}[\mathbf{E}^{\#}, \bar{x}]: \bar{x} \in A^{<\omega}\}$$

satisfy the Δ_1^1-Comprehension Schema with parameters and hence a fortiori the Δ_1^1-Comprehension Schema without parameters. This result should be contrasted with the fact that the class $\mathbf{SEC}[\mathbf{E}]$ of the "boldface" recursive in $\mathbf{E}$ relations is not in general a model of the Δ_1^1-Comprehension Schema. We see therefore that the above schemata for indexed families are finer "lightface" versions of the Δ_1^1-Comprehension Schema for a class of relations. The main theorem in this section is a characterization of the indexed family $\mathcal{R} = \{\mathrm{SEC}[\mathbf{E}, \bar{x}]: \bar{x} \in A^{<\omega}\}$ on an acceptable structure $\mathfrak{A}$ in terms of these schemata. Moreover, the assumption that the structure is acceptable is used only in one direction, so that we actually establish a result for arbitrary structures.

3.3. <u>Theorem</u>. Let $\mathfrak{A} = \langle A, R_1, \ldots, R_n, f_1, \ldots, f_m, c_1, \ldots, c_\ell\rangle$ be a structure such that $\omega \subseteq A$.

i) The indexed family $\mathcal{R} = \{\mathrm{SEC}[\mathbf{E}, \bar{x}]: \bar{x} \in A^{<\omega}\}$ of the recursive in $\mathbf{E}$ relations is contained in any indexed family which is a model of the Δ_1^1-Comprehension Schema without parameters, i.e., if $\mathcal{A} = \{\Lambda(\bar{x}): \bar{x} \in A^{<\omega}\}$ is any other such family, then $\mathrm{SEC}[\mathbf{E}, \bar{x}] \subseteq \Lambda(\bar{x})$ for every $\bar{x} \in A^{<\omega}$.

ii) If in addition the structure $\mathfrak{A}$ is acceptable, then the indexed family $\mathcal{R} = \{\mathrm{SEC}[\mathbf{E}, \bar{x}]: \bar{x} \in A^{<\omega}\}$ is the smallest model of the Δ_1^1-Comprehension Schema without (or with) parameters. ⊣

The above Theorem 3.3 provides a model theoretic characterization of recursion in $\mathbf{E}$ which is the analog of the characterization for recursion in $\mathbf{E}^{\#}$ given by Theorem 2.3. However, one difference between these two results is that here we deal with indexed families, while Theorem 2.3 refers to classes of relations. We will return to this point in the next section, where we will restore the analogy by proving a model theoretic characterization of the indexed family $\mathcal{H} = \{\mathrm{SEC}[\mathbf{E}^{\#}, \bar{x}]: \bar{x} \in A^{<\omega}\}$.

In the rest of this section we will complete the picture for recursion in $\mathbf{E}$ and make a few comments about the proofs of the results. The next theorem is an invariant definability characterization of recursion in $\mathbf{E}$ on acceptable structures and it should be compared with the characterization of recursion in $\mathbf{E}^{\#}$ provided by Theorem 2.4.

3.4. <u>Theorem</u>. Let $\mathfrak{A} = \langle A, R_1, \ldots, R_n, f_1, \ldots, f_m, c_1, \ldots, c_\ell\rangle$ be a structure such that $\omega \subseteq A$ and let $\bar{z}$ be a finite sequence from A.

i) If R is a relation recursive in $\mathbf{E}$ from $\bar{z}$, then R is Δ^1_1 invariantly definable from $\bar{z}$ over all indexed families which are models of the Δ^1_1-Comprehension Schema without (or with) parameters, i.e., there are formulas $\varphi(Y, \bar{u}, \bar{v})$, $\psi(Y, \bar{u}, \bar{v})$ of the language $\mathcal{L}^{\mathfrak{A}}$ such that for every indexed family $\mathcal{A} = \{\Lambda(\bar{x}) : \bar{x} \in A^{<\omega}\}$ which is a model of the Δ^1_1-Comprehension Schema without (or with) parameters

$$R(\bar{y}) \iff (\exists Y \in \Lambda(\bar{y}, \bar{z}))\varphi(Y, \bar{y}, \bar{z}) \iff (\forall Y \in \Lambda(\bar{y}, \bar{z}))\psi(Y, \bar{y}, \bar{z}) .$$

ii) If in addition the structure $\mathfrak{A}$ is acceptable, then a relation R is recursive in $\mathbf{E}$ from $\bar{z}$ if and only if it is Δ^1_1 invariantly definable from $\bar{z}$ over all indexed families which are models of the Δ^1_1-Comprehension Schema without (or with) parameters. ⊣

3.5. In the special case of recursion in the type $(n + 2)$ object ${}^{n+2}\mathbf{E}$ the preceding characterizations in terms of the Δ^1_1-Comprehension Schema are due to MacQueen [1972]. However, in his proofs MacQueen [1972] used in a crucial way the fact that the notion of wellfoundedness is first-order on the higher types structure. This of course is not true for arbitrary structures and therefore a different technical argument is needed in the abstract setting.

In establishing the preceding Theorems 3.3 and 3.4 we approach recursion in $\mathbf{E}$ as a branch of the general theory of inductive definability, according to the program introduced by Moschovakis [1977] and developed in Kechris-Moschovakis [1977]. One of the main technical tools we use in our proofs is the second stage comparison theorem for recursion in normal functionals of Kolaitis [1979]. This result extends the corresponding one of Moschovakis [1969c[, [EIAS] for positive elementary induction and provides the necessary link between recursion in $\mathbf{E}$ and second order definability. Another technical tool we use in the proofs is a new normal form theorem for functional induction the details of which will appear elsewhere. We should mention here that this theorem gives analogs for functional induction of the Kleene master recursion even when no coding machinery is available.

Both Moschovakis [1969c], [EIAS] and MacQueen [1972] considered only Δ^1_1-Comprehension Schemata with parameters. Here we have made a distinction between the Δ^1_1-Comprehension Schema without parameters and the one with parameters. The reason for this distinction is that in general the first is weaker than the second, although they have the same smallest model. In fact, even over the integers it is not very hard to construct by a diagonal argument a non-hyperarithmetic real α such that the class $\Delta = \mathrm{HYP} \cup \{\alpha\}$ is a model of the Δ^1_1-Comprehension Schema without parameters, but it is not a model of the Δ^1_1-Comprehension Schema with parameters.

Finally, we should point out that recursion in $\mathbf{E}$ on an abstract structure does not possess a hierarchy similar to the $\langle H_\alpha, \alpha < \omega_1^{ck} \rangle$ one for the hyperarithmetic sets. The preceding Theorems 3.3 and 3.4 are proved by establishing first a second order hierarchy result for recursion in $\mathbf{E}$ which is the "lightface" analog without parameters of Moschovakis' Theorem 2.6 in the previous section.

Let $\mathfrak{A} = \langle A, R_1, \ldots, R_n, f_1, \ldots, f_m, c_1, \ldots, c_\ell \rangle$ be a structure such that $\omega \subseteq A$, let $\mathcal{A} = \{\Lambda(\bar{x}): \bar{x} \in A^{<\omega}\}$ be an indexed family on $\mathfrak{A}$ and let $\bar{z}$ be a finite sequence from A. We say that a relation $R \subseteq A^k$ is Σ^1_1-definable from $\bar{z}$ with basis the indexed family $\mathcal{A}$ if there is a formula $\varphi(Y, \bar{u}, \bar{v})$ of the language $\mathcal{L}^{\mathfrak{A}}$ such that

$$R(\bar{y}) \iff (\exists Y \in \Lambda(\bar{y}, \bar{z}))\varphi(Y, \bar{y}, \bar{z}) \iff (\exists Y)\varphi(Y, \bar{y}, \bar{z}) .$$

We say that $R \subseteq A^k$ is Δ^1_1-definable from $\bar{z}$ with basis the indexed family $\mathcal{A}$ if both R and $A^k - R$ are Σ^1_1-definable from $\bar{z}$ with basis $\mathcal{A}$.

3.6. Theorem. Let $\mathfrak{A} = \langle A, R_1, \ldots, R_n, f_1, \ldots, f_m, c_1, \ldots, c_\ell \rangle$ be a structure such that $\omega \subseteq A$. For each $\bar{z} \in A^{<\omega}$ and for each ordinal ξ define the class of relations $\mathcal{D}^\xi(\mathbf{E}, \bar{z})$ by the induction

$$\mathcal{D}^0(\mathbf{E}, \bar{z}) = \text{the class of first order definable from } \bar{z} \text{ relations on } \mathfrak{A},$$
$$\mathcal{D}^\xi(\mathbf{E}, \bar{z}) = \text{the class of the } \Delta^1_1\text{-definable relations from } \bar{z} \text{ with basis the indexed family } \mathcal{D}^{<\xi}(\mathbf{E}) = \{ \bigcup_{\eta<\xi} \mathcal{D}^\eta(\mathbf{E}, \bar{x}): \bar{x} \in A^{<\omega}\} .$$

Then the indexed family $\mathcal{R} = \{\mathrm{SEC}[\mathbf{E}, \bar{x}]: \bar{x} \in A^{<\omega}\}$ is contained in the indexed family $\mathcal{D}(\mathbf{E}) = \{\bigcup_\xi \mathcal{D}^\xi(\mathbf{E}, \bar{x}): \bar{x} \in A^{<\omega}\}$. If in addition the structure $\mathfrak{A}$ is acceptable, then for every $\bar{z} \in A^{<\omega}$

$$\mathrm{SEC}[\mathbf{E}, \bar{z}] = \bigcup_\xi \mathcal{D}^\xi(\mathbf{E}, \bar{z}) .$$

⊣

§4. On the Difference between Recursion in $\mathbf{E}$ and Recursion in $\mathbf{E}^\#$

4.1. The results of Moschovakis in §2 characterized the class $\underset{\sim}{\mathbf{SEC}}[\mathbf{E}^\#]$ of the "boldface" recursive in $\mathbf{E}^\#$ relations, while in §3 we characterized the indexed family $\mathcal{R} = \{\mathrm{SEC}[\mathbf{E}, \bar{x}]: \bar{x} \in A^{<\omega}\}$. Here we restore the analogy by obtaining results about the indexed family $\mathcal{H} = \{\mathrm{SEC}[\mathbf{E}^\#, \bar{x}]: \bar{x} \in A^{<\omega}\}$. These results imply the "boldface" characterizations of Moschovakis and shed more light on the difference between recursion in $\mathbf{E}$ and recursion in $\mathbf{E}^\#$ on an abstract structure.

4.2. Let $\mathcal{A} = \{\Lambda(\bar{x}): \bar{x} \in A^{<\omega}\}$ be an indexed family on the structure $\mathfrak{A}$. We say that $\mathcal{A}$ is a model of the $\Delta^1_1(\exists)$-Comprehension Schema without parameters if for any relation R on A, any $\bar{x} \in A^{<\omega}$ and any formulas $\varphi(Y, \bar{u}, y, \bar{v})$, $\psi(Y, \bar{u}, y, \bar{v})$ of the language $\mathcal{L}^{\mathfrak{A}}$ of the structure $\mathfrak{A}$

if $(R(\bar{w}) \iff (\exists y)(\exists Y \in \Lambda(\bar{x},y,\bar{w}))\varphi(Y,\bar{x},y,\bar{w}) \iff (\forall y)(\forall Y \in \Lambda(\bar{x},y,\bar{w}))\psi(Y,\bar{x},y,\bar{w}))$,

then $$R \in \Lambda(\bar{x}) .$$

In a similar way we can define also the *$\Delta^1_1(\exists)$-Comprehension Schema with parameters*.

If $\mathfrak{U}$ is an acceptable structure, then the indexed family $\mathcal{H}$ is a model of the $\Lambda^1_1(\exists)$ - Comprehension Schema with parameters and hence it is also a model of the one without parameters. On the other hand, the indexed family $\mathcal{R}$ is not always a model of the $\Delta^1_1(\exists)$-Comprehension Schema without parameters.

4.3. *Theorem*. Let $\mathfrak{U} = \langle A, R_1, \ldots, R_n, f_1, \ldots, f_m, c_1, \ldots, c_\ell \rangle$ be a structure such that $\omega \subseteq A$.

i) The indexed family $\mathcal{H} = \{SEC[\mathbf{E}^\#, \bar{x}]: \bar{x} \in A^{<\omega}\}$ of the recursive in $\mathbf{E}^\#$ relations is contained in any indexed family which is a model of the $\Delta^1_1(\exists)$-Comprehension Schema without parameters, i.e., if $\mathcal{L} = \{\Lambda(\bar{x}): \bar{x} \in A^{<\omega}\}$ is any other such family, then $SEC[\mathbf{E}^\#, \bar{x}] \subseteq \Lambda(\bar{x})$ for every $\bar{x} \in A^{<\omega}$.

ii) If in addition the structure $\mathfrak{U}$ is acceptable, then the indexed family $\mathcal{H} = \{SEC[\mathbf{E}^\#, \bar{x}]: \bar{x} \in A^{<\omega}\}$ is the smallest model of the $\Delta^1_1(\exists)$-Comprehension Schema without (or with) parameters. $\dashv$

We obtain next an invariant definability characterization of the indexed family $\mathcal{H} = \{SEC[\mathbf{E}^\#, \bar{x}]: \bar{x} \in A^{<\omega}\}$.

4.4. *Theorem*. Let $\mathfrak{U} = \langle A, R_1, \ldots, R_n, f_1, \ldots, f_m, c_1, \ldots, c_\ell \rangle$ be a structure such that $\omega \subseteq A$ and let $\bar{z}$ be a finite sequence from A.

i) If R is a relation recursive in $\mathbf{E}^\#$ from $\bar{z}$, then R is *$\Lambda^1_1(\exists)$ invariantly definable from $\bar{z}$* over all indexed families which are models of the $\Delta^1_1(\exists)$-Comprehension Schema without (or with) parameters, i.e., there are formulas $\varphi(Y, \bar{u}, y, \bar{v})$, $\psi(Y, \bar{u}, y, \bar{v})$ of the language $\mathcal{L}^{\mathfrak{U}}$ such that for every indexed family $\mathcal{L} = \{\Lambda(\bar{x}): \bar{x} \in A^{<\omega}\}$ which is a model of the $\Delta^1_1(\exists)$-Comprehension Schema without (or with) parameters

$$R(\bar{w}) \iff (\exists y)(\exists Y \in \Lambda(\bar{z}, y, \bar{w}))\varphi(Y, \bar{z}, y, \bar{w}) \iff (\forall y)(\forall Y \in \Lambda(\bar{z}, y, \bar{w}))\psi(Y, \bar{z}, y, \bar{w}).$$

ii) If in addition the structure $\mathfrak{U}$ is acceptable, then a relation R is recursive in $\mathbf{E}$ from $\bar{z}$ if and only if it is $\Delta^1_1(\exists)$ invariantly definable from $\bar{z}$ over all indexed families which are models of the $\Delta^1_1(\exists)$-Comprehension Schema without (or with) parameters. $\dashv$

4.5. It is quite easy to see that the above Theorems 4.3 and 4.4 about the indexed family $\mathcal{H} = \{SEC[\mathbf{E}^\#, \bar{x}]: \bar{x} \in A^{<\omega}\}$ imply immediately the Moschovakis characterizations 2.3 and 2.4 of the class $\underset{\sim}{SEC}[\mathbf{E}^\#]$. In addition to these results we can obtain also a hierarchy theorem for the indexed family $\mathcal{H}$, which gives as a corollary Moschovakis' hierarchy Theorem 2.6 for the class $\underset{\sim}{\text{SEC}}[E^{\#}]$.

Let $\mathfrak{A} = \langle A, R_1, \ldots, R_n, f_1, \ldots, f_m, c_1, \ldots, c_\ell \rangle$ be a structure such that $\omega \subseteq A$, let $\mathcal{L} = \{\Lambda(\bar{x}) : \bar{x} \in A^{<\omega}\}$ be an indexed family on $\mathfrak{A}$ and let $\bar{z}$ be a finite sequence from A. We say that a relation $R \subseteq A^k$ is $\Sigma^1_1(\exists)$-definable from $\bar{z}$ with basis the indexed family $\mathcal{L}$ if there is a formula $\varphi(Y, \bar{u}, y, \bar{v})$ of the language $\mathcal{L}^{\mathfrak{A}}$ such that

$$R(\bar{w}) \iff (\exists y)(\exists Y \in \Lambda(\bar{z}, y, \bar{w}))\varphi(Y, \bar{z}, y, \bar{w}) \iff (\exists y)(\exists Y)\varphi(Y, \bar{z}, y, \bar{w}) .$$

We say that $R \subseteq A^k$ is $\Delta^1_1(\exists)$-definable from $\bar{z}$ with basis the indexed family $\mathcal{L}$ if both R and $A^k - R$ are $\Sigma^1_1(\exists)$-definable from $\bar{z}$ with basis $\mathcal{L}$.

4.6. Theorem. Let $\mathfrak{A} = \langle A, R_1, \ldots, R_n, f_1, \ldots, f_m, c_1, \ldots, c_\ell \rangle$ be a structure such that $\omega \subseteq A$. For each $\bar{z} \in A^{<\omega}$ and for each ordinal ξ define the class of relations $\mathcal{D}^{\xi}(E^{\#}, \bar{z})$ by the induction

$\mathcal{D}^{0}(E^{\#}, \bar{z})$ = the class of first order definable from $\bar{z}$ relations on $\mathfrak{A}$.
$\mathcal{D}^{\xi}(E^{\#}, \bar{z})$ = the class of the $\Delta^1_1(\exists)$-definable relations from $\bar{z}$ with basis the indexed family $\mathcal{D}^{<\xi}(E^{\#}) = \{ \bigcup_{\eta<\xi} \mathcal{D}^{\eta}(E^{\#}, \bar{x}) : \bar{x} \in A^{<\omega}\}$.

Then the indexed family $\mathcal{R} = \{\text{SEC}[E^{\#}, \bar{x}] : \bar{x} \in A^{<\omega}\}$ is contained in the indexed family $\mathcal{D}(E^{\#}) = \{\bigcup_{\xi} \mathcal{D}^{\xi}(E^{\#}, \bar{x}) : \bar{x} \in A^{<\omega}\}$. If in addition the structure $\mathfrak{A}$ is acceptable, then for every $\bar{z} \in A^{<\omega}$

$$\text{SEC}[E^{\#}, \bar{z}] = \bigcup_{\xi} \mathcal{D}^{\xi}(E^{\#}, \bar{z}) . \qquad \dashv$$

4.7. The preceding Theorems 3.3, 3.4, 3.6 and 4.3, 4.4, 4.6 contribute to the comparison of recursion in **E** with recursion in $\mathbf{E}^{\#}$ and provide a way to "measure" the different strength of existential first-order quantification embodied by each of these theories.

These results show also that the Schemata of Δ^1_1-Comprehension and $\Delta^1_1(\exists)$-Comprehension for indexed families succeed in distinguishing recursion in **E** from recursion in $\mathbf{E}^{\#}$. One might consider in addition the "boldface" version of the $\Delta^1_1(\exists)$-Comprehension Schema for classes of relations and hope to characterize the class $\underset{\sim}{\text{SEC}}[\mathbf{E}]$ of the "boldface" recursive in **E** relations. It turns out however that the Schemata of Δ^1_1 and $\Delta^1_1(\exists)$-Comprehension for classes of relations coincide. In fact, if a class of relations Δ is a model of the Δ^1_1-Comprehension Schema, then it is also a model of the $\Delta^1_1(\exists)$-Comprehension Schema, because it satisfies the equivalence

$$(\exists y)(\exists Y \in \Delta)\varphi(Y, y, \bar{z}) \iff (\exists Y' \in \Delta)(\exists Y \in \Delta)(\exists y)(Y' = \{y\} \ \&\ \varphi(Y, y, \bar{z})) .$$

The above remarks make clear why it is necessary to consider "lightface" Δ^1_1-Comprehension Schemata for indexed families in order to characterize recursion in **E**. Moreover, they suggest the following

Problem. Is there a model theoretic or invariant definability characterization of the class $\mathbf{SEC}[\mathbf{E}] = \cup\{\mathrm{SEC}[\mathbf{E}, \bar{x}]: \bar{x} \in A^{<\omega}\}$ of the "boldface" recursive in **E** relations on an acceptable structure?

4.8. We conclude this paper by mentioning some further model theoretic characterizations of the indexed families

$$\mathcal{R} = \{\mathrm{SEC}[\mathbf{E}, \bar{x}]: \bar{x} \in A^{<\omega}\} \text{ and } \mathcal{H} = \{\mathrm{SEC}[\mathbf{E}^{\#}, \bar{x}]: \bar{x} \in A^{<\omega}\} .$$

Kreisel [1965] characterized the collection HYP of the hyperarithmetic relations on the integers as the smallest class of relations which is closed under recursive operations and satisfies the Schema of Σ^1_1-Choice. This result does not generalize to recursion in $\mathbf{E}^{\#}$ on abstract structures. However, Moschovakis [1969c], [EIAS] showed that the class $\mathbf{SEC}[\mathbf{E}^{\#}]$ of the "boldface" recursive in $\mathbf{E}^{\#}$ relations is the smallest model of the Schemata of Δ^0_∞-Comprehension and Σ^1_1-Collection. In Kolaitis [1979] we introduced "lightface" versions of these schemata and characterized the indexed family $\mathcal{R} = \{\mathrm{SEC}[\mathbf{E}, \bar{x}]: \bar{x} \in A^{<\omega}\}$. This characterization can be obtained now as a direct consequence of the preceding Theorem 3.3.

We say that an indexed family $\mathcal{L} = \{\Lambda(\bar{x}): \bar{x} \in A^{<\omega}\}$ on a structure $\mathfrak{A}$ is a model of the Δ^0_∞-Comprehension Schema if for any $\bar{x} \in A^{<\omega}$, any formula $\varphi(\bar{u}, \bar{y}, Y)$ of the language $\mathcal{L}^{\mathfrak{A}}$ and any relation S in $\Lambda(\bar{x})$ there is a relation R in $\Lambda(\bar{x})$ such that

$$R(\bar{y}) \iff \varphi(\bar{x}, \bar{y}, S) .$$

We say that the indexed family $\mathcal{L}$ satisfies the Σ^1_1-Collection Schema without parameters if for any $\bar{x} \in A^{<\omega}$ and any formula $\varphi(\bar{u}, \bar{y}, Y)$ of the language $\mathcal{L}^{\mathfrak{A}}$ we have that

$$(\forall \bar{y})(\exists Y \in \Lambda(\bar{x}, \bar{y}))\varphi(\bar{x}, \bar{y}, Y) \iff (\exists W \in \Lambda(\bar{x}))(\forall \bar{y})(\exists n \in \omega)\varphi(\bar{x}, \bar{y}, W_{\bar{y},n}) .$$

In a similar way we define also the Σ^1_1-Collection Schema with parameters.

4.9. Theorem. Let $\mathfrak{A} = \langle A, R_1, \ldots, R_n, f_1, \ldots, f_m, c_1, \ldots, c_\ell\rangle$ be a structure such that $\omega \subseteq A$.

i) The indexed family $\mathcal{R} = \{SEC[E, \bar{x}]: \bar{x} \in A^{<\omega}\}$ is contained in any indexed family which is a model of the Schemata of Δ^0_∞-Comprehension and Σ^1_1-Collection without parameters.

ii) If in addition the structure $\mathfrak{A}$ is acceptable, then the indexed family $\mathcal{R}$ is the smallest model of the Schemata of Δ^0_∞-Comprehension and Σ^1_1-Collection without (or with) parameters. ⊣

The above Theorem 4.9 is proved in Kolaitis [1979], but as we pointed out before it follows also from the preceding Theorem 3.3. Moreover, we can obtain a new characterization of the indexed family $\mathcal{H} = \{SEC[E^\#, \bar{x}]: \bar{x} \in A^{<\omega}\}$ by introducing the following schema.

We say that the indexed family $\mathcal{A} = \{\Lambda(\bar{x}): \bar{x} \in A^{<\omega}\}$ on $\mathfrak{A}$ is a model of the <u>$\Sigma^1_1(\exists)$-Collection Schema without parameters</u> if for any $\bar{x} \in A^{<\omega}$ and any formula $\varphi(\bar{u}, y, Y)$ of the language $\mathcal{L}^{\mathfrak{A}}$ we have that

$$(\exists y)(\exists Y \in \Lambda(\bar{x}, y))\varphi(\bar{x}, y, Y) \iff (\exists W \in \Lambda(\bar{x}))(\exists y)(\exists n \in \omega)\varphi(\bar{x}, y, W_{y,n}) .$$

In a similar way we can define the <u>$\Sigma^1_1(\exists)$-Collection Schema with parameters</u>.

4.10. <u>Theorem</u>. Let $\mathfrak{A} = \langle A, R_1, \ldots, R_n, f_1, \ldots, f_m, c_1, \ldots, c_\ell \rangle$ be a structure such that $\omega \subseteq A$

i) The indexed family $\mathcal{H} = \{SEC[E^\#, \bar{x}]: \bar{x} \in A^{<\omega}\}$ is contained in any indexed family which is a model of the Schemata of Δ^0_∞-Comprehension, Σ^1_1-Collection and $\Sigma^1_1(\exists)$-Collection.

ii) If in addition the structure $\mathfrak{A}$ is acceptable, then the indexed family $\mathcal{H}$ is the smallest model of the Schemata of Δ^0_∞-Comprehension, Σ^1_1-Collection and $\Sigma^1_1(\exists)$-Collection. ⊣

4.11. As a final remark we mention that the theorems of this paper about recursion in E and recursion in $E^\#$ can be extended to characterizations of recursion in generalized quantifiers. In particular, we can obtain model theoretic and invariant definability characterizations of recursion in the functionals F_{Q_1}, $F^{\wedge}_{Q_2}$, $F^\#_{Q_3}$, where Q_1, Q_2, Q_3 are generalized quantifiers. The precise statements and the proofs of these results will appear elsewhere.

REFERENCES

A. Grzegorczyk, A. Mostowski and C. Ryll-Nardzewski [1958], The classical and the ω-complete arithmetic, Journal of Symbolic Logic 23 (1958), 188-206.

A. S. Kechris and Y. N. Moschovakis [1977], Recursion in higher types, J. Barwise (ed.), Handbook of Mathematical Logic, 681-737, North-Holland (1977).

S. C. Kleene [1959a], Recursive functionals and quantifiers of finite type 1, Trans. Amer. Math. Soc. 91 (1959), 1-52.

S. C. Kleene [1959b], Quantification of number theoretic functions, Compositio Math. 14 (1959), 23-40.

Ph. G. Kolaitis [1978], On recursion in **E** and semi-Spector classes, A. S. Kechris-Y. N. Moschovakis (eds.), Cabal Seminar 76-77, 209-243, Lecture Notes in Mathematics 689, Springer-Verlag (1978).

Ph. G. Kolaitis [1979], Recursion in a quantifier vs. elementary induction, Journal of Symbolic Logic 44 (1979), 235-259.

G. Kreisel [1961], Set theoretic problems suggested by the notion of potential totality, Infinitistic Methods, 103-140, Pergamon (1961).

G. Kreisel [1965], The Axiom of choice and the class of hyperarithmetic functions, Indagationes Mathematicae 24 (1962), 307-319.

D. B. MacQueen [1972], Post's problem for recursion in higher types, Ph.D. Thesis, Massachusetts Institute of Technology (1972).

Y. N. Moschovakis [1967], Hyperanalytic predicates, Trans. Amer. Math. Soc. 129 (1967), 249-282.

Y. N. Moschovakis [1969a], Abstract first order computability I, Trans. Amer. Math. Soc. 138 (1969), 427-464.

Y. N. Moschovakis [1969b], Abstract first order computability II, Trans. Amer. Math. Soc. 138 (1969), 465-504.

Y. N. Moschovakis [1969c], Abstract computability and invariant definability, Journal of Symbolic Logic 34 (1969), 605-633.

Y. N. Moschovakis [EIAS], Elementary Induction on Abstract Structures, North Holland (1974).

Y. N. Moschovakis [1977], On the basic notions in the theory of induction, Butts and Hintikka (eds.), Logic, Foundations of Mathematics and Computability Theory, 207-236, Reidel (1977).

D. Normann [1978], Set Recursion, J. E. Fendstad, R. O. Gandy, G. E. Sacks (eds.), Generalized recursion theory II, 303-320, North Holland (1978).

C. Spector [1961], Inductively defined sets of natural numbers, Infinitistic methods, 97-102, Pergamon (1961).

$L_{\infty\omega_1}$-ELEMENTARILY EQUIVALENT

MODELS OF POWER ω_1

David W. Kueker*
Department of Mathematics
University of Maryland
College Park, MD 20742

0. INTRODUCTION.

A well-known result of D. Scott says that any two countable models which are $L_{\infty\omega}$-elementarily equivalent are isomorphic. M. Morley showed that the natural generalization of this result to uncountable cardinalities fails by constructing two models of power ω_1 which are $L_{\infty\omega_1}$-elementarily equivalent but not isomorphic (in fact, neither is embeddable in the other). Nevertheless, $L_{\infty\omega_1}$-elementary equivalence is a natural equivalence between models, as the standard back-and-forth criterion indicates. In this paper we investigate how much alike two $L_{\infty\omega_1}$-elementarily equivalent models of power ω_1 must be. Our main results give a characterization of $L_{\infty\omega_1}$-elementary equivalence of models of power ω_1 in terms of how the models are built up from below by countable models. This characterization seems intuitively more meaningful, and seems to yield more information, than the back-and-forth criterion alone. We derive, as a consequence, sufficient conditions for a model of power ω_1 to be embeddable in every model $L_{\infty\omega_1}$-elementarily equivalent to it. The specific examples of ω_1-like linear orders and free algebras are also dealt with in our framework.

Our results generalize to arbitrary uncountable regular κ in place of ω_1. These, and related material, will be considered in a future paper of the author.

We assume throughout that the underlying language L has just countably many non-logical symbols. The logic $L_{\infty\omega_1}$ allows conjunctions $\wedge\Phi$ and disjunctions $\vee\Phi$ of arbitrary sets Φ of formulas, and allows the simultaneous universal $\forall\vec{x}\varphi$ and existential $\exists\vec{x}\varphi$ quantification of a countable sequence $\vec{x} = \langle x_i \rangle_{i\in\omega}$ of variables. For its basic properties see [1, 8].

*Research partially supported by the NSF under Grant MCS 77-03993.

1. CHARACTERIZING INFINITARILY EQUIVALENT MODELS.

We view models of power ω_1 as built up from below by countable models. That is, if M is the union of a chain $\{M_\xi\}_{\xi<\omega_1}$ of countable submodels we think of the chain as constructing M. The sorts of characterizations we are after say that M and N are equivalent iff they are built up from similar chains of countable submodels.

As an example, we rephrase a result about $L_{\infty\omega}$ from [9]. First recall that a chain $\{M_\xi\}_{\xi<\omega_1}$ is <u>smooth</u> if $M_\eta = \bigcup_{\xi<\eta} M_\xi$ for every limit η.

PROPOSITION: If M and N have power ω_1 then $M \equiv_{\infty\omega} N$ iff M and N can be written as the unions of smooth chains $\{M_\xi\}_{\xi<\omega_1}$ and $\{N_\xi\}_{\xi<\omega_1}$ of countable submodels such that $M_\xi \cong N_\xi$ for all $\xi < \omega_1$.

It may not be immediately obvious why two models satisfying the criterion in the Proposition need not also be $L_{\infty\omega_1}$-elementarily equivalent. The following example is instructive.

EXAMPLE: Let the only non-logical symbol of L be a unary predicate P. Let M be the model of power ω_1 in which P is interpreted by a countably infinite set, and let N be the model of power ω_1 in which the interpretation of P is uncountable and co-uncountable. Then $M \equiv_{\infty\omega} N$, since they can be written as the unions of smooth chains $\{M_\xi\}_{\xi<\omega_1}$ and $\{N_\xi\}_{\xi<\omega_1}$ of countable models in each of which P is interpreted as an infinite and co-infinite set, hence $M_\xi \cong N_\xi$.

M and N are not $L_{\infty\omega_1}$-elementarily equivalent since "P is countable" is expressible by a sentence of $L_{\infty\omega_1}$. How is this reflected by the chains $\{M_\xi\}_{\xi<\omega_1}$ and $\{N_\xi\}_{\xi<\omega_1}$? For all sufficiently large ξ we will have $P^M = P^{M_\xi}$, so $M_{\xi+1}$ will contain no new points satisfying P; on the other hand $P^{N_\xi} \subsetneq P^N$ for all ξ so for arbitrarily large ξ $N_{\xi+1}$ will add new points satisfying P. Hence there are arbitrarily large ξ such that

$$(M_{\xi+1}, M_\xi) \not\cong (N_{\xi+1}, N_\xi),$$

that is, although $M_\xi \cong N_\xi$ and $M_{\xi+1} \cong N_{\xi+1}$ no isomorphism of M_ξ with N_ξ can be extended to an isomorphism of $M_{\xi+1}$ with $N_{\xi+1}$.

This example suggests that if M and N both have power ω_1 and

$M \equiv_{\infty\omega_1} N$ then M and N can be written as the unions of chains $\{M_\xi\}_{\xi<\omega_1}$ and $\{N_\xi\}_{\xi<\omega_1}$ of countable submodels such that

$$(M_\xi, M_\nu) \cong (N_\xi, N_\nu) \quad \text{whenever} \quad \nu < \xi < \omega_1.$$

This is true, but the condition is not sufficient. We need to know that "most" isomorphisms of M_ν with N_ν extend to isomorphisms of M_ξ with N_ξ whenever $\nu < \xi$. This is so we can take an isomorphism of M_{ξ_0} with N_{ξ_0}, extend it to an isomorphism of M_{ξ_1} with N_{ξ_1} for any $\xi_1 > \xi_0$, extend this extension to an isomorphism of M_{ξ_2} with N_{ξ_2} for any $\xi_2 > \xi_1$, etc. The precise statement corresponding to this is as follows.

THEOREM 1.1: Assume M and N both have power ω_1. Then $M \equiv_{\infty\omega_1} N$ iff M and N can be written as the unions of chains $\{M_\xi\}_{\xi<\omega_1}$ and $\{N_\xi\}_{\xi<\omega_1}$ of countable submodels such that

$$(M_\xi, M_\nu) \cong (N_\xi, N_\nu) \quad \text{whenever} \quad \nu < \xi < \omega_1,$$

and in fact there are non-empty sets G_ξ of isomorphisms of M_ξ with N_ξ, for all $\xi < \omega_1$, such that any isomorphism in G_ν extends to one in G_ξ for any $\nu < \xi < \omega_1$.

Note that the chains $\{M_\xi\}_{\xi<\omega_1}$ and $\{N_\xi\}_{\xi<\omega_1}$ are not assumed to to be smooth. If they were both smooth then we could conclude that $M \cong N$ provided G_ξ consisted of all isomorphisms of M_ξ and N_ξ, contradicting some known examples (cf. section 3).

The sufficiency of the condition is immediate from the standard back-and-forth criterion (see below). The hard direction is necessity, that any two $L_{\infty\omega_1}$-elementarily equivalent models of power ω_1 must be built up this way.

At the end of this section we will compare the condition in this theorem with the back-and-forth criterion and see what additional information this gives.

The statement of Theorem 1.1 would be neater if we could avoid reference to the sets G_ξ. This can be done in the case where G_ξ can be taken to be all isomorphisms of M_ξ and N_ξ. Such cases are fairly common and include some of the most important examples.

DEFINITION: (i) A chain $\{M_\xi\}_{\xi<\omega_1}$ of models is <u>good</u> if for all $\xi < \nu < \omega_1$ every automorphism of M_ξ extends to an automorphism of M_ν.

(ii) A model M of power ω_1 is <u>decomposable</u> if it can be written as the union of a smooth good chain of countable models.

THEOREM 1.2: Assume M and N both have power ω_1 and M is decomposable. Then the following are equivalent:

(i) $M \equiv_{\infty\omega_1} N$;

(ii) M and N can be written as the unions of chains $\{M_\xi\}_{\xi<\omega_1}$ and $\{N_\xi\}_{\xi<\omega_1}$ of countable models such that

$$(M_\xi, M_\nu) \cong (N_\xi, N_\nu) \quad \text{for all} \quad \nu < \xi < \omega_1 ,$$

and $\{M_\xi\}_{\xi<\omega_1}$ is a good chain;

(iii) M and N can be written as the unions of chains $\{M_\xi\}_{\xi<\omega_1}$ and $\{N_\xi\}_{\xi<\omega_1}$ of countable models such that

$$(M_\xi, M_\nu) \cong (N_\xi, N_\nu) \quad \text{for all} \quad \nu < \xi < \omega_1 ,$$

where the models M_ξ can be required to come from any smooth chain of countable models whose union is M.

In this paper we will only prove Theorem 1.2 and will leave the more involved case of Theorem 1.1 to a subsequent paper. The following two important examples show the applicability of Theorem 1.2.

EXAMPLE 1: Let $M = (M, \leq)$ be an ω_1-like linear order. Then M is decomposable since it can be written as the union of a smooth chain of countable initial segments, and any chain of initial segments is clearly good. Using (iii) of Theorem 1.2 we see easily that $M \equiv_{\infty\omega_1} N$ iff N is ω_1-like and there are increasing sequences $\langle m_\xi \rangle_{\xi<\omega_1}$ and $\langle n_\xi \rangle_{\xi<\omega_1}$ such that

$$(M^{<m_\xi}, m_\nu) \cong (N^{<n_\xi}, n_\nu) \quad \text{for all} \quad \nu < \xi < \omega_1 ,$$

where $M^{<m_\xi}$ is the submodel of M whose universe is $\{x \in M : x < m_\xi\}$.

EXAMPLE 2: Let $\mathbb{K}$ be some class of algebras having $\mathbb{K}$-free algebras on κ-generators for all non-zero $\kappa \leq \omega_1$ (see [3] for definitions). If N_0 and N_1 are both $\mathbb{K}$-free we write $N_0 \mid N_1$ if $N_0 \subseteq N_1$ and a set of free generators for N_0 can be extended to a set of free generators for N_1. Let M be K-free on ω_1-generators. Then M is decomposable since it can be written as the union of a smooth chain $\{M_\xi\}_{\xi<\omega_1}$ of countable free subalgebras such that $M_\nu \mid M_\xi$ whenever $\nu < \xi < \omega_1$, and clearly any chain under $\mid$ is good (if $\langle m_\xi\rangle_{\xi<\omega_1}$ freely generate M, let M_ξ be the subalgebra generated by $\langle m_\nu\rangle_{\nu<\xi}$). It follows from Theorem 1.2 that, if N has power ω_1, then $M \equiv_{\infty\omega_1} N$ iff N can be written as the union of a chain under $\mid$ of countable K-free algebras. This characterization was first obtained directly and announced by the author in [7]. Note that if K is the class of abelian groups, then $N_0 \mid N_1$ holds iff N_1 is free and N_0 is a direct summand of N_1; a characterization of abelian groups which are $L_{\infty\omega_1}$-elementarily equivalent to free abelian groups was also given in [2], see also the discussion below.

REMARKS: (1) Results similar to Theorems 1.1 and 1.2 hold even without the assumption that N has power ω_1. They will be discussed in a future paper.

(2) As mentioned in the introduction, versions of Theorems 1.1 and 1.2 hold for arbitrary uncountable regular κ in place of ω_1, and they also will be given in a future paper.

(3) The back-and-forth criterion for $L_{\infty\omega_1}$-elementary equivalence can be expressed as follows (see [8]): $M \equiv_{\infty\omega_1} N$ iff there is a non-empty set G of isomorphisms between countable submodels of M and N such that for any $g \in G$ and any countable $M_0 \subseteq M$, $N_0 \subseteq N$ there is some $g' \in G$ such that $g \subseteq g'$, $M_0 \subseteq \mathrm{dom}(g')$, and $N_0 \subseteq \mathrm{ran}(g')$. Thus the condition in Theorem 1.1 clearly implies $M \equiv_{\infty\omega_1} N$ (let $G = \bigcup_{\xi<\omega_1} G_\xi$). What Theorem 1.1 adds to the back-and-forth characterization is the additional information that the domains and ranges of the partial isomorphisms in G form chains of corresponding submodels.

To appreciate what this means, let us look at Example 2, the K-free algebras. It easily follows from the back-and-forth criterion that $N \equiv_{\infty\omega_1} M$ (where M is K-free on ω_1-generators) iff N is the ω_1-union of a set S of countable free subalgebras which is ω-directed under $\mid$ (i.e., every countable $N_0 \subseteq N$ is contained in some N_1 in S, and given finitely many $N_1,\dots,N_k$ in S there is some N' in S with $N_i \mid N'$ for all $i = 1,\dots,k$). Theorem 1.2 implies that, for N of power ω_1, the collection S can be taken as a chain under $\mid$ (more generally, that S is ω_1-directed under $\mid$). I see no way of getting this stronger form in general without considerable additional argument along the lines of the proof of Theorem 1.2. The stronger conclusion that S can be taken as a chain under $|$ is used in section 3 to show that the K-free algebra on ω_1-generators can be embedded in every algebra $L_{\infty\omega_1}$-elementarily equivalent to it. The characterization for abelian groups found in [2] is the weaker version.

2. THE PROOF OF THEOREM 1.2.

Clearly (iii) implies (ii) since M is decomposable and hence can be written as the union of a smooth good chain of countable models.

If (ii) holds then every isomorphism of M_ξ with N_ξ extends to an isomorphism of M_ν with N_ν for any $\nu > \xi$, since $\{M_\xi\}$ is a good chain. Thus the back-and-forth criterion yields (i).

The remainder of this section is devoted to showing that (i) implies (iii).

Let M have power ω_1 and let $\{M_\xi\}_{\xi<\omega_1}$ be any smooth chain of countable models whose union is M. For any $N \equiv_{\infty\omega_1} M$ of power ω_1 we need to find some I cofinal in ω_1 and a chain $\{N_\xi\}_{\xi\in I}$ whose union is N and such that

$$(M_\xi, M_\nu) \cong (N_\xi, N_\nu) \quad \text{whenever} \quad \nu, \xi \in I, \quad \nu < \xi.$$

If $N \equiv_{\infty\omega_1} M$ we want to define what it means for a countable $N^0 \subseteq N$ to be ξ-like in N; the intention being that N^0 is ξ-like in N if N^0 inside N looks like M_ξ inside M. Since almost all countable extensions of M_ξ inside M are M_μ's for $\mu > \xi$, we want N^0 to be ξ-like in N if almost all countable extensions N' of N^0 inside N are such that

$$(N', N^0) \cong (M_\mu, M_\xi) \quad \text{for} \quad \mu > \xi.$$

We therefore define a game $G(N, N^0, \xi)$ as follows: players (I) and (II) play alternately, each picking an element of N and a countable ordinal at every move; after ω moves we form the submodel N' of N generated by the elements of N chosen and the supremum μ of all the ordinals chosen; (II) wins if

$$(N', N^0) \cong (M_\mu, M_\xi).$$

We define N^0 is <u>ξ-like</u> in N to mean that (II) has a winning strategy for the game $G(N, N^0, \xi)$.

More concisely we could write that N^0 is ξ-like in N iff

$$(\forall x_0 \in N\ \forall \xi_0 < \omega_1\ \exists x_1 \in N\ \exists \xi_1 < \omega_1 \ldots)_{k<\omega}[(N \upharpoonright \{x_k : k \in \omega\}, N^0) \cong (M_{\bigcup_{k<\omega} \xi_k}, M_\xi)].$$

Certainly M_ξ is ξ-like in M. There will normally be other ξ-like submodels of M, but the following holds:

(1) If M^n is μ_n-like in M, for all $n \in \omega$, then there are arbitrarily large countable μ such that

$$(M_\mu, M^n) \cong (M_\mu, M_{\mu_n}) \quad \text{for every } n \in \omega.$$

PROOF: The second player has a winning strategy in each game $G(M, M^n, \mu_n)$. We will play all these games simultaneously with (II) using his winning strategies. Furthermore, we will have (I) play in the following way:

(i) in each game (I) plays every $x \in M$ and every ordinal ξ played in any of the other games;

(ii) in each game (I) plays every element of M_ξ for every ξ played;

(iii) in each game (I) plays large enough ordinals ξ so that every $x \in M$ played belongs to M_ξ for some ξ played.

Now (i) guarantees that we end up with the same submodel M' and the same ordinal μ in each game $G(M, M^n, \mu_n)$; (ii) and (iii) guarantee that $M' = M_\mu$. Since (II) plays his winning strategies we are guaranteed that

$$(M_\mu, M^n) \cong (M_\mu, M_{\mu_n}) \quad \text{for all } n \in \omega. \qquad \dashv$$

What makes being ξ-like suitable for our purposes is that it is preserved under $L_{\infty\omega_1}$-elementary equivalence.

(2) Assume that $(M, \vec{m}^0) \equiv_{\infty\omega_1} (N, \vec{n}^0)$ where $\vec{m}^0$ and $\vec{n}^0$ list the universes of countable submodels M^0 and N^0. Then M^0 is ξ-like in M iff N^0 is ξ-like in N.

PROOF: Say, for example, that M^0 is ξ-like in M. Then (II) can win $G(N, N^0, \xi)$ by taking the roles of both players in $G(M, M^0, \xi)$. In $G(N, N^0, \xi)$ let (I) choose $y_0 \in N$ and a countable ξ_0. Then in $G(M, M^0, \xi)$ we have (I) choose the same ξ_0 and some $x_0 \in M$ such that

$$(M, \vec{m}^0, x_0) \equiv_{\infty\omega_1} (N, \vec{n}^0, y_0);$$

(II) then uses the winning strategy in $G(M, M^0, \xi)$ to choose $x_1 \in M$ and ξ_1. (II)'s strategy in $G(N, N^0, \xi)$ is to pick the same ξ_1 and some $y_1 \in N$ so that

$$(M,\vec{m}^0,x_0,x_1) \equiv_{\infty\omega_1} (N,\vec{n}^0,y_0,y_1).$$

Thus, (II) always moves so that the plays in the two games are $L_{\infty\omega_1}$-elementarily equivalent. Since (II) wins $G(M,M^0,\xi)$ he will also win $G(N,N^0,\xi)$. (This argument is essentially in [6] where Keisler shows that certain sorts of game sentences are preserved by $L_{\infty\omega_1}$-elementary equivalence.) ⊣

With (2) we find that the following version of (1) holds in every $N \equiv_{\infty\omega_1} M$.

(3) Assume that $M \equiv_{\infty\omega_1} N$, and let N^n be μ_n-like in N for each $n \in \omega$. Then for any $y \in N$ and any countable μ' there are countable $\mu \geq \mu'$ and $N' \subseteq N$ such that N' is μ-like in N and

$$(N',N^n) \cong (M_\mu,M_{\mu_n}) \quad \text{for all} \quad n \in \omega.$$

PROOF: If $M = N$ then (3) is true by (1). For $M \equiv_{\infty\omega_1} N$ we use (2) to reduce the question back to M again. So, let $\vec{y}^n$ enumerate the universe of N^n, for each $n \in \omega$. We can find $\vec{x}^n$ and x in M such that

$$(M,x,\vec{x}^i)_{i<\omega} \equiv_{\infty\omega_1} (N,y,\vec{y}^i)_{i<\omega}.$$

If M^n is the submodel of M whose universe is listed by $\vec{x}^n$, then M^n is μ_n-like in M by (2). So by (1) there is some $\mu \geq \mu'$ such that $x \in M_\mu$ and

$$(M_\mu,M^n) \cong (M_\mu,M_{\mu_n}) \quad \text{for all} \quad n \in \omega.$$

If $\vec{x}^*$ lists all elements of M_μ then we can find $\vec{y}^*$ in N such that

$$(M,x,\vec{x}^*,\vec{x}^i)_{i<\omega} \equiv_{\infty\omega_1} (N,y,\vec{y}^*,\vec{y}^i)_{i<\omega}.$$

Then $\vec{y}^*$ lists the universe of some μ-like $N' \subseteq N$ which is clearly as desired. ⊣

Finally, let $N \equiv_{\infty\omega_1} M$ have power ω_1. By using (3) ω_1 times we can find a cofinal $I \subseteq \omega_1$ and a chain $\{N_\xi\}_{\xi\in I}$ of submodels of N such that N_ξ is ξ-like in N for each $\xi \in I$, N is the union of the chain, and

$$(N_\xi, N_\nu) \cong (M_\xi, M_\nu) \quad \text{whenever} \quad \xi, \nu \in I, \quad \nu < \xi.$$

This completes the proof of Theorem 1.2.

3. EMBEDDING AND CATEGORICITY.

In this section we first use the characterization from section 1 to obtain a simple sufficient condition for M to be embeddable in every model $L_{\infty\omega_1}$-elementarily equivalent to it; in fact the embedding will be $L_{\infty\omega_1}$-elementary for formulas with just finitely many free variables.

DEFINITION: (i) A function f with domain $\subseteq M$ and range $\subseteq N$ is <u>weak (∞,ω_1)-elementary on M into N</u> if

$$(M, a_0, \ldots, a_n) \equiv_{\infty\omega_1} (N, f(a_0), \ldots, f(a_n))$$

for any finite number of points $a_0, \ldots, a_n$ from the domain of f.

(ii) M of power ω_1 is (∞,ω_1)-<u>homogeneous</u> if M can be written as the union of a smooth chain $\{M_\xi\}_{\xi<\omega_1}$ of countable models such that any function f mapping M_ξ into M which is weak (∞,ω_1)-elementary on M can be extended to a function mapping all of M into M which is weak (∞,ω_1)-elementary on M.

We leave to the reader the verification that any ω_1-like dense linear order is (∞,ω_1)-homogeneous and also that any K-free algebra on ω_1 generators is (∞,ω_1)-homogeneous.

THEOREM 3.1: Let M of power ω_1 be (∞,ω_1)-homogeneous, and let $M \equiv_{\infty\omega_1} N$. Then M can be embedded in N by a function which is weak (∞,ω_1)-elementary on M into N; in fact N is the ω_1-union of such images of M.

PROOF: We assume that N also has power ω_1, for simplicity. Let $\{M_\xi\}_{\xi<\omega_1}$, $\{N_\xi\}_{\xi<\omega_1}$ and G_ξ, $\xi < \omega_1$, be as given by Theorem 1.1. In fact the proof of Theorem 1.1 shows (just as in 1.2) that the M_ξ's can be taken from any smooth chain whose union is M; since M is (∞,ω_1)-homogeneous we may assume the M_ξ's all come from a smooth chain whose models all have the property given in the definition of (∞,ω_1)-homogeneity. For limit ordinals λ let $M_{(\lambda)} = \bigcup_{\xi<\lambda} M_\xi$. Then $M_{(\lambda)} \subseteq M_\lambda$, but every $M_{(\lambda)}$ still comes from the chain showing that M is (∞,ω_1)-homogeneous. Thus $\{M_{(\lambda)}\}_{\text{limit}\,\lambda<\omega_1}$ is a smooth chain whose union is M and having the property given in the definition of

(∞,ω_1)-homogeneity.

Let $G_{(\lambda)}$ be the set of all isomorphisms of $M_{(\lambda)}$ into N_λ which are weak (∞,ω_1)-elementary on M into N. If $h_n \in G_{(\lambda_n)}$ for all $n \in \omega$ and $h_n \subseteq h_{n+1}$ for all n, then $\bigcup_{n\in\omega} h_n \in G_{(\lambda)}$ for $\lambda = \bigcup_{n\in\omega} \lambda_n$. Similarly, we see that $G_{(\omega)} \neq 0$. We will show that any $h_0 \in G_{(\lambda)}$ can be extended to some $h_1 \in G_{(\eta)}$ for some $\eta > \lambda$. This will establish the theorem since we will then be able to embed M into N by a function which is the union of functions from $G_{(\lambda)}$'s, hence is weak (∞,ω_1)-elementary on M into N.

So, let $h_0 \in G_{(\lambda)}$ be given. Pick any $g_0 \in G_\lambda$. Then $g_0^{-1} \circ h_0$ maps $M_{(\lambda)}$ into M_λ and is weak (∞,ω_1)-elementary on M, hence it can be extended to some $\bar{f}$ defined on all of M which is weak (∞,ω_1)-elementary on M. Since the $M_{(\lambda)}$'s form a smooth chain there is some $\eta > \lambda$ such that the restriction f of $\bar{f}$ to $M_{(\eta)}$ maps $M_{(\eta)}$ into itself. Let $g_1 \in G_\eta$ be such that $g_0 \subseteq g_1$ and let $h_1 = g_1 \circ f$. Then h_1 is as desired. ⊣

COROLLARY: If M is K-free on ω_1-generators then M can be embedded in any $N \equiv_{\infty\omega_1} M$ by a function which is weak (∞,ω_1)-elementary on M into N; in fact every countable subalgebra of N is contained in such an image of M.

Finally, we mention a theorem which (for special sorts of M) exactly characterizes when there is some N of power ω_1 such that $M \equiv_{\infty\omega_1} N$ and $M \not\cong N$. The proof of this, and several related results, will be given in a subsequent paper.

DEFINITION: (i) M of power ω_1 is <u>freely decomposable</u> if M can be written as the union of a smooth good chain $\{M_\xi\}_{\xi<\omega_1}$ of countable models such that

$$(M_\xi, M_\nu) \cong (M_1, M_0) \quad \text{whenever} \quad \nu < \xi < \omega_1.$$

(ii) Given a chain $\{M_\xi\}_{\xi<\omega_1}$ as in (i) we write $N^0 \subseteq^* N^1$ if

$$(N^1, N^0) \cong (M_1, M_0).$$

(iii) $N^0 \subseteq M$ is <u>good</u> if $N^0 \subseteq^* M_\xi$ holds for all sufficiently large $\xi < \omega_1$.

THEOREM 3.2: Let M of power ω_1 be freely decomposable. Then the following are equivalent:

(i) there is some N of power ω_1 such that $M \equiv_{\infty\omega_1} N$ and $M \not\cong N$;

(ii) there is some chain of length ω under $\subseteq^*$ of countable good submodels of M whose union is not a good submodel of M.

EXAMPLE 1: Any ω_1-like dense linear order is freely decomposable, and (ii) holds. We leave the verification of this to the reader. The conclusion that there are $L_{\infty\omega_1}$-elementarily equivalent non-isomorphic ω_1-like dense linear orders was previously known; see [11].

EXAMPLE 2: Any K-free algebra on ω_1 generators is freely decomposable. If K is the class of groups or of abelian groups then (ii) holds.

Let M be K-free on ω_1 generators. Then $N^0 \subseteq^* N^1$ means just that $N^0 \mid N^1$, N^0 is not finitely generated, and N^1 is not finitely generated over N^0. And $N^0 \subseteq M$ is good iff $N^0 \mid M$ and N^0 is K-free on ω generators. It is then clear that M is freely decomposable.

It is easy to show that (ii) of Theorem 3.2 holds if K is the class of groups or of abelian groups; for groups, e.g., it is implicit in Higman's definition of $U_{\lambda+1}$ in [4, p.288].

The consequence that there are abelian groups of power ω_1 which are $L_{\infty\omega_1}$-elementarily equivalent to free abelian groups but not free was first obtained by Eklof [2]. The corresponding conclusion for groups has also been obtained by Mekler [10].

Some of the deepest results about algebras infinitarily equivalent to free abelian groups are due to S. Shelah. Many of these results are proved using games similar to the ones we use, and are known to hold for more general classes K than just abelian groups. As an example, see the recent paper [5].

REFERENCES

[1] M.A. Dickman, Large Infinitary Languages, North-Holland Publ. Co., 1975.

[2] P.C. Eklof, Infinitary equivalence of abelian groups, Fund. Math. 81 (1974), 305-314.

[3] G. Grätzer, Universal Algebra, Van Nostrand Co., 1968.

[4] G. Higman, Almost free groups, Proc. London Math. Soc. 3 (1951), 284-290.

[5] W. Hodges, For singular λ, λ-free implies free, to appear in Algebra Universalis.

[6] H.J. Keisler, Formulas with linearly ordered quantifiers, in: J. Barwise (ed.), Syntax and Semantics of Infinitary Languages, Springer-Verlag (1968), 96-130.

[7] D.W. Kueker, Free and almost free algebras, abstract 701-02-5, Notices Amer. Math. Soc. 20 (1973), A-31.

[8] D.W. Kueker, Back-and-forth arguments and infinitary logics, in: D.W. Kueker (ed.) Infinitary Logic: In Memoriam Carol Karp, Springer-Verlag (1975), 17-71.

[9] D.W. Kueker, Countable approximations and Löwenheim-Skolem theorems, Ann. Math. Logic 11 (1977), 57-103.

[10] A.H. Mekler, How to construct almost free groups, to appear.

[11] M. Nadel and J. Stavi, $L_{\infty\lambda}$-equivalence, isomorphism and potential isomorphism, Trans. Amer. Math. Soc. 236 (1978), 51-74.

ON RECURSIVE LINEAR ORDERINGS

Manuel Lerman[1]
The University of Connecticut
Storrs, Ct. 06268 U.S.A.

We consider questions about recursive suborderings of recursive linear orderings, and about representations of sets by recursive linear orderings. These questions were raised by Rosenstein [R] in a draft of his book (all references to Rosenstein's book are to Chapter 16 of that book). Our motivation to try to answer these questions came from a seminar presentation of R. Watnick in which these questions were raised.

Let ω be the order-type of N, the set of positive integers; let ω^* be the order-type of the set of negative integers; and let η be the order-type of the set Q of rational numbers. Order-types are added by concatenation, e.g., $\omega^* + \omega$ is the order-type of the set Z of integers. Order-types are multiplied on the right, e.g., $\omega \cdot \eta$ is the order-type obtained by replacing each point in Q with a copy of N.

It is easily shown [R] that any infinite recursive linear ordering must have a recursive subset with one of the following order-types; ω, ω^*, $\omega + \omega^*$, $\omega + (\omega^* + \omega) \cdot \eta + \omega^*$. Rosenstein asks if all these order-types are necessary. Clearly ω and ω^* are necessary, and Tennenbaum has constructed a recursive linearly ordered set of order-type $\omega + \omega^*$ which has no recursive subset of order-type ω or ω^*. In Section 1, we construct a recursive linearly ordered set which has no recursive subset of order-type ω, ω^*, or $\omega + \omega^*$, thus showing that all the above order-types are necessary.

Let α be an order-type of an infinite linear ordering, and let A be a set of natural numbers. An <u>α-representation</u> of A is a linear ordering of order-type $\alpha + a_0 + \alpha + a_1 + \ldots$ where $\{a_i\}$ is a listing of the elements of A (repetitions are allowed in this listing). If $\{a_i\}$ is a one-one enumeration of the elements of A in order of magnitude, then $\alpha + a_0 + \alpha + a_1 + \ldots$ is an <u>α-representation</u> of A <u>in order of magnitude</u>. Rosenstein [R] asks for characterizations of those sets which have recursive $\omega^* + \omega$-representations in order of magnitude, and of those sets which have recursive η-representations in order of magnitude. We show in Section 2 that A has a recursive $\omega^* + \omega$-representation in order of magnitude if and only if $A \in \Sigma_3^0$. η-representations are studied in Section 3. Let $M(\mathcal{L}_0^*) = \{A : A$ has a recursive η-representation in order of magnitude$\}$. Rosenstein [R] notes that $\Sigma_2^0 \subseteq M(\mathcal{L}_0^*) \subseteq \Delta_3^0$; and Fellner [F] has shown that $\Pi_2^0 \subseteq M(\mathcal{L}_0^*)$. We show that $M(\mathcal{L}_0^*) \neq \Delta_3^0$. We also classify the degrees of those sets which have recursive η-representations.

---

(1) Research partially supported by the National Science Foundation under Grant #MCS 78-01849.

1. RECURSIVE SUBORDERINGS OF RECURSIVE ORDERINGS

Rosenstein [R] shows that any recursive linear ordering must have a recursive subordering of order-type ω, ω^*, $\omega + \omega^*$, or $\omega + (\omega^*+\omega)\cdot\eta + \omega^*$ and asks if all these order-types are necessary. The positive and negative integers respectively show that ω and ω^* are necessary. Tennenbaum has shown that $\omega + \omega^*$ is necessary (a proof appears in [R]). We now show that all the order-types in the list are necessary.

<u>DEFINITION 1.1</u>: A <u>string</u> is a finite sequence of integers. Given strings σ and τ, $\sigma * \tau$ is the string obtained by concatenation, and $lh(\sigma)$ is the length of σ, or the cardinality of the domain of σ. $\mathscr{S}$ will denote the set of all strings, and $\mathscr{S}_n$ will denote all strings of integers $<n$.

<u>THEOREM 1.2</u>: There is a recursive linear ordering having no recursively enumerable suborderings of order-types ω, ω^*, or $\omega + \omega^*$.

<u>Proof</u>: We will construct a recursive linear ordering L whose universe is N. Let $\{W_e : e \in N\}$ be a recursive enumeration of all the recursively enumerable sets, with $W_e^s = \{x : x$ has appeared in W_e by stage $s\}$. For each $e \in N$, we guarantee that if W_e is infinite, then there is an $x \in W_e$ which has infinitely many predecessors and infinitely many successors in L. Thus the elements of W_e form a recursively enumerable subset of L whose induced ordering is not ω, ω^*, or $\omega + \omega^*$. Since every recursively enumerable subordering of L has universe W_e for some e, this will suffice to prove the theorem.

We begin by describing the process through which the requirements R_0 and R_1 corresponding to the sets W_0 and W_1 respectively are satisfied. The construction begins by designating 0 as the smallest element of L and 1 as the largest element of L. R_0 establishes the constraint that all remaining elements of L be enumerated in $(0,1)$. If no element of $N-\{0,1\}$ appears in W_0, then R_0 does not act again; W_0 is then finite. Suppose that a first element, say x_0, of $N-\{0,1\}$ appears in W_0^s. Then R_0 establishes two intervals, $(0,x_0)$ and $(x_0,1)$. R_0 dictates that all elements of N are to be enumerated in $(0,x_0)$ (except those elements already enumerated) until, and if, an element of W_0 appears in this interval. If such an element appears, then R_0 changes the constraint it imposes to the interval $(x_0,1)$ etc., alternating between the intervals and changing intervals each time a new element of W_0 appears in the currently designated interval. Thus if W_0 is not finite, then x_0 will have infinitely many predecessors and infinitely many successors in L.

At each stage beyond the first stage, R_1 reacts to the interval designated by R_0. R_1 designates an interval within the interval designated by R_0, and plays a strategy similar to that played by R_0 within each of the possible intervals designated by R_0. Thus if W_1 is infinite, R_1 will succeed in forcing an element $x_1 \in W_1$ to have infinitely many predecessors and infinitely many successors in L.

We now describe the construction.

Stage 0: Place $0 < 1$ in L.

Stage s+1: We define a finite sequence $\sigma_{s+1} \in \mathcal{S}_3$ of length s+1 which selects the designated intervals. The definition proceeds by induction on $j \leq s$ and by cases. Let $t(s,j)$ be the greatest stage $t \leq s$ such that $t > j$ and $\sigma_t[j] = \sigma_{s+1}[j]$ ($\sigma_t[j]$ is the sequence of length j which is extended by σ) if such a t exists, and $t(s,j) = 0$ otherwise.

Case 1: $t(s,j) = 0$. Let $\sigma_{s+1}(j) = 0$ and $I(\sigma_{s+1}[j]) = I(\sigma_{s+1}[j-1])$. ($I(\sigma_{s+1}[j])$ is the interval designated by the requirement R_j corresponding to W_j at stage s+1, assuming that $\sigma_{s+1}[j-1]$ selects the designated intervals for R_i, $i < j$.)

Case 2: $t(s,j) \neq 0$ and $\sigma_{t(s,j)}(j) = 0$. If $I(\sigma_{s+1}[j-1]) \cap (W_j^{s+1} - W_j^{t(s,j)}) \neq \emptyset$, fix the least x in this intersection. Let $I(\sigma_{s+1}[j-1]) = (a,b)$. Set $I(\sigma_{s+1}[j-1] * 1) = (a,x)$ and $I(\sigma_{s+1}[j-1] * 2) = (x,b)$ and let $\sigma_{s+1}(j) = 1$. Otherwise, let $\sigma_{s+1}(j) = 0$.

Case 3: $t(s,j) \neq 0$ and $\sigma_{t(s,j)}(j) \neq 0$. Let $\sigma_{t(s,j)}(j) = k \in \{1,2\}$. If $I(\sigma_{s+1}[j-1] * k) \cap (W_j^{s+1} - W_j^{t(s,j)}) \neq \emptyset$, let $\sigma_{s+1}(j) = 3 - k$. Otherwise, let $\sigma_{s+1}(j) = k$.

In all cases, place s+2 into L as the least element of $I(\sigma_{s+1}[s])$.

This completes the construction. Note that for all n, there are only finitely many $\sigma \in \mathcal{S}_3$ such that $lh(\sigma) = n$. Thus it is easily verified by induction that $N - \cup\{I_\sigma : lh(\sigma) = n\}$ is finite for all n. Hence if W_j is infinite, then there is a $\sigma \in \mathcal{S}_3$ such that $lh(\sigma) = j$, $\{s : \sigma_s[j] = \sigma\}$ is infinite, and $W_j \cap I(\sigma)$ is infinite. By Case 2, $\sigma_s(j) = 0$ for only finitely many s, and so by Case 3, $\{s : \sigma_s[j] = \sigma * 1\}$ and $\{s : \sigma_s[j] = \sigma * 2\}$ are both infinite. Hence there is an $x \in W_j$ which has infinitely many predecessors and infinitely many successors in L.

2. $\omega^* + \omega$ -REPRESENTATIONS

DEFINITION 2.1: Let L be a linear ordering. The spectrum of L is $M(L) = \{n \in N : L$ has a maximal finite interval of cardinality $n\}$. If $\mathcal{L}$ is a set of linear orderings, then $M(\mathcal{L}) = \{M(L) : L \in \mathcal{L}\}$.

Rosenstein [R] investigates $M(\mathcal{L})$ for various classes $\mathcal{L}$ of recursive linear orderings. It is easily checked that if L is a recursive linear ordering, then M(L) is a Σ_3^0 set.

In this section, we investigate $\omega^* + \omega$-representations. Let $\mathcal{L}$ be the class of all recursive $\omega^* + \omega$-representations, and let $\mathcal{L}_0$ be the class of all

$\omega^* + \omega$-representations in order of magnitude. Rosenstein shows that $M(\mathcal{L}) = \Sigma_3^0$. Fellner shows that $A \subset M(\mathcal{L}_0)$ for every $A \subseteq N$ such that $x \in A$ can be expressed as a conjunction of Σ_2^0 and Π_2^0 formulas. We will show that $M(\mathcal{L}_0) = \Sigma_3^0$. We begin with a discussion of Σ_3^0 sets.

Let A be a Σ_3^0 set. Then there is a recursive function $f: N \to N^3$ such that for each $n \in N$,

(1) if $n \in A$ then there is exactly one m such that $\text{rng}(f) \cap \langle n,m\rangle \times N = \langle n,m\rangle \times N$; and

(2) if $n \notin A$ then for all m, $\text{rng}(f) \cap \langle n,m\rangle \times N$ is finite.

Thus each $n \in N$ is assigned a copy $N \times N$ of the plane, f fills in points on each copy of the plane, filling in an initial part of each column and at most one full column on any copy of the plane, and $n \in A$ if and only if the nth copy of the plane has a complete column filled in by f.

Fix a one-one recursive correspondence, $\{\langle n_k, m_k\rangle : k \in N\}$ of N with N^2. We say that $\langle n_i, m_i\rangle$ <u>has higher priority than</u> $\langle n_j, m_j\rangle$ if $i < j$.

Each $\sigma \in \mathcal{S}_2$ will be used to guess at which of a certain sequence of columns is finite, and which is full. $\sigma(k)$ will make this guess for $\langle n_k, m_k\rangle$ corresponding to column m_k on copy n_k of the plane. $\sigma(k) = 0$ will be the guess that f enumerates the whole column, and $\sigma(k) = 1$ will be the guess that f enumerates only finitely much of column m_k on copy n_k of the plane. The inclusion ordering is placed on $\mathcal{S}_2$, producing a tree. A priority ordering is also placed on $\mathcal{S}_2$, σ having <u>higher priority than</u> τ if $\sigma \subset \tau$ or if $\sigma(x) < \tau(x)$ for the least x such that $\sigma(x) \neq \tau(x)$.

The enumeration $f: N \to N^3$ induced by the Σ_3^0 set A will determine which string (i.e., the branch on the tree $\mathcal{S}_2$) is to be selected at stage σ of the construction. A string σ is <u>dormant</u> at stage $s+1$ unless all $\tau \subset \sigma$ are discharged at the end of stage s, in which case σ is <u>active</u> at stage $s+1$. While σ is active, say beginning at stage $s+1$, each $k < lh(\sigma)$ such that $\sigma(k) = 0$ receives a check for σ at stage $t > s$ if f enumerates an element in column $\langle n_k, m_k\rangle$ at stage t. σ <u>requires attention</u> at stage $s+1$ if each $k < lh(\sigma)$ such that $\sigma(k) = 0$, has received a check for σ which has not been cancelled. Existing checks for σ, as well as the discharged state of σ may be <u>cancelled</u> at stage t because some τ of higher priority than σ requires attention at stage t, in which case σ must wait again until it becomes active and then begin accumulating checks anew. The σ of highest priority which requires attention at stage t, say σ_t, is the one which determines the action of the construction at stage t. For each $n \in N$, let γ_n be the string of highest priority such that $lh(\gamma_n) = n$ and $\gamma_n \subseteq \sigma_t$ for infinitely many t. It is easily shown that γ_n is defined for each $n \in N$ and that if $i < j$ then $\gamma_i \subset \gamma_j$. $\{\gamma_n\}$ codes A in the following sense:

(3) $x \in A \leftrightarrow \exists k \exists m(\gamma_{k+1}(k) = 0 \;\&\; \langle n_k, m_k\rangle = \langle x, m\rangle)$.

Let $\Gamma = \cup\{\gamma_n : n \in N\}$. The priority ordering of $\mathcal{S}_2$ can be extended to include Γ by letting σ have higher priority than Γ if and only if σ has higher priority than γ_n for all but finitely many n. It then follows that

(4) σ has higher priority than Γ if and only if σ has higher priority than σ_s for all but finitely many s.

The above procedure to select σ_t from A yielding (3) and (4) is a standard procedure of recursion theory. This procedure is now used to prove:

THEOREM 2.2: If $A \in \Sigma_3^0$ then A has a recursive $\omega^* + \omega$-representation in order of magnitude.

Proof: Let $\{a_i : i \in N\}$ be an enumeration of A in order of magnitude. Since A can be modified finitely without effecting its representability, we may assume that $0 \notin A$. Furthermore, we may assume that A is infinite, else the theorem follows easily. The ordering L which we build will have order-type $\omega + \Sigma\{a_i + (\omega^* + \omega) : i \in N\}$. ω^* can be recursively placed at the beginning of L to get the desired representation.

During the construction, n-tuples will be assigned to strings for $k \in N$, and such assignments may be cancelled. Assignments of n-tuples will have the property that if $\tau \subset \sigma \in \mathcal{S}_2$, $k < lh(\tau)$ and an n-tuple is assigned to σ for k at stage s, then that same n-tuple is assigned to τ for k at stage s.

The construction proceeds as follows:

Stage 0: ϕ is discharged. Assign the a_o-tuple $<0,1,\ldots,a_o-1>$ to ϕ, and place $0 \prec 1 \prec \ldots \prec a_o - 1$ into L, where $\prec$ is the ordering of L.

Stage s+1: Choose σ_{s+1} as described above, and let $k + 1 = lh(\sigma_{s+1})$. There are three steps.

Cancellation: Cancel all assignments of n-tuples to those strings τ having lower priority than σ_{s+1}.

Interval Maximization: For each $\delta \subset \sigma_{s+1}$ and each n-tuple $<b_o,\ldots,b_{n-1}>$ assigned to δ for $lh(\delta)$, place a new integer into L immediately preceding b_o and another new integer into L immediately following b_{n-1}.

Realization: Let $\sigma_{s+1}^- = \sigma_{s+1}[k-1]$. All n-tuples assigned to σ_{s+1}^- for some j are assigned to σ_{s+1} for the same j. If $\sigma_{s+1}(k) = 1$ or if $n_k \leq a_o$, proceed to the next stage. Suppose that $\sigma_{s+1}(k) = 1$ and $n_k > a_o$. Let $c_o,\ldots,c_{n_k-1}$ be the least n_k integers not yet placed in L, with $c_o < c_1 < \ldots < c_{n_k-1}$; assign the n_k-tuple $<c_o,\ldots,c_{n_k-1}>$ to σ_{s+1} for k. These integers are inserted into L as an interval of L in order of magnitude as follows. Fix the least $r < k$, if any, such that

(i) $\sigma_{s+1}(r) = 0$.

(ii) $n_r > n_k$.

(iii) $\forall v < k\ (\sigma_{s+1}(v) = 0 \rightarrow n_v \geq n_r$ or $n_v \leq n_k)$.

Let $\langle d_0,\ldots,d_{n_r-1}\rangle$ be the n_r-tuple assigned to σ_{s+1} for r. Insert $c_0,\ldots,c_{n_k-1}$ into L immediately preceding d_0. If no such r exists, $c_0,\ldots,c_{n_k-1}$ are inserted at the end of L.

This completes the construction. L is readily seen to be recursive. We conclude the proof of the theorem with some lemmas.

LEMMA 2.3: $\sigma \neq \phi$ is assigned an n-tuple for r at all sufficiently large stages if and only if $\sigma \neq \phi$ has higher priority than Γ, $\sigma(r) = 0$, and $n_r > a_0$.

Proof: If σ does not have higher priority than Γ, then by (4), all assignments of n-tuples to σ are cancelled at infinitely many stages. Conversely, if σ has higher priority than Γ and $\sigma \neq \phi$, then by (4) there must be a last stage t at which $\sigma_t = \sigma$; and for all $s > t$, σ_s has higher priority than σ. σ is assigned an n_r-tuple at stage t exactly if $n_r > a_0$ and this assignment is never cancelled at any stage $s > t$.

We also note that the a_0-tuple assigned to ϕ at stage 0 is never cancelled.

LEMMA 2.4: Suppose that $\gamma \subset \Gamma$ and that the n-tuple $\langle d_0,\ldots,d_{n-1}\rangle$ is assigned to Γ for r and never cancelled. Then $\{d_0,\ldots,d_{n-1}\}$ is a maximal finite interval of L.

Proof: Let $\langle d_0,\ldots,d_{n-1}\rangle$ be assigned to $\gamma \subset \Gamma$ at stage t. Since this assignment is never cancelled and $\gamma \subset \Gamma$, γ has higher priority than σ_s for all $s > t$ and $\gamma \subset \sigma_s$ for infinitely many s. The lemma now follows from the interval maximization process.

LEMMA 2.5: (i) The order-type of $\{x \in L : x \prec 0\}$ is ω. (ii) Given any $x \in L$ there are $d \in L$ and $\gamma \subset \Gamma$ such that $x \prec d$ and d is an element of an m-tuple which is assigned to γ for some r and never cancelled.

Proof: (i) is immediate from the interval maximization process and the requirement that new intervals are added only for $n_k > a_0$. We now verify (ii). By (3) and since A is infinite, it follows that there are infinitely many $\gamma \subset \Gamma$ such that if $k = lh(\gamma)$ then $\sigma(k-1) = 0$ and $n_{k-1} > n_r$ for all $r < k - 1$ such that $\sigma(r) = 0$. Hence there are infinitely many stages s at which the realization process attaches an interval to the end of the part of L constructed before stage s, and this interval is assigned to some $\gamma \subset \Gamma$ and never cancelled. d can be chosen for x as an element of such an interval.

LEMMA 2.6: Suppose that $u,v \in A$, $u < v$, and $A \cap \{x : u < x < v\} = \phi$. Then there are unique maximal intervals $\{b_0,\ldots,b_{u-1}\}$ and $\{c_0,\ldots,c_{v-1}\}$ of L of lengths u and v respectively. Furthermore, $\{x \in L : b_{n-1} \prec x \prec c_o\}$ has order-type $\omega^* + \omega$.

Proof: By Lemma 2.3 and since for each $n \in A$ there is exactly one m such that the column corresponding to $\langle n,m\rangle$ is full, $\{b_i : i < u\}$ and $\{c_i : i < v\}$ as above exist and are unique. Fix $d_o, d_1 \in \{x \in L : b_{n-1} \prec x \prec c_o\}$ with $d_o \prec d_1$, and let $[d_o,d_1] = \{x \in L : d_o \preceq x \preceq d_1\}$. Fix a stage s_o such that d_o and d_1 have been placed in L by the end of stage s_o. Let $R(s) = \{x \in [d_o,d_1]$: for some σ and r, x is in an interval assigned to σ for r by the end of stage s, but this assignment is not cancelled before the end of stage s}. By Lemma 2.3, if $x \in R(s)$ is in an interval assigned to σ for r, then σ has lower priority than Γ so this interval is eventually cancelled; hence there is a $t > s$ such that $R(s) \cap R(t) = \phi$. If a new interval is inserted into $[d_o,d_1]$ at some stage $t > s_o$ and is assigned to τ, then τ has lower priority than σ for all σ for which there is an $x \in R(s_o)$ in an interval assigned to σ. Thus there is an $s_1 > s_o$ such that $R(s_1) = \phi$. Thus neither the interval maximization nor the realization processes insert new elements of L into $[d_o,d_1]$ after stage s_1, so $[d_o,d_1]$ is finite. The lemma now follows from Lemma 2.4 and the interval maximization process.

The theorem now follows easily from the initial comments in the proof and Lemmas 2.4-2.6.

The following corollary is immediate from Theorem 2.2 and the facts that $\mathcal{L}_o \subseteq \mathcal{L}$ and $M(\mathcal{L}) \subseteq \Sigma_3^o$.

COROLLARY 2.7: $M(\mathcal{L}_o) = M(\mathcal{L}) = \Sigma_3^o$.

3. η- REPRESENTATIONS

We now turn to η-representations. Since each interval of order-type η can be cut to represent 0 and 1 infinitely often, questions about η-representations are only interesting for sets $A \subseteq N - \{0,1\}$. Thus we study $M^*(L) = M(L) - \{0,1\}$ and $M^*(\mathcal{L}) = \bigcup\{M^*(L) : L \in \mathcal{L}\}$.

Again we concentrate on two classes of representations. Let $\mathcal{L}^*$ be the class of recursive η-representations and let $\mathcal{L}_o^*$ be the class of recursive η-representations in order of magnitude. The classification of η-spectra for these classes of η-representations is still incomplete. The following facts are known (see [R]):

(1) $M(\mathcal{L}_o^*) \subseteq M(\mathcal{L}^*) \subseteq \Sigma_3^o$.

(2) $M(\mathcal{L}_o^*) \subseteq \Delta_3^o$.

(3) $\Sigma_2^o \subseteq M(\mathcal{L}_o^*)$.

Fellner [F] has also shown that:

(4) $\Pi_2^o \subseteq M(\mathcal{L}_o^*)$.

Our first result shows that the arithmetical hierarchy cannot be used to classify the spectra of $\mathcal{L}_0^*$ or $\mathcal{L}^*$.

<u>THEOREM 3.1</u>: There is an $A \in \Delta_3^0$ such that $A \subseteq N - \{0,1\}$ and $A \neq M^*(L)$ for any $L \in \mathcal{L}^*$.

<u>Proof</u>: Fix a recursive enumeration $\{L_i : i \in N\}$ of all partial recursive linear orderings. Let D be a complete Δ_3^0 set. We will use a D oracle to locate maximal finite intervals of the L_i and keep the cardinalities of some of these intervals out of the Δ_3^0 set A which is constructed. This will be our basic strategy to guarantee that L_i does not represent A. We will also need to make A infinite. D cannot tell us whether L_i has a maximal finite interval of length n. It can locate finite intervals, however, and expand them until they become maximal (the expansion continues forever if the interval is not part of a maximal finite interval). In the latter case, $L_i \notin \mathcal{L}^*$. Each L_i will be ordered by $<_i$, and we define $[a,b]_i = \{x \in L_i : a \leq_i x \leq_i b\}$.

The D oracle can answer the following questions:

(5) Is L_i total?

(6) Does L_i have an interval of cardinality n?

(7) Is $[a,b]_i$ an interval of L_i of cardinality n?

(8) Is $[a,b]_i$ a maximal finite interval of L_i?

(9) Is $[a',b']_i$ an interval of L_i of cardinality $n+1$ which extends the interval $[a,b]_i$ of L_i of cardinality n?

We will construct a set A recursively in D which is not η-represented by any L_i. Since A is recursive in D, $A \in \Delta_3^0$. At each stage of the construction, intervals will be <u>assigned</u> to finitely many L_i. Such assignments may be <u>cancelled</u> at later stages. L_i may be <u>discharged</u> at a stage; this will mean that L_i does not η-represent A. The construction proceeds by induction on $\{s : s \geq 1\}$.

<u>Stage s</u>: There are four basic steps in the construction.

<u>Discharging</u>: For each $i < s$ to which an interval $[a_i^s, b_i^s]_i$ is assigned at the end of stage $s-1$, we ask whether the interval $[a_i^s, b_i^s]_i$ is a maximal finite interval of L_i of cardinality s (use (7) and (8)). If the answer is no for all i, place $s \in A$ and go to the Realization step. If the answer is yes for some i, let $i(s)$ be the least such i. Place $s \in \mathbb{N} - A$, discharge $L_{i(s)}$, and go to the Cancellation step.

<u>Cancellation</u>: Cancel the assignment of $[a_i^s, b_i^s]_i$ to L_i for all $i \geq i(s)$. Go to the Expansion step.

Realization: Fix the least i such that L_i is not discharged and no interval is currently assigned to L_i. If L_i is not total (use (5)), discharge L_i and go to the Expansion step. Suppose that L_i is total. Use (6) to ask if L_i has a finite interval of cardinality $s+1$. If the answer is no, discharge L_i and go to the Expansion step. If the answer is yes, use (7) and (8) to fix a finite interval $[a_i^s, b_i^s]_i$ of L_i recursively in D which is not maximal and has cardinality s, and assign this interval to L_i. Go to the Expansion step.

Expansion: For each $i \leq s$ such that an interval is currently assigned to L_i, the assigned interval will have cardinality s but will not be maximal. Use (9) to find, recursively in D, an interval $[a_i^{s+1}, b_i^{s+1}]_i \supseteq [a_i^s, b_i^s]_i$ of L_i of cardinality $s+1$. This interval is now assigned to L_i in place of $[a_i^s, b_i^s]_i$. Go to the next stage.

This completes the construction. The construction is recursive in D, so $A \in \Delta_3^o$. Suppose that $i(s)$ is defined. Then Realization is not followed at stage s. Hence by Cancellation, if $i(s+1)$ is defined, then $i(s+1) < i(s)$. There must then be infinitely many stages s such that $i(s)$ is not defined; for each such s, $s \in A$ so A is infinite.

We now show that for all $i \in N$, L_i does not η-represent A. Fix $i \in N$. First suppose that L_i is discharged. Let s be the first stage at which L_i is discharged. Then either L_i is not total or L_i η-represents a finite set or L_i does not η-represent any set or $s \notin A$ and L_i has a maximal interval of cardinality s. In all these cases, L_i cannot η-represent A. Suppose that L_i is not discharged. Let $J = \{j \leq i : L_j$ is discharged$\}$. Then there is a stage s_o such that for all $j \in J$, L_j is discharged at the end of stage s_o. Fix $t \geq s_o$ such that an interval $[a_i^{t+1}, b_i^{t+1}]_i$ is assigned to L_i at stage t. Note that Realization will produce such a stage t. Then by choice of $t \geq s_o$ and Expansion if $v > u > t$ then a finite interval $[a_i^v, b_i^v]_i \supseteq [a_i^u, b_i^u]_i$ of L_i of cardinality $v+1$ will be assigned to L_i at stage $v+1$. Thus $\cup\{[a_i^v, b_i^v]_i : v > t\}$ is an interval of L_i having order-type ω, ω^*, or $\omega^* + \omega$, so L_i is not an η-representation of any set. This completes the proof of the theorem.

Although the arithmetical hierarchy cannot be used to classify $M^*(\mathcal{L}^*)$, the Turing degrees of sets in $M^*(\mathcal{L}^*)$ can be neatly classified.

DEFINITION 3.2: For any set $A \subseteq N$, let $\underline{A}$ denote the Turing degree of A. If $\mathcal{C}$ is a class of subsets of N, then $\underline{\mathcal{C}} = \{\underline{C} : C \in \mathcal{C}\}$.

THEOREM 3.3: $\underline{M^*(\mathcal{L}^*)} = \underline{\Sigma_3^o}$.

Proof: Fix $A \in \Sigma_3^o$. Let $A^* = \{2n : n \in A\} \cup \{2n+1 : n \in N\}$. Clearly $\underline{A} = \underline{A^*}$.

We will construct an η-representation L of A^*. As in Section 2, the fact that $A \in \Sigma_3^o$ provides a recursive one-one correspondence between ordered pairs $\langle n,m\rangle \in N^2$ and columns $\{i\} \times N \subseteq N^2$ (let $\{i(n,m)\} \times N$ correspond to $\langle n,m\rangle$) and a recursive enumeration of elements in the columns such that for each $i \in N$ either all or finitely much of $\{i\} \times N$ is enumerated, and:

(10) $n \in A \to \exists m$(all of column $\{i(n,m)\} \times N$ is enumerated).

(11) $n \notin A \to \forall m$(finitely much of column $\{i(n,m)\} \times N$ is enumerated).

For each $i \in N$ we will uniformly recursively construct an interval I_i. L will be a recursive linear ordering η-representing A^* defined by $L = \{I_i : i \in N\}$ ordered by $I_o < I_1 < \dots$. $I_{i(n,m)}$ will have order-type $\eta+n+\eta$ if either n is odd or if n is even and all of column $\{i(n,m)\} \times N$ is enumerated; and $I_{i(n,m)}$ will have order-type $\eta+(n+1)+\eta$ otherwise. By (10) and (11), L will then η-represent A^*.

$I_{i(n,m)}$ is easily constructed. If n is odd, directly recursively construct an interval of the desired order-type. If n is even, begin by specifying an interval of $I_{i(n,m)}$ of cardinality n. At each stage, this interval is placed inside a designated interval of $I_{i(n,m)}$ of cardinality $n+1$, with care taken to guarantee density outside this interval. Whenever a new element is enumerated in column $\{i(n,m)\} \times N$, the $n+1^{st}$ element of the designated interval is separated off into the dense part, and a new candidate for an $n+1^{st}$ element of the interval is specified. The original interval together with this new element now becomes the designated interval. If all elements of column $\{i(n,m)\} \times N$ are enumerated, then all candidates for the $n+1^{st}$ element of the interval will eventually be separated off into a dense part of $L_{i(n,m)}$, so $L_{i(n,m)}$ will have order-type $\eta+n+\eta$. And if only finitely many elements of column $\{i(n,m)\} \times N$ are enumerated, there will be a last candidate for the $n+1^{st}$ element of the interval, so $L_{i(n,m)}$ will have order-type $\eta+(n+1)+\eta$. This completes the proof of the theorem.

We note that the proof of Theorem 3.3 can easily be modified to show that if $A \in \Sigma_3^o$ and A is not immune, then there is an η-representation of A.

The characterization of $M^*(\mathcal{L}_o^*)$ at this point is unsatisfactory. It seems to us that the proofs showing $\Sigma_2^o \subseteq M^*(\mathcal{L}_o^*)$ and $\Pi_2^o \subseteq M^*(\mathcal{L}_o^*)$ can be combined and extended to construct a recursive η-representation of any Δ_3^o set L for which there are column representations for both A and $N - A$ and a Δ_2^o function f such that if $n \in A$ (or $n \in N - A$) then there is a column corresponding to $\langle n,m\rangle$ all of which is enumerated, and $m \leq f(n)$. We have not attempted to prove this fact here since, by itself, it still does not yield a satisfactory classification of $M^*(\mathcal{L}_o^*)$. We conjecture, however, that the above procedure can be carried out.

Other questions which we have considered briefly but not pursued are questions about the classification of $M^*(\mathcal{L}_1^*)$, where $\mathcal{L}_1^*$ is some class of η-representations other than $\mathcal{L}^*$ or $\mathcal{L}_o^*$. We briefly looked at one such class, $\mathcal{L}_1^*$ = class of linear orderings such that for all n, there is at most one maximal interval of cardinality n. In that case, it seemed likely that one could show that $M^*(\mathcal{L}_1^*) = M^*(\mathcal{L}^*)$, but we have not carried out such a proof.

REFERENCES

[F] Fellner, S., Recursiveness and finite axiomatizability of linear orderings, Ph.D. Thesis, Rutgers Univ., 1976.

[R] Rosenstein, J., Linear Orderings, Academic Press (to appear).

The Complexity of Types in Field Theory

Angus Macintyre*

§0. Introduction. While studying primes in nonstandard models of arithmetic [M1], I proved that the residue class field of a nonstandard prime is never recursive. (Some cases of this were anticipated by Tennenbaum, but never published). From this setting I extracted the following conjecture:

Conjecture. Let K be an infinite recursive field, with K recursively saturated. Then K is algebraically closed of infinite transcendence degree.

This conjecture has resisted proof, but the main result of the present paper is a close approximation. Namely:

Theorem 1. Let K be an infinite recursive field, with K recursively saturated. Then G_K, the absolute Galois group of K, is procyclic, and $\tilde{K}$ is got from K by adjoining roots of unity. K is not formally real. If K is not algebraically closed, then K is neither henselian nor pseudoalgebraically closed. If K has positive characteristic, and is not algebraically closed, then the algebraic closure of the prime field in K is finite, and K is perfect.

It will emerge that the proof is a cousin of my proof that ω-stable fields are finite or algebraically closed [M2]. I use an effective version of Duret's proof [D1] that p.a.c. fields, if not algebraically closed, have the independence property. I also show that there is no stable counterexample to my conjecture.

The title of the paper calls attention to the complexity of arbitrary type spaces of fields, save the finite or algebraically closed ones. From the proof of the main theorem, I obtain:

Theorem 2. Let K be an infinite field, not algebraically closed. Let T = Th(K), and suppose that for each n $S_n(T)$ has cardinal $< 2^{\aleph_0}$. Then G_K is procyclic, and all the other conclusions in the main theorem hold.

For applications to arithmetic, a relativized version of the main theorem is proved, where recursive and recursively saturated are replaced by analogues at arbitrary Turing degrees.

I am grateful to Denef, van den Dries, Duret, Poizat and Wilkie for various suggestions.

1. The Tennenbaum phenomenon. Tennenbaum [T] first implicitly called attention to the conflict between recursive and recursively saturated. A nonstandard model M of arithmetic is weakly recursively saturated (definition below). From this it follows,

*Supported by NSF

by an argument involving representation of recursively inseparable r.e. sets, that neither the $\oplus$ nor the $\odot$ of M is recursive.

It will be convenient to put Tennenbaum's argument in a general setting.

Let L be a first-order language, recursively arithmetized. Let C be a subset of $P(\omega)$, and let $\mathcal{D}$ be a set of L-formulas. Let M be an L-structure. I say M is $(C,\mathcal{D})$-saturated if the following holds:

If $\Sigma(v_1,\ldots,v_n, w_1,\ldots,w_m) \subseteq \mathcal{D}$, and the set of codes of Σ is in C, and $a_1,\ldots,a_m \in M$, and $\Sigma(v_1,\ldots,v_n, \vec{a})$ is finitely satisfiable in M, then $\Sigma(v_1,\ldots,v_n, \vec{a})$ is satisfiable in M.

The most familiar instance is when C is the class of recursive sets, and $\mathcal{D}$ is the set of all L-formulas. Then $(C,\mathcal{D})$-saturation is called recursive saturation.

Let $\mathcal{D}_n$ be the set of all Σ_n L-formulas. Let C be the class of recursive sets. Then $(C,\mathcal{D}_n)$-saturation is called recursive Σ_n-saturation. If M is recursively Σ_n-saturated for all n, M is called weakly recursively saturated. The most notable weakly recursively saturated structures are the nonstandard models of Peano arithmetic.

A useful observation, depending on a trick of Craig [C] is that recursively enumerable saturation is no stronger than recursive saturation. For my purposes it is useful to record the general version, Lemma 1 below.

For the purposes of this paper, I say C is closed under Turing reducibility if whenever $A_1,\ldots,A_n \in C$ and A is Turing reducible to $\{<a_1,\ldots,a_n> : a_i \in A_i\}$ then $A \in C$.

Lemma 1. Suppose $\mathcal{D}$ is closed under $\wedge$. Suppose C_1 is closed under Turing reducibility. Suppose that for each E_2 in C_2 there is an E_1 in C_1 so that

$$E_2 = \{x \in \mathbb{N} : (\exists y)(<x,y> \in E_1)\}$$

(where $<,>$ is a fixed recursive pairing function). Suppose M is $(C_1,\mathcal{D})$-saturated. Then M is $(C_2,\mathcal{D})$-saturated.

Proof: Let $\Sigma(v_1,\ldots,v_n, w_1,\ldots,w_m) \subseteq \mathcal{D}$, and suppose E_2 in C_2 is the set of codes of Σ. Choose E_1 in C_1 so that $E_2 = \{x \in \mathbb{N} : (\exists y)(<x,y> \in E_1)\}$. Let Σ' consist of all formulas

$$\underbrace{\Phi \wedge \Phi \wedge \ldots \wedge \Phi}_{k \text{ times}}$$

where $\Phi \in \Sigma$, Φ has Gödel number ℓ and $<\ell,k> \in E_1$.

Then for any $a_1,\ldots,a_n$ in M, $\Sigma'(\vec{a}, w_1,\ldots,w_m)$ is satisfiable iff $\Sigma(\vec{a}, w_1,\ldots,w_m)$ is. But the set of codes of Σ' is in C_1. The result follows. □

Now let E be a set of relations and operations on N. I say M is E-codable if M is countable and there is a bijection $f : M \leftrightarrow \mathbb{N}$ which maps the basic relations and operations of M to elements of E. (For the purposes of this paper I disregard finite M).

When E is the set of recursive relations and operations, and M is E-codable, I simply say M is recursive.

Definition. E has the Rosser property if there are disjoint sets A_1 and A_2 in E which cannot be separated by a set in E whose complement is also in E.

For example, the set of recursively enumerable sets has the Rosser property, because of the existence of recursively inseparable r.e. sets.

The following lemma gives the abstract version of Tennenbaum's original argument.

Lemma 2. Suppose E has the Rosser property, and is closed under Turing reducibility. Suppose $\mathcal{D}$ is closed under $\wedge$ and contains all Σ_1 formulas. Let $v_1,\ldots,v_n$, $w_1,\ldots,w_m$ be distinct variables. Let

$$k \longmapsto \psi_k(\vec{v}, \vec{w}),$$

$$k \longmapsto \theta_k(\vec{v}, \vec{w})$$

be recursive maps from $\mathbb{N}$ to the set of Σ_1 formulas. Suppose M is such that

i) there is a tuple $\vec{a}$ from M so that for each pair (U,V) of disjoint finite subsets of $\mathbb{N}$ there is a $\vec{b}$ in M such that

$$M \models \psi_k(\vec{a}, \vec{b}) \qquad (k \in U)$$

$$M \models \theta_\ell(\vec{a}, \vec{b}) \qquad (\ell \in V);$$

ii) $M \models (\forall\vec{w})[\psi_k(\vec{a}, \vec{w}) \leftrightarrow \neg\, \theta_k(\vec{a}, \vec{w})]$;

iii) M is $(E_s,\mathcal{D})$-saturated, where E_s consists of the sets in E.

Then M is not E-coded.

Proof: Select $\vec{a}$ as in (ii). Fix A_1, $A_2 \in E$ witnessing the Rosser property. Consider the type

$$\psi_k(\vec{a}, \vec{w}) \qquad (k \in A_1)$$

$$\theta_\ell(\vec{a}, \vec{w}) \qquad (\ell \in A_2)$$

By (i) this is consistent over M.
By (iii) and Lemma 1, the above type is realized in M, say by $\vec{b}$. Let

$B_1 = \{k : M = \psi_k(\vec{a}, \vec{b})\}$

$B_2 = \{\ell : M = \theta_\ell(\vec{a}, \vec{b})\}$.

$B_1 \cap B_2 = \emptyset$, $B_1 \cup B_2 = N$ by (ii).

If M is E-coded, then both B_1 and B_2 are r.e. in a member of E. So

B_1 and B_2 are in $\mathcal{E}$, since they are complementary and $\mathcal{E}$ is closed under Turing reducibility. But then B_1 separates A_1 and A_2, contradiction. □

In Tennenbaum's original argument

$$\psi_k(w) \text{ is } p_k | v$$

and

$$\theta_k(w) \text{ is } \neg(p_k | v)$$

both of which are (equivalent to) Σ_1 formulas relative to P.

2. Application to fields 2.1. I now give the first application of Lemma 2.

For the rest of the paper, I assume $\mathcal{E}$ is fixed, closed under Turing reducibility, and with the Rosser property. $\mathcal{E}_s$ is the restriction of $\mathcal{E}$ to sets.

Let K be a field, with multiplicative group K^*. Let p be a prime. One considers (as in [M2]) the filtration of definable subgroups

$$K^* \supseteq K^{*p} \supseteq K^{*p^2} \supseteq \ldots \supseteq K^{*p^n} \supseteq \ldots,$$

and the corresponding projective system

$$K^*/1 \twoheadleftarrow K^*/K^{*p} \twoheadleftarrow K^*/K^{*p^2} \twoheadleftarrow \ldots .$$

Suppose now $K^* \neq K^p$, and let $\alpha \in K^*$, $\alpha \notin K^{*p}$. One pulls back the distinct cosets

$K^{*p}\cdot 1$ and $K^{*p}\cdot\alpha$ through the projective system to get the "trees"

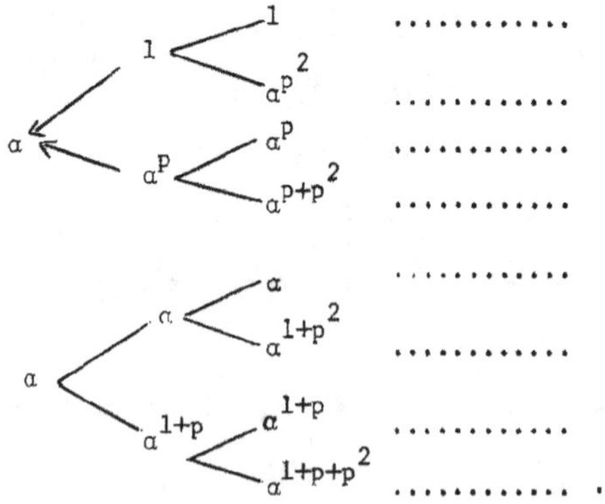

Given $\alpha^k \cdot K^{*p^j}$, it has natural "pullbacks" $\alpha^k \cdot K^{*p^{j+1}}$ and $\alpha^{k+p^j} \cdot K^{*p^{j+1}}$. The former will be called the left pullback, and the latter the right pullback.

Now let A and B be disjoint sets of integers. I want to concoct a partial type $\Sigma_{A,B}(v,a)$ which "says":

The path of v in the projective system is one of the branches in the above tree, and for $n \in A$ the branch goes left at the nth stage and for $m \in B$ the branch goes right at the mth stage

This is easily expressed. Let $f(n)(v,w)$ say: (for $n \geq 2$)

$$\bigwedge_{1\le j\le n}\left(\bigvee_{\substack{c_0,\ldots,c_{j-2}\\ c_r\in\{0,1\}}} v\cdot w^{-\left(\sum_{r=0}^{j-1} c_r\cdot p^r\right)} \text{ is a } p^{j\text{th}} \text{ power}\right)$$

$$\wedge\left(\bigvee_{\substack{c_0,\ldots,c_{j-2}\\ c_r\in\{0,1\}}} v\cdot w^{-\left(\sum_{r=0}^{j-2} c_r\cdot p^r\right)} \text{ is a } p^{j\text{th}} \text{ power}\right).$$

Let $g(n)(v,w)$ say

$$\bigwedge_{1\le j\le n}\left(\bigvee_{\substack{c_0,\ldots,c_{j-1}\\ c_r\in\{0,1\}}} v\cdot w^{-\left(\sum_{r=0}^{j-2} c_r\cdot p^r\right)} \text{ is a } p^{j\text{th}} \text{ power}\right)$$

$$\wedge\left(\bigvee_{\substack{c_0,\ldots,c_{j-2}\\ c_r\in\{0,1\}}} v\cdot w^{-\left(\sum_{r=0}^{j-2} c_r\cdot p^r\right)+p^{j-1}} \text{ is a } p^{j\text{th}} \text{ power}\right).$$

So $f(n)(v,\alpha)$ says that up to the nth level in the projective system the branch of v agrees with one of those in the tree above, and at the nth level the branch is on a left node. $g(n)(v,\alpha)$ says that up to the nth level the branch of v is in the tree, and at the nth level is on a right node. (As will be seen, one has to worry about "fusion" of left and right).

Suppose now that for each n, the elements

$$\alpha^{\left(\sum_{r=0}^{n-1} c_r\cdot p^r\right)} \quad (c_r\in\{0,1\})$$

have distinct cosets modulo K^{*p^n} (i.e. there is no fusion in the tree).

Then evidently for each pair U, V of disjoint finite sets of integers (≥ 2) there is β in K so that

$$K \models f(n)(\alpha,\beta) \qquad n\in U$$
$$K \models g(n)(\alpha,\beta) \qquad n\in V.$$

Then since f and g are recursive one may conclude by Lemma 2 that either K is not E-coded, or K is not (E_s, Σ_1)-saturated.

So if K is E-coded, and (E_s, Σ_1)-saturated, there is fusion. It follows easily that for some $n > 1$, $\alpha^{p^{n-1}}$ is a $p^{n\text{th}}$ power, i.e. $\alpha^{p^{n-1}} = y^{p^n}$ some y, so $(y^p/\alpha)^{p^{n-1}} = 1$, so $y^p/\alpha = \beta$, β a $p^{n-1\text{th}}$ root of 1.

So the coset of α mod K^{*p} is that of β, a p^{n-1}th root of 1.

We now have our first step towards Theorem 1.

Lemma 3. Suppose K is E-coded and (E_s, Σ_1)-saturated. Then for each prime p, the cosets of K^{*p} in K^* are represented by p^nth roots of 1 for various n.

Corollary. If K is E-coded and (E_s, Σ_1)-saturated, K is perfect.

Note: Lemma 3 easily gives the result (not a trivial consequence of Tennenbaum's Theorem) that if M is a nonstandard model of P then its field of fractions K is not a recursive field. The key idea to be added to Lemma 3 is that K is necessarily weakly recursively saturated. This is a consequence of [R] and overspill.

2.2. A reader familiar with [M2] will have guessed that I must now get information about Artin-Schreier extensions. So I suppose K is E-coded, and (E_s, Σ_1)- saturated. Let K be of characteristic $p > 0$. Let $\tau(x) = x^p - x$. We consider the filtration (of $\mathbb{F}_p$ spaces).

$$K \supset \tau(K) \supset \tau^2(K) \supset \dots .$$

An argument analogous to that for Lemma 3 gives us:

Lemma 4. If K is E-coded and (E_s, Σ_1)-saturated then every coset of $\tau(K)$ in K is occupied by a root of some equation $\tau^j(y) = 0$ (so every coset is occupied by a root of unity).

2.3. Lemmas 3 and 4 are very powerful, in combination with some Galois theory. First we need

Lemma 5. Suppose K is E-coded and (E_s, Σ_1)-saturated. Let L be a finite extension of K. Then L is E-coded and (E_s, Σ_1)-saturated.

Proof: K is perfect, so $L = K(\gamma)$, some γ. Let f be the minimum polynomial of γ over K. Let C be the set of coefficients of f. The L is Σ_0 interpretable in K using C ([M2]). So L can be presented recursively in a presentation of K. So L is E-coded. Then (E, Σ_1)-saturation follows similarly.

Corollary: Suppose K is E-coded and (E_s, Σ_1)-saturated. Then for every finite extension L of K

(a) L is perfect;

(b) for every prime p, the cosets of L^{*p} in L are represented by p^n roots of 1;

(c) If K has characteristic p, the cosets of $\tau(L)$ in L are represented by roots of $\tau^j(y) = 0$ (for various y).

Now we apply a Galois-theoretic argument reminiscent of [M2].

Lemma 6. Suppose K is E-coded and (E_s, Σ_1)-saturated. Then $\tilde{K}$ is obtained from K by adjoining roots of 1.

Proof: K is perfect. Let L be the field got from K by adjoining all roots of 1. By [M2] it suffices to show that every finite extension of L has divisible L^* and is closed under Artin-Schreier extensions.

Let L_1 be a finite extension of L. Let p be a prime. Let $\alpha \in L_1^*$. In $K(\alpha)$, there is a $p^{j\text{th}}$ root β of unity so that $\alpha/\beta \in (K(\alpha))^{*p}$. But $\beta^{1/p} \in L_1$. So $\alpha \in L_1^{*}$. So L_1^* is divisible. A similar argument works for Artin-Schreier extensions, and proves the lemma. □

Corollary. Under the hypotheses of the lemma, G_K is abelian.

Proof: Cyclotomic extension are abelian. □

I now sharpen the corollary. Let Abs(K) be the relative algebraic closure of the prime field in K. The inclusion $\text{Abs}(K) \to K$ is regular, and so the restriction $G_K \to G_{\text{Abs}(K)}$ is surjective. But, under the hypotheses of Lemma 6, $\tilde{K} = \widetilde{K \cdot \text{Abs}(K)}$. So the restriction $G_K \to G_{\text{Abs}(K)}$ is an isomorphism. If K has positive characteristic, $G_{\text{Abs}(K)}$ is procyclic [A].

If K has characteristic 0, $G_{\text{Abs}(K)}$ is a closed subgroup of G_Q, and so by [G], is procyclic, if abelian. This proves:

Lemma 7. Suppose K is $\mathcal{E}$-coded and $(\mathcal{E}_s, \Sigma_1)$-saturated. Then G_K is procyclic, and $\tilde{K}$ is obtained from K by adjoining roots of 1.

This is significant progress, but many K survive the elimination process. That is, there are many elementarily inequivalent K with G_K procyclic and $\tilde{K}$ obtained from K by adjoining roots of 1. Unfortunately, I do not know the general structure of such examples. For now I record what I do know, and later eliminate these known possibilities:

Example 1. K real closed.

Example 2. Let K be an algebraic extension of Q_p, henselian with respect to the unique extension of the p-adic valuation, and having divisible value group. K is an example, and its residue class field may be any algebraic extension of $\mathbb{F}_p$.

Example 3. K any algebraic extension of $\mathbb{F}_p$.

Example 4. If K is infinite in Example 3, K is p.a.c. [J-K] by Weil's Riemann Hypothesis. Now I exhibit p.a.c. examples in characteristic 0. I prefer to do so in terms of nonstandard number theory, but it can routinely be done in terms of ultraproducts of finite fields [A]. (Van den Dries gave me essential help on this example).

Let Q^{ab} be the maximal abelian extension of Q. By Kronecker's Theorem [Ri] Q^{ab} is the union of all cyclotomic extensions of Q. Let $\underline{\Gamma} = \text{Gal}(Q^{ab}|Q)$. Then $\underline{\Gamma} \cong \Pi\, U_p$ [Ri], where U_p is the group of p-adic units, and the product is over all primes. But [Ri] $U_p \cong Z/(p-1) \times \hat{Z}_p$ $(p \neq 2)$, and $U_2 \cong Z/2 \times Z_2$. So there is a continuous epi $\Gamma \twoheadrightarrow \Pi\, Z_p \cong \hat{Z}$. Let Δ be the kernel. Let F be the fixed field of Δ in Q^{ab}. Then $G(\underline{F}|Q) \cong \hat{Z}$. Using Cebotarev's Theorem in the style of [A] or [M1], and exploiting the fact that F/Q is procyclic, one easily obtains an ultrapower Z^* of Z, and a nonstandard prime q in Z^* such that the unique extension of Z^*/q of dimension n is $F_n \otimes_Q Z^*/q$, where F_n is the unique subfield of F of dimension n over Q. Z^*/q is p.a.c. [M1], and since each F_n is included in a cyclotomic extension of Q, it follows that Z^*/q is pseudofinite [A] and

$$\widetilde{(Z^*/q)} = Z^*/q \otimes_Q F,$$

obtained from Z^*/q by adjoining elements of Q^{ab}, and so by adjoining roots of 1.

This concludes my list of examples, and I now show that none of these can provide a counterexample to my conjecture.

First, I consider formally real K.

Lemma 8. Suppose K is $\mathcal{E}$-coded and $(\mathcal{E}_s, \Sigma_1)$-saturated. Then K is not formally real.

Proof: The idea is related to that used in [D2] to show that formally real fields are unstable. Firstly, on Q the order > is definable by

$$(*) \qquad x > 0 \Leftrightarrow (\exists u,v,w,t)(uvwt \neq 0 \wedge x = u^2 + v^2 + w^2 + t^2).$$

Note that for x in Q, this equivalence holds no matter in which formally real K one interprets the right hand side.

To get the Tennenbaum phenomenon, the idea is this. One cuts the rational interval [0,1] into the left box [0, 1/2) and the right box (1/2, 1) and forgets about 1/2. Then one cuts these boxes unto left and right again, and so on. There is complete independence between the left-right decisions at each stage. (Smorynski told me he also had used this idea to show that there are no recursive recursively saturated real closed fields).

Formally, one defines

$\psi_o(v)$ as (the natural formula, using * above, expressing)

$0 < v < 1/2$

$\theta_o(v)$ as: $1/2 < v < 1$;

$\psi_{k+1}(v)$ as:

$0 < v < 1/2^{k+2}$

$\vee\ 2/2^{k+2} < v < 3/2^{k+2}$

$\vee\ 4/2^{k+2} < v < 5/2^{k+2}$

....

$\vee\ \dfrac{2^{k+2}-2}{2^{k+2}} < v < \dfrac{2^{k+2}-1}{2^{k+2}}$

and $\theta_{k+1}(v)$ as

$1/2^{k+2} < v < 2/2^{k+2}$

$\vee$ ....

$\vee \frac{2^{k+2}-1}{2^{k+2}} < v < 1$

Each E_ℓ and θ_ℓ is Σ_1, and one easily verifies that Lemma 2 applies. □.

Next I outline the treatment of the henselian case.

Lemma 9. Suppose K is henselian with respect to a nontrivial valuation, E-coded and and (E_s, Σ_1)-saturated. Then K is algebraically closed.

Proof: Suppose K is henselian with respect to v, with residue class field F and value group Γ.

My objective is to define the relation $v(x) > 0$ by a Σ_1-formula of field theory. (I allow parameters in the definition). If this can be done, one proceeds rather as in the proof of Lemma 8. The details are routine, and I give only the main idea which is used to apply Lemma 2.

Suppose we have t with $v(t) > 0$, and α,β with $v(\alpha) = v(\beta) = 0$, and $v(\alpha-\beta) = 0$. Our sequence of left-right choices for x are:

Stage 0: Left: $v(x-\alpha) > 0$

Right: $v(x-\beta) > 0$

Stage n+1: Restrict x to satisfy the condition:

There are $\lambda_0,\ldots,\lambda_{n-1} \in \{\alpha,\beta\}$ such that

$$v(x - \sum_{j=0}^{n-1} \lambda_j \cdot t^j) > v(t^{n-1}).$$

Then the left condition is:

for some $\lambda_0,\ldots,\lambda_{n-1} \in \{\alpha,\beta\}$

$$v(x - \sum_{j=0}^{n-1} \lambda_j \cdot t^j - \alpha \cdot t^n) > v(t^n).$$

The right condition is:

for some $\lambda_0,\ldots,\lambda_{n-1} \in \{\alpha,\beta\}$

$$v(x - \sum_{j=0}^{n-1} \lambda_j \cdot t^j - \beta \cdot t^n) > v(t^n)$$

It is a routine exercise in valuation theory to verify that the obvious formalizations of the stage n left and right conditions enable one to apply Lemma 2. Indeed the various sums $\sum_{j=0}^{n-1} \lambda_j t^j$ used above witness the finite satisfiability conditions needed in Lemma 2.

It is important for what follows to observe that the preceding argument does not require quite as much as a Σ_1 definition of $v(x) > 0$. It is clearly enough to have a Σ_1 $V(x)$ (perhaps with parameters) so that

$$K \models v(x) > 0 \rightarrow V(x)$$

and

$$K \models v(x) < 0 \rightarrow \neg V(x) \qquad (*)$$

(The point is that we assume nothing about $V(x)$ when $v(x) = 0$.)

Now I suppose that there is no Σ_1 V satisfying the above conditions, and I will deduce that K is algebraically closed, thereby proving the lemma.

First note that $\underline{\Gamma}$ is divisible, since the cosets of K^{*n} in K^* are occupied by roots of 1. (v is trivial on roots of 1.)

Now I analyze the residue class field F. There are unfortunately three cases.

<u>Case 1</u>. Characteristic $F = 0$.

<u>Case 2</u>. Characteristic F = characteristic $K = p > 0$

<u>Case 3</u>. Characteristic $F = p > 0$, but characteristic $K = 0$.

<u>Case 1</u>. I claim F^* is divisible. Suppose not. Then there is $\alpha \in K$, $v(\alpha) = 0$, and an integer n so that there is no y in K with $v(y^n - \alpha) > 0$. Now define $W(x)$ as

$$(\exists y)(y^n = \alpha + x^n).$$

Clearly $v(x) > 0 \Rightarrow \neg W(x)$.
Conversely, by Hensel's Lemma

$$v(x) < 0 \Rightarrow W(x)$$

For let $g(y) = y^n - \alpha - x^n$.

Then
$$v(g'(y)) = v(ny^{n-1})$$
$$= (n-1)v(y),$$
and
$$v(g(y)) = 0.$$

Now define $V(x)$ as

$$x = 0 \ \vee \ (x \neq 0 \ \wedge \ W(x^{-1})).$$

Then V satisfies (*), contradiction.

So F^* is divisible. The same argument applies to F_1^* for any finite extension F_1 of F. For F_1 will be the residue class field of a finite extension K_1 of K, and K_1 will be E-coded, etc.

I conclude by Galois theory that F is algebraically closed. Now I use this to show that K is algebraically closed. It suffices to show that L^* is divisible, for all finite extensions L^* of K. But this is easy, by Hensel's Lemma, using the fact that $\underline{\Gamma}$ is divisible.

Case 2. F is perfect, since K is. The same argument as in Case 1 shows that F^* is n-divisible, for n prime to p. Then Hensel's Lemma shows that L^* is n-divisible, for n prime to p, and L any finite extension of K. L^* is of course perfect. To get K algebraically closed, I could show that each L is closed under Artin-Schreier extensions. However, there is another method, related to Duret's [D1]. I will show in Lemma 10 below that Abs(L), the algebraic closure of F_p in L, must be finite if K is E-coded and (E_s, Σ_1)-saturated, and not algebraically closed. But the above divisibility conditions on L imply that some Abs(L) is infinite. So K is algebraically closed.

Case 3. The same argument as in the preceding cases shows that L^* is n-divisible, for n prime to p and L a finite extension of K. Now suppose some L^* is not p-divisible. It follows from Lemma 3 that L^* contains all p^{th} roots of 1. By the n-divisibility, G_L is a pro-p-group, and unless G_L is 1 G_L is $\mathbb{Z}_p$. Each extension of L is Kummer, and the unique extension of dimension p^j is obtained by adjoining a suitable $p^{\ell^{th}}$ root of 1.

I now show that L is algebraically closed, whence K is, since K is not formally real. It suffices to show that Abs(L) is algebraically closed. Now, v on Abs(L) is an extension of the p-adic valuation on Q, and Abs(L) is henselian. So Abs(L) naturally contains the algebraic p-adic field $Q_p \cap \widetilde{Q}$, which is elementarily equivalent to Q_p, and algebraically maximal [K]. All finite extensions of $Q_p \cap \widetilde{Q}$ are algebraically maximal too. It follows that Abs(L) is algebraically maximal.

So, since the value group of Abs(L) is divisible, any extension of v to a proper finite extension L_1 of L must extend the residue class field of L. But [Ri] if M is a finite extension of Q_p and α is a primitive $p^{\ell^{th}}$ root of 1, $M(\alpha)$ is a totally ramified extension of M, so that the residue class field of M is not extended. The same holds for $Q_p \cap \widetilde{Q}$ of course. So the $p^{\ell^{th}}$ cyclotomic extension of L is immediate, so is L. So $L = \widetilde{L}$. □

I now turn to the last supplement to Lemma 3. It is important to recall that the relevant hypothesis, that K be p.a.c. [J-K], was the one I originally encountered in [M1]. The treatment I give is based on work of Duret [D1], and replaces my original ad hoc analysis of pseudofinite fields.

Duret proved that if K is perfect and p.a.c. but not algebraically closed then Th(K) has the independence property (so Th(K) is not stable). Since any infinite algebraic extension of F_p is p.a.c. [A], it follows from the details of [D1] that if K has positive characteristic and Abs(K) is infinite but not algebraically closed then Th(K) has the independence property.

For my purposes, it suffices to observe that Duret actually proved something stronger, namely the independence property using existential formulas. A careful inspection of his proof shows the following for each perfect p.a.c. field K which is not algebraically closed:

There is an existential formula $\Phi(\vec{x}, \vec{u}, \vec{v}, t)$, tuples $\vec{u}_o, \vec{u}_1, \vec{v}_o$ from K and an infinite subset A of K such that

i) $K \models (\forall x, t)\ (\neg\, \Phi(\vec{x}, \vec{u}_o, \vec{v}_o, t) \vee \neg\Phi(\vec{x}, \vec{u}_1, \vec{v}_o, t))$;

ii) for each pair X,Y of disjoint finite subsets of A, there is $\vec{\gamma}$ in K such that

$$t \in X \Rightarrow K \models \Phi(\vec{\gamma}, \vec{u}_o, \vec{v}_o, t)$$

$$t \in Y \Rightarrow K \models \Phi(\vec{\gamma}, \vec{u}_1, \vec{v}_o, t).$$

Further, if K has characteristic 0, A can be chosen as $\mathbb{N}$, and if K has positive characteristic p then A can be taken as any infinite set linearly independent over $\mathbb{F}_p$.

The final useful observation is that one can replace the hypothesis K is p.a.c. by K has a not algebraically closed p.a.c. subfield K_o which is relatively algebraically closed in K. The only change to be made above is that in characteristic p A must be a subset of K_o.

Now I easily prove:

Lemma 10. Suppose K has a p.a.c. subfield K_o which is not algebraically closed but is relatively algebraically closed in K. If K is E-coded then K is not (E_s, Σ_1)-saturated.

Proof: I use Φ, $\vec{u}_o, \vec{u}_1, \vec{v}_o$ A as above. In characteristic 0, A is $\mathbb{N}$. In characteristic p, a counterexample to the lemma must be of positive transcendence degree (by saturation). Let w be transcendental and let $A = \{w^n : n \in \mathbb{N}\}$. Now it is obvious how to separate the inseparable, if K is E-coded and (E_s, Σ_1)-saturated. □

Corollary. If K has positive characteristic, is not algebraically closed and is both E-coded and (E_s, Σ_1)-saturated, then Abs(K) is finite.

Proof: If Abs(K) is infinite, Abs(K) is p.a.c. [A]. Then use Lemma 10. □

This concludes the proof of Theorem 1. I regret as much as the reader the above ad hoc discussion.

I make one final remark about any counterexample K to my conjecture. K cannot be stable. For by the above, for some prime q and some finite extension L of K, L^*/L^{*q} is finite and the cosets are occupied by roots of 1, so by [M2] one gets a symmetric irreflexive relation definable on an infinite subset of L. So L, whence K, is unstable. It is important to note that the preceding corollary lets me bypass Artin-Schreier extensions.

3. Towards Vaught's Conjecture for Fields.

The original motivation for this work was from recursive model theory. But there are interesting implications for pure model theory.

Theorem 2. Let K be an infinite field which is not algebraically closed. Let $T = Th(K)$. Suppose that for each n $S_n(T)$ has cardinal $< 2^{\aleph_0}$ (equivalently, $S_n(T)$ is countable). Then G_K is procyclic, and $\widetilde{K}$ is obtained from K by adjoining roots of 1. K is not formally real, henselian nor p.a.c. If K has positive characteristic, K is perfect and $Abs(K)$ is finite.

Corollary. Same result, but with the third sentence replaced by T has $< 2^{\aleph_0}$ non-isomorphic countable models.

The corollary is of course immediate from the theorem. To prove the theorem one observes that each time we applied Lemma 3 above we actually constructed $2^{\aleph_0}$ n-types for some n. This is essentially obvious, but note that e.g. in the case of p.a.c. fields it was important to get A as $\mathbb{N}$ or $\{w^n : n \in \mathbb{N}\}$ (i.e. "finitely generated").

4. Concluding Remarks. How is one to prove the conjecture, and the obvious related one that some $S_n(T)$ has cardinal $2^{\aleph_0}$? Somehow one has to convert the known instability of a counterexample into a kind of independence property. Note however that $Th(\mathbb{R})$ does not have the independence property (this was communicated to me by Poizat). In the case of $\mathbb{R}$ we got by with a "Cantor decomposition" of $\mathbb{Q}$. I believe it worth while to look for an analogue for the order one gets via the symmetric irreflexive relation mentioned at the end of 2. Prima facie, this seems to involve additive Ramsey combinatorics in the style of [Mi].

References

(A) J. Ax, The elementary theory of finite fields, Ann. of Math. (2) 88 (1968), 239-271.

(C). W. Craig, On axiomatizability within a system, J. S. L. 18 (1953), 30-32.

(D1) J. L. Duret, Les corps pseudofinis ont la propriété d' indépendance. C. R. Acad. Sc. Paris 290 (1980), 981-3.

(D2) ________, Instabilité des corps formellement reéls, Canad. Math. Bull 20 (3), 1977.

(G) W-D Geyer, Unendliche algebraische Zahlkörper, über denen jede Gleichung auflösbar von beschränkter stufe ist, Journal of Number Theory 1 (1970), 346-374.

(J-K) M. Jarden and U. Kiehne, The elementary theory of algebraic fields of finite corank, Inventiones Math. 30 (1975), 275-294.

(K) I. Kaplansky, Maximal fields with valuation, Duke Math. Journal, 9 (1942), 303-321.

(M1) A. Macintyre, Residue fields of models of IP, to appear in Proceedings of 1979 Hannover I. C. L. H. P. S. meeting.

(M2) A. Macintyre, On ω_1-categorical theories of fields, Fund. Math. LXXI (1971), 1-25.

(Mi) K. Milliken, Hindman's theorem and groups, Journal of Combinatorial Theory A 25 (1978), 175-180.

(R) J. Robinson, Definability and decision problems in arithmetic, J. S. L. 14 (1949), 98-114.

(Ri) P. Ribenboim, L' Arithmetique des Corps, Hermann, Paris, 1972.

(T) S. Tennenbaum, unpublished, c. 1958.

The topos of types

by M. Makkai*

Introduction

In this paper we introduce and study a construction associating a certain new topos, the prime completion, with any coherent topos. Bearing in mind the close connection of coherent toposes with (finitary first order) theories (see MR), we can also say that we associate a new topos, called the topos of types in this context, with any theory.

In the main part of the paper, the terminology will be categorical; in this introduction, we make some remarks clarifying connections with ordinary model theoretical concepts.

For us, a theory $T = (T,F)$ consists of a fragment F of $L_{\omega\omega}$ over a possibly many sorted language (allowing possibly empty sorts), closed under $\wedge$, $\vee$ and $\exists$ (but not necessarily $\neg$), and a set T of axioms of the form $\forall \vec{x}(\varphi(\vec{x}) \rightarrow \psi(\vec{x}))$ with φ and ψ in F. Given a model M of T, and a finite tuple $\vec{a}$ of elements of M (of various sorts), the type $\vec{a}$ in M is $t^M(\vec{a}) \underset{\text{df}}{=} \{\varphi(\vec{x}) \in F : M \models \varphi[\vec{a}]\}$; a (complete) type of T is any set $p(\vec{x})$ of formulas such that there is $M \models T$ and $\vec{a}$ in M such that $p(\vec{x}) = t^M(\vec{a})$. (There also is a straightforward 'syntactical' definition of a type; this will be translated into the notion of prime filter in Section 1.) This notion naturally generalizes the one usually considered in relation with complete theories (by definition, with F as the full logic $L_{\omega\omega}$). An F-elementary map between models of T is one that preserves the truth of formulas in F; the category of models of T, $\mathrm{Mod}\,T$, is the one whose objects are the models of T, and whose morphisms are the F-elementary maps.

One of the basic constructions of categorical logic is that of the classifying topos $\mathcal{E}(T)$ of the theory T; see MR, and concerning arbitrary fragments, Section 5 in [10]. Just as $\mathcal{E}(T)$, the topos of types $\widetilde{P}(T)$ of T is a syntactical construction in the sense that it is directly made up of formulas in $L_{\infty\omega}$ 'coming from T' (although not just from its underlying fragment F). Although our definition of $\widetilde{P}(T)$ below does not present it explicitly as a syntactical construction, the elaborations in Chapter 8, MR, if applied to $\widetilde{P}(T)$, would show the syntactical nature of $\widetilde{P}(T)$.

---

* This research was supported by a grant of the Natural Sciences and Engineering Research Council of Canada.

The topos of types is not primarily a tool for solving problems already stated in model theory; rather, just as other constructions of categorical logic, it is a conceptual tool meant to enable us to formulate precisely certain natural intuitive questions, as well as to put in a conceptually satisfactory form results that would without it sound rather technical. To indicate the need for structures like the topos of types, we point out that if we want to state precisely assertions such as the category of models of a theory determines the theory, or at least certain syntactical aspects of the theory, then it is reasonable to define an abstract structure embodying these syntactical aspects that will be invariant under renaming of symbols and other trivial notational changes, since notational features cannot possibly be recaptured from the category of models.

The definition of $\tilde{P}(T)$ is most simply put by using the terminology of sheaf theory. First of all, we construct the category P of types of T. The objects of P are the types of T. To describe the morphisms, let X and Y (for simplicity) be two sorts of T, $p: = p(x)$, $q: = q(y)$ be two types (x,y variables of sort X, Y, respectively), $A_i(x)$, $B_i(x,y)$ ($i = 1,2$) formulas (always in F) such that T proves that B_i defines a function β_i from the extension of $A_i(x)$, to the extension of $q(y)$ ($=\bigcap\{B: B(y) \in q\}$). We say that B_1 and B_2 define the same <u>germ of functions</u> $p \longrightarrow q$ if $A_i(x) \in p$ and for some $A(x) \in p$ with $T \models \forall x(A(x) \to A_1(x) \wedge A_2(x))$, we have that T proves that β_1 and β_2 restricted to A are the same. A morphism $p \longrightarrow q$ of P is defined to be a germ of definable functions $p \to q$, also with tuples of variables in place of x and y. Given a germ $\tilde{f}: p \to q$ (the equivalence class of f) we say that $\tilde{f}$ is a <u>cover</u> if q is (set theoretically) the smallest type q': $q'(y)$ over Y such that f defines a germ $p \to q'$. We make P into a site by endowing it with the Grothendieck topology generated by all the (single) covers. Finally, $\tilde{P}(T)$ is defined as the category of sheaves over the site P.

Besides the definition, there are two other descriptions of the topos of types. One is a universal property defining it; it is one similar to the universal property of the classifying topos ('the most general topos valued model') but it is more involved; see Theorem 1.1. The final description is through a representation theorem, Theorem 2.3, which says the following. Let K be a subcategory of Mod T. A functor F from K to SET, the category of sets, is said to have the <u>finite support property</u> (f.s.p.) if the following holds: whenever M is a model in K and $a \in F(M)$, then there is a finite set $\{x_1,\ldots,x_n\}$ of elements of M (a <u>support</u> of a) such that: for any N in K and any two F-elementary maps in K, $M \underset{h}{\overset{g}{\rightrightarrows}} N$, if for all $i = 1,\ldots,n$ we have $g(x_i) = h(x_i)$, then $(F(g))(a) = (F(h))(a)$. Let λ be a fixed cardinal in which sufficiently many special models (see CK) of T exist, and let K be the full subcategory of Mod T consisting of the special models of power λ. Then Theorem 2.3 says that $\tilde{P}(T)$ is equivalent to the category f.s.p.(K, SET), the full subcategory of the category (K, SET) of all functors $K \to \text{SET}$ (with natural transformations as morphisms) consisting of the functors with the f.s.p.

We remark that this gives a purely semantical description of the topos of types. (For $\mathcal{E}(T)$, the classifying topos, we do not know of any 'semantical' description.) In fact, this description may serve as an introduction of the notion of topos to the model theorist. One starts by observing that the formulas, and also the types, of the theory naturally give rise to certain functors $\mathrm{Mod}\,T \longrightarrow \mathrm{SET}$, and by restriction, functors $K \longrightarrow \mathrm{SET}$; call these functors coming from types <u>standard</u>; next one observes that the standard functors have the f.s.p.; one wonders if the f.s.p. is to any extent characteristic of standard functors; the answer is 'yes but not quite'; the functors $K \longrightarrow \mathrm{SET}$ with the f.s.p. form the subtopos of (K,SET) generated by the standard functors.

We arrived at the notion of the topos of types through our studies on M. Barr's full embedding theorem [2] on exact categories, itself a generalization of B. Mitchell's full embedding theorem on Abelian categories. We found that the topos of types can be used to show the existence of full and in fact continuous embeddings of certain coherent toposes into functor categories, considerably generalizing Barr's theorem. We then found the simply defined class of prime generated coherent toposes all of which have such embeddings; we wrote out a direct proof of this fact avoiding the prime completion in [10]. Here we restate the true state of affairs by proving a rather technical but general theorem (2.6) concerning the canonical embedding of $\mathcal{E}(T)$ into $\widetilde{P}(T)$; this, together with 2.3, gives us in Section 3 a proper generalization of the theorem on coherent prime generated toposes (Theorem 3.2).

Observe that f.s.p.(K,SET) is not defined for an abstract category K: in particular, from the equivalence of two categories K, K' of models one cannot conclude that $\widetilde{P}(T) \simeq$ f.s.p.(K,SET) is equivalent to $\widetilde{P}(T') \simeq$ f.s.p.(K',SET). On the other hand, as André Joyal observed, there is another property of 'standard' functors $\mathrm{Mod}\,T \longrightarrow \mathrm{SET}$ (see above), namely that they are <u>upcontinuous</u>, i.e., they preserve directed colimits (ascending unions) in $\mathrm{Mod}\,T$; one has the category of all upcontinuous functors, $\mathrm{Upcon}(K,\mathrm{SET})$, defined for any abstract category K. After some special results obtained by the present author in the same direction, Daniel Lascar, using 'generalized' stability theory [6], succeeded in proving an interesting theorem saying that for a certain large class of theories T, it is true that every upcontinuous functor $\mathrm{Mod}\,T \longrightarrow \mathrm{SET}$ has the f.s.p. (see also Section 3 below); this class includes those theories that he calls G-trivial. We deduce from Lascar's theorem that the topos of types of a G-trivial theory T can be recovered from the category of models of T (Theorem 3.3 and its proof).

Perhaps the most interesting conclusion emerging from Lascar's work is Theorem 3.4: for a G-trivial $\aleph_0$-categorical theory T, the classifying topos $\mathcal{E}(T)$ (hence, 'for all practical purposes', the theory T itself) is determined by $\mathrm{Mod}\,T$, in the sense that for any other (not necessarily G-trivial) theory T', if Mod T' ≃ Mod T, then $\mathcal{E}(T') \simeq \mathcal{E}(T)$. G-trivial $\aleph_0$-categorical theories include the theories of equality on an infinite set, of dense linear orders without end points, of infinite dimensional vector spaces over a finite field, and many others: the theorem is non-trivial already for these special cases. We deduce the above result from Lascar's by a simple application of a theorem in MR.

Throughout this paper, we freely use the terminology of (Grothendieck) topos theory. Except occasional explicit references to the original source, SGA4, what is contained in MR suffices. We also use the connections between logic and topos theory established in MR.

A piece of new terminology is the notion of a <u>regular site</u>. A regular site is one whose underlying category is finitely complete, whose topology is subcanonical and is generated by finite covering families and in which every morphism f can be factored as $f = h \circ g$ where g is a cover (i.e. it forms by itself a covering of its codomain), and h is a monomorphism.

In MR, p.166, we define a theory $T_{\mathcal{C}}$ associated with an arbitrary site $\mathcal{C}$; in particular, a model of $T_{\mathcal{C}}$ becomes the same as a continuous functor $\mathcal{C}$ into SET. We emphasize that by a continuous functor we mean one that is left exact and preserves coverings (this terminology is at variance with that of SGA4); also, any (Grothendieck) topos is understood to be the site with the canonical topology. We also talk about a model of $\mathcal{C}$, meaning a continuous functor with domain $\mathcal{C}$.

The presentation in Section 1 was inspired by the paper [5]. In Section 3 of [5], the authors introduce the category of existential types; that is the same as the category of types $\mathcal{P}$ as described above. However, the topology on $\mathcal{P}$ introduced in [5] is different from the one we use. Originally (in the Fall of 1977, and independently of [5]), we constructed the topos of types in the way sketched at the end of Section 1 (although at that time we did not have 1.1 and 2.3 in their present forms). The presentation given here perhaps has the advantage of consisting of steps that are natural from the point of view of topos theory, more so than with the original presentation.

§1. <u>The prime completion</u>.

Let $\mathcal{D}$ be a site. An object X of $\mathcal{D}$ is called a <u>prime object</u> (or a <u>prime</u>) if the following holds: whenever $\{X_i \longrightarrow X: i \in I\}$ is a covering in $\mathcal{D}$, there is $i \in I$ such that the singleton $\{X_i \longrightarrow X\}$ is a covering as well (we also say: the morphism $X_i \longrightarrow X$ is a cover). $\mathcal{D}$ is <u>prime-generated</u> if every object X of $\mathcal{D}$ has a covering $\{X_i \longrightarrow X: i \in I\}$ with each X_i being a prime. A (Grothendieck) topos is <u>prime-generated</u> if it is as a site with the canonical topology, i.e. if it has a family of generators consisting of prime objects. In a topos, an object is prime just in case it (its maximal subobject) is not the supremum of its proper subobjects.

If $\widetilde{\mathcal{D}}$ is the category of sheaves over $\mathcal{D}$, $\underset{\sim}{\varepsilon}: \mathcal{D} \to \widetilde{\mathcal{D}}$ is the canonical functor, i.e. the composite of the Yoneda embedding $\mathcal{D} \to \hat{\mathcal{D}}$ followed by the associated sheaf-functor $\underset{\sim}{a}: \hat{\mathcal{D}} \to \widetilde{\mathcal{D}}$, and X is a prime object in $\mathcal{D}$, then $\underset{\sim}{\varepsilon}(X)$ is a prime object in $\widetilde{\mathcal{D}}$ (this is because every covering $\{E_i \to \underset{\sim}{\varepsilon}X: i \in I\}$ can be refined to a covering $\{\underset{\sim}{\varepsilon}X_{ij} \to \underset{\sim}{\varepsilon}E_i \to \underset{\sim}{\varepsilon}X: j \in J_i, i \in I\}$ with $\underset{\sim}{\varepsilon}X_{ij} \to \underset{\sim}{\varepsilon}X$ being $\underset{\sim}{\varepsilon}(f)$ for some f in $\mathcal{D}$; see 1.3.8 (i), p. 35 in MR). It follows that for a prime-generated site $\mathcal{D}$, the category of sheaves over $\mathcal{D}$, $\widetilde{\mathcal{D}}$, is a prime-generated topos.

A site $\mathcal{P}$ in which every object is a prime is called a <u>prime site</u>; in this case the topology on $\mathcal{P}$ is generated by coverings that are singletons, and the class C of covers in $\mathcal{P}$ has the following properties: (i) all isomorphisms are in C, (ii) C is closed under composition, (iii) if $p \longrightarrow q$ belongs to C, $r \longrightarrow q$ is any morphism, then there is $(s \longrightarrow r) \in C$ and $s \longrightarrow p$ such that

$$\begin{array}{ccc} p & \longrightarrow & q \\ \uparrow & & \uparrow \\ s & \longrightarrow & r \end{array}$$

commutes, and (iv) if the composite $p \longrightarrow q \longrightarrow r$ belongs to C, then so does $q \longrightarrow r$.

Conversely, if $\mathcal{P}$ is a category, C is a class of morphisms with the above properties (i) - (iv), then C generates a Grothendieck topology in which $\{p_i \longrightarrow p: i \in I\}$ is a covering just in case $(p_i \to p) \in C$ for some $i \in I$, hence $\mathcal{P}$ with this topology is a prime site.

If $\mathcal{D}$ is a prime-generated site, $\mathcal{P}$ is the full subcategory of $\mathcal{D}$ consisting of the prime objects, and $\mathcal{P}$ is regarded a site with the induced topology, then $\mathcal{P}$ is a prime site and the categories of sheaves over $\mathcal{D}$ and $\mathcal{P}$ are canonically equivalent ('Lemme de comparaison', [SGA4, I, p. 288]). In particular, if $\mathcal{E}$ is a prime-generated topos, and $\mathcal{P}_{\mathcal{E}}$ is the full subcategory of the primes of $\mathcal{E}$ with the induced topology, then $\mathcal{E} \simeq \widetilde{\mathcal{P}_{\mathcal{E}}}$.

From now on, we fix a site $\mathcal{C}$, and assume that it is a regular site (see the Introduction). All entities we introduce below will depend on $\mathcal{C}$, although the dependence might not be mentioned explicitly.

Let X be an object in $\mathcal{C}$. A <u>prime filter on</u> X is a set p of subobjects of X with the following properties: (i) 1_X (the maximal subobject of X) $\in p$; (ii) $A \in p$ and $B \in p$ imply $A \wedge B \in p$, (iii) $A \in p$ and $A \leq B$ imply $B \in p$, and (iv) $A \in p$ and $\{A_i \leq A: i \in I\}$ is a covering in $\mathcal{C}$ (with $A_i \in \mathrm{Sub}_{\mathcal{C}}(X)$) imply that $A_i \in p$ for some $i \in I$. [An equivalent definition would be: p is a prime filter on X iff for some $h: Sub_C(X) \to \mathbf{2}$, with $\mathbf{2}$ the two-element Boolean algebra, such that h is (left exact and) continuous with respect to the topology on $Sub_C(X)$ induced by that on C and the obvious topology on $\mathbf{2}$, we have $p = \{A \in Sub_C(X): h(A) = 1\}$]. The set of prime filters on X is denoted by $P(X)$. To stress the object X, we also write (p,X) for p. Given a morphism $f: A \longrightarrow Y$ with $A \in Sub_C(X)$ [we ambiguously use the same notation for a subobject and the domain object of a representative monomorphism of it] and a prime filter $p \in P(X)$ with $A \in p$, $f(p)$ denotes the set $\{B \in Sub_C(Y): f^{-1}(B) \in p\}$ ($f^{-1}(B)$ is a subobject of A, hence, in the natural way, a subobject of X as well; it is in the latter sense that we use the notation here). It is immediately seen that $f(p)$ is a prime filter on Y.

Intuitively, we will deal with a (prime) filter as if it were a (formal) intersection of the subobjects contained in it. This is important to keep in mind to understand our definitions; also, in the prime completion (see below), this 'becomes' literally true.

Let $M: C \to E$ be a model of C in a topos E (i.e., M is continuous with respect to the topology on C and the canonical topology on E, also, M is left exact). For a prime filter (p,X) of C, $M(p,X)$ denotes the subobject $\bigwedge^{(E)}\{M(A): A \in p\}$ of $M(X)$ in E ($\bigwedge^{(E)}$ denotes intersection (g.l.b.) in $Sub_E(M(X))$). We call M a p-<u>model</u> if the following holds: whenever $p \in P(X)$, $A \in p$, $f: A \longrightarrow Y$, then $Im(M(f)|M(p,X)) = M(f(p),Y)$.

The main result of this section is the construction of a "generic p-model of C in a prime-generated topos". The morphisms between prime-generated toposes are taken to be the continuous functors that preserve intersections of arbitrary families of subobjects of any fixed object in the domain topos; let $\bigwedge(E_1, E_2)$ denote the category of all such functors $E_1 \longrightarrow E_2$, a full subcategory of the category (E_1, E_2) of all functors $E_1 \longrightarrow E_2$. With this in mind, a "generic p-model of C in a prime-generated topos" is a p-model $M_0: C \longrightarrow \widetilde{P}$ in a prime-generated topos $\widetilde{P}$ such that for any p-model $M: C \longrightarrow E$ in a prime-generated topos E there is $\widetilde{M} \in |\bigwedge(\widetilde{P},E)|$, unique up to a unique isomorphism, such that

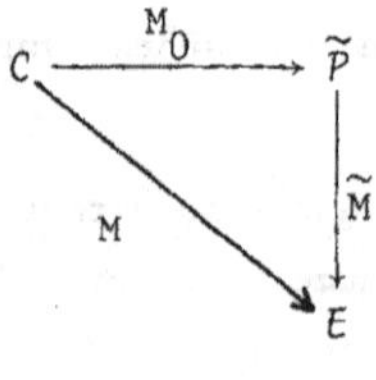

commutes. Formulated more sharply, this means that with the category $p(C,E)$ of all p-models $C \to E$ (a full subcategory of (C,E)), the functor

$$\bigwedge(\widetilde{P},E) \longrightarrow (C,E)$$

defined by composition with M_0 factors through $p(C,E) \xrightarrow{\text{incl.}} (C,E)$ and gives an equivalence

$$\Lambda(\widetilde{P},E) \xrightarrow{\sim} p(C,E).$$

The topos $\widetilde{P}$, also called the *prime completion* of C, will be the category of sheaves over P, the *prime site of prime filters of* C, described as follows.

Fix a prime filter (p,X) and an object $Y \in |C|$. Two morphisms f, f' ($f: A \to Y$, $f': A' \to Y$; $A, A' \in p$) are *equivalent* if for some $A'' \in p$, $A'' \leq A \wedge A'$, we have $f|A'' = f'|A''$. A *germ* (of morphisms) $(p,X) \to Y$ is an equivalence class of the above equivalence relation; the germ represented by $f: X \to Y$ is denoted by $\tilde{f}$. If in addition $q \in P(Y)$, then the germ $\tilde{f}: (p,X) \to Y$ is a *morphism* $\tilde{f}: (p,X) \to (q,Y)$ if for all $B \in q$ we have $f^{-1}(B) \in p$; it is easy to see that this definition does not depend but on $\tilde{f}$. The *category of prime filters of* C, P, is the category whose objects are the prime filters of C, and whose morphisms are as indicated. Composition of morphisms is defined in the obvious way: for $\tilde{f}: (p,X) \longrightarrow (q,Y)$ with $f: A \longrightarrow Y$, and $\tilde{g}: (q,Y) \longrightarrow (r,Z)$, with $g: B \longrightarrow Z$, we define $\tilde{g} \circ \tilde{f} = \widetilde{g \circ f'}$, where f' comes from the following factorization:

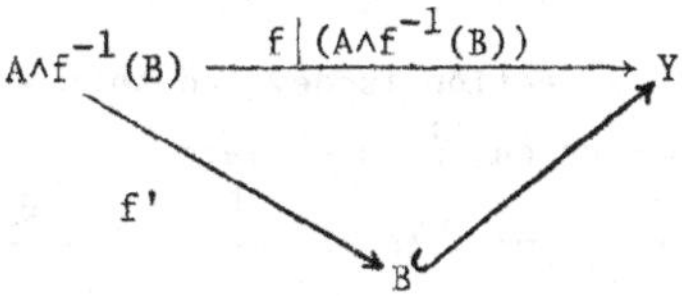

Again, it is easy to see that this definition is legitimate. For later reference, note that the definition of composition makes sense if, instead, $\tilde{g}: (q,Y) \longrightarrow Z$ (without r), and then it gives a germ $\tilde{g} \circ \tilde{f}: (p,X) \longrightarrow Z$; also $\tilde{f}: (p,X) \longrightarrow Y$ and $g: Y \longrightarrow Z$ naturally combine to give $g \circ \tilde{f}: (p,X) \longrightarrow Z$.

It would be simpler to define P by only considering germs $\tilde{f}: (p,X) \longrightarrow Y$ derived from morphisms $f: X \longrightarrow Y$ with 'total' domain X. However, this is not sufficient (the proof of $\hat{(-)}$ preserving equalizers breaks down, see (1.6) below). On the other hand, it is often possible to 'pretend' that we have the simpler definition. In fact, let $p \in P(X)$ and $A \in p$, and let A also denote the domain object of a monomorphism representing A. Then the set $p' = \{B \in p: B \leq A\}$, with each $B \leq A$ understood, in the natural way, as a subobject of the object A, is easily seen to be a prime filter over A; moreover the morphism $(p',A) \xrightarrow{\tilde{i}} (p,X)$ with $i: A \longrightarrow X$ the structure morphism of the subobject $A \in \mathrm{Sub}_C(X)$ is an isomorphism (because of our extended definition of morphisms!). So, if one is given a morphism $(p,X) \xrightarrow{\tilde{f}} Y$ with $A \in p$, $f: A \longrightarrow Y$, then by passing to the object (p',A) of P isomorphic to (p,X), we are in the situation of having a morphism $(p',A) \xrightarrow{\tilde{f}} Y$ with $\mathrm{dom}(f) = A$. This procedure allows simplifying the notation sometimes.

We make P into a site by defining the topology on P to be the one generated by the singletons $\{(p,X) \xrightarrow{f} (f(p),Y)\}$, for all $p \in P(X)$ and $f: A \to Y$ $(A \in p)$, with $f(p)$ defined above; such a morphism will be called a <u>cover</u>. We will prove below that P is a prime site. As usual, $\tilde{P}$ denotes the category of sheaves over P.

We define the functor $(\hat{-}): C \longrightarrow \hat{P}$, with $\hat{P}$ the category of presheaves over P, as follows. We assign to $X \in |C|$ the presheaf $\hat{X}$ such that $\hat{X}(p,Y) = \operatorname{Hom}((p,Y),X) =$ the set of all germs $(p,Y) \to X$. For $\tilde{f}: (p',Y') \to (p,Y)$, we define $\hat{X}(\tilde{f})$ to be the map $\tilde{g} \longmapsto \tilde{g} \circ \tilde{f}$ $(\tilde{g}: (p,Y) \to X)$. To $X \xrightarrow{f} X'$ in C, we assign the natural transformation $\hat{X} \xrightarrow{\hat{f}} \hat{X}'$ such that $\hat{f}_{(p,Y)}$ is the map $\tilde{g} \longmapsto f \circ \tilde{g}$ $(g: (p,Y) \longrightarrow X)$. Let $\underset{\sim}{a}: \hat{P} \longrightarrow \tilde{P}$ be the associated sheaf-functor; let $M_0 = C \longrightarrow \tilde{P}$ be the composite $M_0 = \underset{\sim}{a} \circ (\hat{-})$.

<u>Theorem 1.1.</u> Suppose C is a regular site. Then $M_0: C \longrightarrow \tilde{P}$ constructed above is a generic p-model of C in the prime-generated topos $\tilde{P}$; composing with M_0 defines an equivalence

$$\bigwedge(\tilde{P},E) \xrightarrow{\sim} p(C,E)$$

for any prime-generated topos E.

Almost all of the rest of this section is devoted to the proof of the theorem; we also obtain additional information on $\tilde{P}$ used later.

As usual, we denote by $\underset{\sim}{\varepsilon}$ the composite $P \xrightarrow{\underset{\sim}{h}} \hat{P} \xrightarrow{\underset{\sim}{a}} \tilde{P}$, where $\underset{\sim}{h}$ is the Yoneda embedding; also, $\underset{\sim}{h}$ will be considered an inclusion, i.e. $\underset{\sim}{h}(p,X)$ will be written (p,X), e.t.c.

Given any model $M: C \longrightarrow E$ in a topos E, we can deduce a functor $\bar{M}: P \longrightarrow E$ as follows. We put $\bar{M}(p,X)$ to be equal to what we called $M(p,X)$ above. For a morphism $(p,X) \xrightarrow{\tilde{f}} (q,Y)$, we define $\bar{M}(\tilde{f})$ as follows. $\tilde{f}$ is the germ represented by some $f: A_0 \to Y$ $(A_0 \in p)$; we have that $B \in q$ implies $f^{-1}(B) \in q$; and we have $\bar{M}(p,X) = \bigwedge\{M(A): A \in p\}$, $\bar{M}(q,Y) = \bigwedge\{M(B): B \in q\}$. It follows that $M(f)|\bar{M}(p,X)$: $\bar{M}(p,X) \to M(Y)$ factors through $\bar{M}(q,Y) \hookrightarrow M(Y)$ and we have a unique morphism, denoted by $\bar{M}(\tilde{f}): \bar{M}(p,X) \to \bar{M}(q,Y)$ such that $M(f)|\bar{M}(p,x)$ equals the composite $\bar{M}(p,X) \xrightarrow{M(\tilde{f})} \bar{M}(q,Y) \hookrightarrow M(Y)$. Note in particular that $\bar{M}(\tilde{f})$ is well-defined, i.e. its definition does not depend but on $\tilde{f}$. We have $\bar{M}(\tilde{g} \circ \tilde{f}) = \bar{M}(\tilde{g}) \circ \bar{M}(\tilde{f})$ whenever $\tilde{g} \circ \tilde{f}$ makes sense, so $\bar{M}$ is indeed a functor.

To say that M is a p-model is equivalent to saying that $\bar{M}$ carries every cover in P into a (canonical) cover in E.

Let $M: C \to SET$ be a model in SET, $X \in |C|$ and $x \in M(X)$. The set $\{A \in \operatorname{Sub}_C(X): x \in M(A)\}$ is called the <u>type</u> of x, and it is denoted by $t_X(x,M)$ (or just $t_X(x)$). It is immediately seen that $t_X(x)$ is a prime filter on X. If

$X \xrightarrow{f} Y$ is a morphism in C, then we have the equality $t_Y(M(f)(x)) = f(t_X(x))$ (for $f(p)$, see above); this is easily seen. A generalization of the fact of $t_X(x)$ being a prime filter is the following useful lemma.

(1.2) Let E be a topos, $M: C \to E$ a model, $X \in |C|$, η a prime in E and $\beta: \eta \to M(X)$ a morphism in E. Then the set $t(\beta) = \{A \in Sub_C(X): \beta$ factors through $M(A) \longrightarrow M(X)\}$ is a prime filter on X. Moreover, β factors through $\bar{M}(t(\beta),X) \longrightarrow M(X)$, and in fact, $t(\beta)$ is the (set-theoretically) largest prime filter on X with this property.

Proof. Suppose $A \in t(\beta)$ and $\{A_i \leq A: i \in I\}$ is a covering in C by subobjects $A_i \in Sub_C(X)$. Then we have a morphism $\eta \to M(A)$ such that β equals the composite $\eta \longrightarrow M(A) \longrightarrow M(X)$. Consider the family $\{(M(A_i) \underset{M(A)}{\times} \eta) \longrightarrow \eta: i \in I\}$; it is a covering since $\{M(A_i) \leq M(A): i \in I\}$ is one. Since η is a prime, there is $i \in I$ such that $(M(A_i) \underset{M(A)}{\times} \eta) \longrightarrow \eta$ is an isomorphism, hence $\beta: = (\eta \to M(A) \longrightarrow M(X))$ factors through $M(A_i) \longrightarrow M(X)$, i.e. $A_i \in t(\beta)$, as required. The other properties of a prime filter are similarly easy to check. The "moreover" part is easily seen. □

(1.3) Let $p \in P(X)$. Then there is a model $M: C \to SET$ with some $x \in M(X)$ such that $t_X(x,M) = p$. In fact, M can be chosen to be $\aleph_0$-saturated.

Proof. This is a good opportunity to point out a more general notion of prime filter that seems suitable for sites that are not regular. In the new sense, a prime filter on $X \in |C|$ is a set p of morphisms in C with the fixed codomain X such that the following are satisfied: (i) every isomorphism $X' \to X$ belongs to p, (ii) if $Y \to X$, $Z \to X$ both belong to p, then so does $Y \underset{X}{\times} Z \to X$, (iii) if the composite $Z \to Y \to X$ belongs to p, then so does $Y \to X$, (iv) if $(Y \to X) \in p$, and $\{Z_i \to Y: i \in I\}$ is a covering of Y, then there is $i \in I$ such that the composite $(Z_i \to Y \to X)$ belongs to p. If every morphism in C factors into a cover followed by a monomorphism (in particular, if C is regular), then the new notion is 'equivalent' to the old one: with p a prime in the new sense, the set p' consisting of the subobjects represented by the monomorphisms in p is a prime in the old sense, and in fact p' determines p. In the context of the present lemma, it is more convenient to work with the 'new' notion; given a prime filter p in the new sense, we want to find a model $M: C \to SET$ such that p equals the set of all morphisms $Y \xrightarrow{f} X$ such that $x \in Im(f)$.

To do so, let us first assume that X is the terminal object in C. Then p is simply a set of objects of C, with certain properties. We define a new topology on the category C, resulting in a finitary (algebraic) site $C^{(p)}$, as follows. The topology of $C^{(p)}$ is the one generated by the covering families of the original site C together with the empty family as a covering of each object $Y \notin p$. We claim that the coverings of $C^{(p)}$ are exactly the following: (i) coverings of C, (ii) any family of morphisms with codomain Y, with $Y \notin p$. It is clear that the ones listed under (i) and (ii) are coverings in $C^{(p)}$. Conversely, it suffices to show that the class of families (i) and (ii) is closed under the closure conditions defining a Grothendieck topology. The condition of stability under pullback, for coverings in (ii), is a consequence of condition (iii) in the definition of prime filter; the condition on composition of coverings, in the case of composing a covering of type (i) with coverings of type (ii), is a consequence of condition (iv); the rest is clear.

As a consequence, it is clear that for any $Y \in p$, $\emptyset \notin \mathrm{Cov}_{C^{(p)}}(Y)$ (by condition (iv) applied to the empty family). Furthermore, by conditions (i) and (ii) in the definition of p being a prime, for any finitely many $Y_1,\ldots,Y_n \in p$ (including $n = 0$) we have $Y_1\times\ldots\times Y_n \in p$, hence $\emptyset \notin \mathrm{Cov}_{C^{(p)}}(Y_1\times\ldots Y_n)$. Clearly, $C^{(p)}$ is a finitary site; by the Joyal-Deligne completeness theorem [MR, 3.5.5, p. 129], for any finitely many $Y_1,\ldots,Y_n \in p$ we have a model $M\colon C^{(p)} \to \mathrm{SET}$ such that $M(Y_1\times\ldots\times Y_n)$ is non-empty, i.e. each $M(Y_i)$ is non-empty. By the compactness theorem, there is a model $M\colon C^{(p)} \to \mathrm{SET}$ such that $M(Y)$ is non-empty for $Y \in p$; of course, $M(Y) = \emptyset$ for $Y \notin p$. We have $M(X) = M(1) = \{x\}$ for some x, and clearly $Y \in p \Leftrightarrow x \in \mathrm{Im}(Y \to 1)$ as promised.

This settles the case $X = 1$; for a general X, we pass to the site C/X defined as follows: the underlying category is the usual 'comma' category C/X; a family

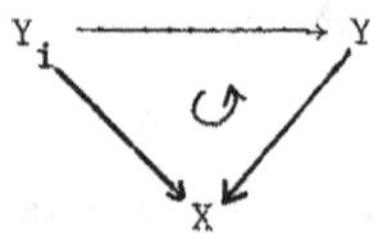

$(i \in I)$

is a covering iff $\{Y_i \to Y\colon i \in I\}$ is a covering in C. The above special case $X = 1$ applied to C/X gives the result for a general X; the straightforward details are omitted.

Since every consistent theory has $\aleph_0$-saturated models [CK, 5.1.4], it is easy to see that in the above argument M can be made $\aleph_0$-saturated [we'll recall this notion in (1.4) below]. □

It is interesting to note that the last lemma remains true for separable (instead of finitary) sites (but with the additional conditions in "regular sites" retained). This fact is related to M. Morley's lemma [11, Lemma 2.2]; the proof is different from the above.

The notion of a p-model is a generalization of the notion of $\aleph_0$-saturation. We have

(1.4) Every $\aleph_0$-saturated model $M: C \to SET$ (i.e., model of T_C, see the Introduction) is a p-model.

Proof. Let $p \in P(X)$, $f: X \to Y$ (for simplicity), $q = f(p)$, $b \in \bar{M}(q,Y) \subset M(Y)$; we want $a \in \bar{M}(p,X)$ such that $b = M(f)(a)$. Consider the following set of formulas with the free variable $\underset{\sim}{x}$ of sort X:

$$\Sigma(\underset{\sim}{x}) := \{f(\underset{\sim}{x}) \approx \underline{b}\} \cup \{\underset{\sim}{A}(\underset{\sim}{x}) : A \in p\}$$

($\underline{b}$ is a new individual constant denoting b; $\underset{\sim}{A}(\underset{\sim}{x})$ is a formula whose canonical interpretation (in C) is the subobject $A \hookrightarrow X$). We claim that every finite subset of $\Sigma(\underset{\sim}{x})$ is satisfiable in M, by a suitable element in $M(X)$ in place of $\underset{\sim}{x}$. Since p is closed under finite intersection, it suffices to show that $f(\underset{\sim}{x}) = \underline{b} \wedge \underset{\sim}{A}(\underset{\sim}{x})$ is satisfiable in M, for any $A \in p$. Let $A \in p$, and let us factorize $f|A$ in the form $A \overset{g}{\twoheadrightarrow} B \hookrightarrow Y$, with g a cover in C (C is regular). Clearly, $A \le f^{-1}(B)$, hence $B \in f(p) = q$. It follows that $M \models \underset{\sim}{B}[b]$ (i.e., $b \in M(B)$). g being a cover, so is $M(g)$; hence there is $a \in M(A) \subset M(X)$ such that $M(g)(a) = M(f)(a) = b$; we have $M \models (f(\underset{\sim}{x}) = \underline{b} \wedge \underset{\sim}{A}(\underset{\sim}{x}))$ [a for $\underset{\sim}{x}$].

M being $\aleph_0$-saturated means that any condition set like $\Sigma(\underset{\sim}{x})$, using altogether finitely many fixed elements of M (b in our case), if finitely satisfiable, is satisfiable in M. If $\Sigma(\underset{\sim}{x})$ is satisfied by $a \in M(X)$, then clearly, a is as required. □

(1.5) P is a prime site.

Proof. It suffices to show that the family C of all covers $(p,X) \xrightarrow{\tilde{f}} (f(p),Y)$ satisfies the four conditions listed at the discussion of prime sites. The first two and the fourth are easy to check. To verify the third one, let $(p,X) \xrightarrow{f} (q,Y)$ be a cover, $q = f(p)$, $(r,Z) \xrightarrow{\tilde{g}} (q,Y)$ any morphism. For simplicity of notation, assume $f: X \to Y$, $g: Z \to Y$. Let, by 1.3 and 1.4, M be a p-model $M: C \to SET$ with $c \in M(Z)$ such that $t_Z(c) = r$, and let $b = M(g)(c) \in \bar{M}(q,Y)$. Since M is a p-model, there is $a \in \bar{M}(p,X) \subset M(X)$ such that $M(f)(a) = b$. Consider the pullback

$$\begin{array}{ccc} X & \xrightarrow{\ f\ } & Y \\ {\scriptstyle g'}\uparrow & & \uparrow{\scriptstyle g} \\ U = X\times_Y Z & \xrightarrow[\ f'\ ]{} & Z\ . \end{array}$$

Since M preserves pullbacks, and $M(f)(a) = M(g)(c) = b$, there is $d \in M(U)$ such that $M(g')(d) = a$ and $M(f')(d) = c$. Let $s = t_U(d)$. We have $(r,Z) = t(c) = f'(t(d)) = f'(s,U)$, hence the germ $\tilde{f}'$ is a cover $(s,U) \to (r,Z)$. Also, since $a \in \bar{M}(p,X)$ and $t(a) = g'(s,U)$, it follows that the germ $\tilde{g}'\colon (s,U) \to X$ defines a morphism $(s,U) \to (p,X)$; and finally, the diagram

$$\begin{array}{ccc} (p,X) & \overset{\tilde{f}}{\twoheadrightarrow} & (q,Y) \\ {\scriptstyle \tilde{g}'}\uparrow & & \uparrow{\scriptstyle \tilde{g}} \\ (s,U) & \underset{\tilde{f}'}{\twoheadrightarrow} & (r,Z) \end{array}$$

clearly commutes. □

(1.6) $M_0\colon \mathcal{C} \to \tilde{\mathcal{P}}$ is continuous.

<u>Proof.</u> The fact that the functor $(\hat{-})\colon \mathcal{C} \to \hat{\mathcal{P}}$ is left exact can easily be seen; we omit the verification. It follows that $M_0\colon \mathcal{C} \xrightarrow{(\hat{-})} \hat{\mathcal{P}} \xrightarrow{\underset{\sim}{a}} \tilde{\mathcal{P}}$ is left exact as well. To show that M_0 preserves coverings, let $\{X_i \xrightarrow[f_i]{} X\colon i \in I\}$ be a covering in $\mathcal{C}$; we have to show that $\{\underset{\sim}{a}\hat{X}_i \xrightarrow{\underset{\sim}{a}\hat{f}_i} \underset{\sim}{a}\hat{X}\colon i \in I\}$ is a (canonical) covering in $\tilde{\mathcal{P}}$. By [SGA4, vol. 1, II.5, pp.251-] this means the following: whenever $q \to \hat{X}$ is a morphism in $\hat{\mathcal{P}}$ with $q \in |\mathcal{P}|$ (we identify $h_{\mathcal{P}}(q)$ with q, for the Yoneda functor $h_{\mathcal{P}}\colon \mathcal{P} \to \hat{\mathcal{P}}$), then the set of morphisms $p \to q$ in $\mathcal{P}$ such that the composite $p \to q \to \hat{X}$ factors through at least one of the morphisms f_i is a covering of q. Note that $\mathcal{P}$ being a prime site (1.5), we should find a single such $p \to q$ which is a cover. A morphism $(q,Y) \to \hat{X}$ is a germ $\tilde{f}\colon (q,Y) \to X$; without loss of generality, assume $f\colon Y \to X$. Consider the pullback:

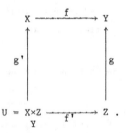

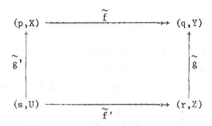

$$\begin{array}{ccc} X_i & \xrightarrow{\;f_i\;} & X \\ {\scriptstyle h}\uparrow & & \uparrow{\scriptstyle f} \\ Y_i = X_i \times_X Y & \xrightarrow[\;g\;]{} & Y \end{array} .$$

By 1.3, let $M: C \to SET$ be a model with some $b \in M(Y)$ such that $t_Y(b) = q$. Since $\{Y_i \to Y: i \in I\}$ is a covering, there is $i \in I$ and $a \in M(Y_i)$ such that $M(g)(a) = b$. It follows that for $p = t_{Y_i}(a)$ we have $(q,Y) = g(p,Y_i)$ and $\tilde{g}: (p,Y_i) \to (q,Y)$ is a cover. Also, clearly the diagram

$$\begin{array}{ccc} X_i & \xrightarrow{\;f_i\;} & X \\ {\scriptstyle \tilde{h}}\uparrow & & \uparrow{\scriptstyle \tilde{f}} \\ (p,Y_i) & \xrightarrow[\;\tilde{g}\;]{} & (q,Y) \end{array}$$

commutes. According to what we said above, and the definitions of $\hat{X}_i$ and $\hat{X}$, this completes the proof. □

(1.7) (i) For $Z \in |C|$, the presheaf $\hat{Z} \in |\hat{P}|$ is a sheaf.

(ii) Every representable presheaf over P is a sheaf.

<u>Proof</u>: (ad(i)). First we show that $\hat{Z}$ is a separated presheaf. Suppose we have the cover $(p,X) \xrightarrow{\tilde{f}} (q,Y)$, and the germs $\tilde{g},\tilde{h}: (q,Y) \rightrightarrows Z$ such that $\tilde{g}\circ\tilde{f} = \tilde{h}\circ\tilde{f}$; assume (without loss) that $g,h: Y \rightrightarrows Z$. Let M be an arbitrary p-model $C \to SET$. Then $\bar{M}(\tilde{g})\circ\bar{M}(\tilde{f}) = \bar{M}(\tilde{h})\circ\bar{M}(\tilde{f})$, and since $\bar{M}(\tilde{f})$ is surjective, $\bar{M}(\tilde{g}) = \bar{M}(\tilde{h})$; in other words, $M(g)|\bar{M}(q,Y) = M(h)|\bar{M}(q,Y)$. In logical language, we can write this as follows: in any $\aleph_0$-saturated model (see 1.4) of $T_C \cup \{\underset{\sim}{A}(\underline{y}): A \in q\}$, the sentence $g(\underline{y}) \approx h(\underline{y})$ is true ($\underline{y}$ a new individual constant of sort Y). It follows that

$$T_C \cup \{\underset{\sim}{A}(\underline{y}): A \in q\} \models g(\underline{y}) \approx h(\underline{y})$$

(if there were a model witnessing the failure of this, there would also be an $\aleph_0$-saturated one). By the compactness theorem, and the fact that q is closed under finite intersection, there is $A \in q$ such that

$$T_C \cup \{\underset{\sim}{A}(\underline{y})\} \models g(\underline{y}) \approx h(\underline{y}) .$$

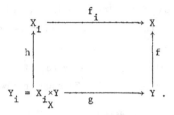

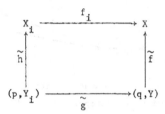

Through the canonical interpretation of the language in C, this means precisely that $g|A = h|A$, hence $\tilde{g} = \tilde{h}$; and this is precisely what is needed for showing that $\hat{X}$ is a separated presheaf.

To show the existence part of the sheaf-property for $\hat{Z}$, assume we have a cover $(p,X) \xrightarrow{\tilde{f}} (q,Y)$, and a germ $\tilde{g}: (p,X) \to Z$ such that whenever $(r,U) \underset{\tilde{h}_2}{\overset{\tilde{h}_1}{\rightrightarrows}} (p,X)$ are such that in the diagram

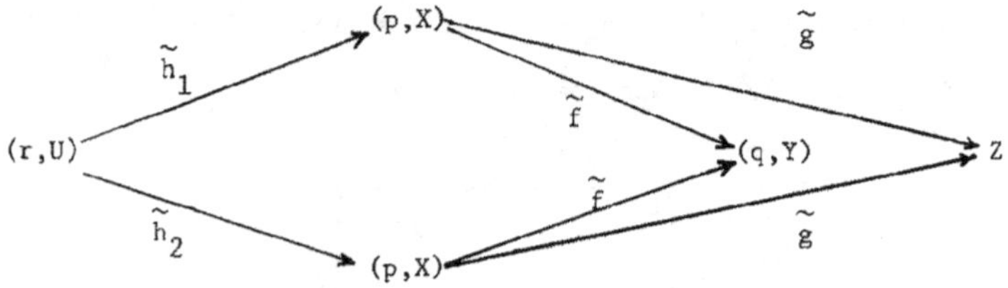

we have $\tilde{f}\circ\tilde{h}_1 = \tilde{f}\circ\tilde{h}_2$, then $\tilde{g}\circ\tilde{h}_1 = \tilde{g}\circ\tilde{h}_2$. We can assume $f: X \to Y$, $g: X \to Z$. Using the canonical language associated with the category C, let's write $p(\underline{x})$ for the set $\{\underset{\sim}{A}(\underline{x}) : A \epsilon p\}$. We claim that

$$T_C \cup p(\underline{x}) \cup p(\underline{x}') \cup \{f\underline{x} \approx f\underline{x}'\} \models g\underline{x} \approx g\underline{x}' \tag{1}$$

Let (M,a,a') be a model of the left-hand-side in (1). Let $U = X\times X$, $r = t_U(\langle a,a'\rangle)$, $h_1, h_2: U \rightrightarrows X$ the two projections. Since $a,a' \in \bar{M}(p,X)$, it follows that the germs $\tilde{h}_1, \tilde{h}_2$ define morphisms $(r,U) \rightrightarrows (p,X)$. Since $M \models (fx \approx fx')[a,a']$, we have that the subobject $[fx \approx fx']$ of $X\times X$ belongs to the set r. Hence, we have $\tilde{f}\circ\tilde{h}_1 = \tilde{f}\circ\tilde{h}_2$. By the starting assumption, this implies $\tilde{g}\circ\tilde{h}_1 = \tilde{g}\circ\tilde{h}_2$; it follows that $M \models (gx \approx gx')[a,a']$. This shows (1).

By compactness applied to (1), there is a single $A \epsilon p$ such that

$$T_C \cup \{\underset{\sim}{A}(\underline{x}) \wedge \underset{\sim}{A}(\underline{x}') \wedge f\underline{x} \approx f\underline{x}'\} \models g\underline{x} \approx g\underline{x}'$$

i.e.

$$T_C \models (\forall x,x' \epsilon X)[\underset{\sim}{A}(x) \wedge \underset{\sim}{A}(x') \wedge fx \approx fx' \Rightarrow gx \approx gx'] \tag{2}$$

Let $B \in S(Y)$ be the subobject $B = \mathrm{Im}(f|A)$; by the assumption that $\tilde{f}$ is a cover, we must have $B \epsilon q$. Define the relation $R \in \mathrm{Sub}_C(Y\times Z)$ by

$$R = [\underset{\sim}{B}(y) \wedge (\exists x \epsilon X)(\underset{\sim}{A}(x) \wedge f(x) \approx y \wedge g(x) \approx z)].$$

By elementary logic, we infer from (2) that R is functional with domain $B \hookrightarrow Y$. Therefore [see MR, 2.4.4, p.89], there is a morphism $B \xrightarrow{h} Z$ in C whose graph is R. Clearly, $(h\circ f)|A = g|A$, hence with the germ $\tilde{h}$: $(q,Y) \to Z$, we have $\tilde{h}\circ\tilde{f} = \tilde{g}$, as required for the sheaf property.

(ad(ii)). This is an easy consequence of part (i). Let $s \in P(Z)$. Then (s,Z) is a sub-presheaf of the sheaf $\hat{Z}$, hence it is separated. To show the existence part of the sheaf-property for (s,Z), we return to the notation introduced above when talking about the existence part of the sheaf-property for $\hat{Z}$; we only have to add that now $\tilde{g}$ is a germ

$$(p,X) \xrightarrow{\tilde{g}} (s,Z).$$

Since in particular we have $(p,X) \xrightarrow{\tilde{g}} Z$, by $\hat{Z}$ being a sheaf, we have a germ $\tilde{h}$: $(q,Y) \to Z$ such that $\tilde{g} = \tilde{h}\circ\tilde{f}$. It follows that $g(p)$ (a prime filter in Z) equals $h(f(q)) = h(q) \in P(Z)$. But clearly, $s \subset g(p)$; hence $s \subset h(q)$ and hence $\tilde{h}$ is actually a germ $(q,Y) \to (s,Z)$, as required. □

In the next remarks and lemma, we discuss what the subobject lattice in $\tilde{P}$ of an object coming from C looks like. First of all, if $p \in P(X)$, the morphism $(p,X) \xrightarrow{\widetilde{\mathrm{Id}}_X} \hat{X}$ in $\hat{P}$ is a monomorphism; with this canonical monomorphism, every prime filter on X is a subobject of $\hat{X}$ in $\hat{P}$. Secondly, let $(q,Y) \xrightarrow{\tilde{f}} \hat{X}$ be a morphism in $\hat{P}$, let $p = f(q) \in P(X)$. We then have that $(q,Y) \xrightarrow{f} \hat{X}$ factors as $(q,Y) \xrightarrow{\tilde{f}} (p,X) \longrightarrow \hat{X}$, with the first morphism being a cover. Next, the associated sheaf-functor $\underset{\sim}{a}$: $\hat{P} \to \tilde{P}$ carries any $p \in P(X)$ into a subobject $\underset{\sim}{a}(p,X)$ of $\underset{\sim}{a}(\hat{X})$, and any $A \in S_C(X)$ into a subobject $M_0(A) = \underset{\sim}{a}(\hat{A})$ of $M_0(X) = \underset{\sim}{a}(\hat{X})$.

We will say of a complete lattice L that it is prime-generated if every element of L is the sup of a family of primes, i.e. elements unequal to the sup of all properly smaller elements.

(1.8) Let $X \in |C|$.

(i) The subobject lattice $S_{\tilde{P}}(\underset{\sim}{a}(\hat{X}))$ is prime-generated; its primes are exactly the subobjects of the form $\underset{\sim}{a}(p,X)$, with $p \in P(X)$.

(ii) $\underset{\sim}{a}(p,X) \leq \underset{\sim}{a}(\hat{A})$ if and only if $A\in p$; here $p \in P(X)$, $A \in S_C(X)$, and $\leq$ is the usual ordering on $S_{\tilde{P}}(\underset{\sim}{a}(\hat{X}))$.

(iii) $\underset{\sim}{a}(p,X) = \bigwedge\{\underset{\sim}{a}(\hat{A}) : A\in p\}$.

(iv) M_0 is a p-model.

Proof. (ad(i)). The family of all objects $\underset{\sim}{a}(q,Y)$ ($q \in P(Y)$, $Y \in |C|$) forms a set of generators for the topos $\tilde{P}$. So, every subobject of $\underset{\sim}{a}(\hat{X})$ is the sup of objects of the form $\mathrm{Im}(\alpha)$, with α: $\underset{\sim}{a}(q,Y) \to q(\hat{X})$. Since (q,Y) and $\hat{X}$ are sheaves (1.7), $\underset{\sim}{a}$ is full and faithful on the hom-set $\mathrm{hom}_{\hat{P}}((q,Y),\hat{X}))$: hence α is of the form $\underset{\sim}{a}(\tilde{f})$, for some $\tilde{f}$: $(q,Y) \to \hat{X}$. Consider the factorization of $\tilde{f}$ as above: $(q,Y) \to (f(q),X) \longrightarrow \hat{X}$; since the first morphism is a cover, $\mathrm{Im}(\alpha) = \underset{\sim}{a}(f(q),X)$. This shows that every subobject of $\underset{\sim}{a}(\hat{X})$ is the sup of subobjects of the form $\underset{\sim}{a}(p,X)$; the latter is clearly a prime of the lattice $S_{\underset{\sim}{P}}(\underset{\sim}{a}(\hat{X}))$. Finally, we also conclude that every prime of $S_{\underset{\sim}{P}}(\underset{\sim}{a}(\hat{X}))$ must be of the form $\underset{\sim}{a}(p,X)$.

(ad(ii)). Note the trivial fact that in $\hat{P}$, in the subobject lattice of $\hat{X}$, $(p,X) \le \hat{A}$ iff $A \in p$. The assertion now follows by 1.7.

(ad(iii). By part (i), $\Sigma \underset{df}{=} \bigwedge\{\underset{\sim}{a}(\hat{A}): A \in p\}$ is the sup of a family of subobjects of $\underset{\sim}{a}(\hat{X})$ of the form $\underset{\sim}{a}(p',X)$. But by part (ii), if $\underset{\sim}{a}(p',X) \le \Sigma$, then $p \subset p'$ (set-theoretically), i.e. $\underset{\sim}{a}(p',X) \le \underset{\sim}{a}(p,X)$; this shows $\Sigma \le \underset{\sim}{a}(p,X)$. The opposite inequality is obvious.

(ad(iv)). By definition, $M_0(p,X) = \bigwedge^{(\widetilde{P})}\{M_0(A): A \in p\} = \bigwedge\{\underset{\sim}{a}(\hat{A}): A \in p\} = \underset{\sim}{a}(p,X)$ by part (iii). For $f: X \to Y$ in C, $M_0(f)|M_0(p)$ is therefore the morphism $\underset{\sim}{a}(p,X) \xrightarrow{\tilde{f}} \underset{\sim}{a}(\hat{Y})$, and its image is $\underset{\sim}{a}(f(p),Y) = M_0(f(p))$, as required for M_0 to be a p-model. □

(1.9) For any $\widetilde{M} \in \bigwedge(\widetilde{P},E)$, the composition $M = \widetilde{M} \circ M_0: C \to E$ is a p-model.

<u>Proof.</u> Direct consequence of 1.8 and the definitions. □

Next, we state some generalities we could not find in the literature in the exact form we need them; nevertheless, they are essentially contained in [SGA4, Expose III, and 4.9, Expose IV], and we did not feel we should supply proofs.

For the moment, P and P' denote arbitrary small sites with not necessarily finitely complete underlying categories. Let $\bar{M}: P \to \widetilde{P}'$ be a functor, let $\dot{M}$ be the composite $P \xrightarrow{\bar{M}} \widetilde{P}' \xrightarrow{i'} \hat{P}'$ (i': inclusion). We deduce the functors

$$\hat{P} \underset{\hat{M}}{\overset{M_{\cdot}}{\rightleftarrows}} \hat{P}'$$

as follows. $M_{\cdot}$ is the composite $\hat{P}' \xrightarrow{\hat{h}'} \hat{\hat{P}}' \xrightarrow{\text{composition with } \dot{M}} \hat{P}$. ($\hat{h}'$: Yoneda embedding); in particular,

$$(M_{\cdot}(G))(p) = \mathrm{Hom}_{\hat{P}'}(\dot{M}(p),G) \qquad (p \in |P|).$$

$\hat{M}$ is a left adjoint of $M_{\cdot}$; the description of $\hat{M}$ is as follows. As usual, $\eta \downarrow \dot{M}$ denotes the comma-category [II.6 in CWM]; in particular, its objects are pairs $\langle \eta \xrightarrow{\alpha} \dot{M}p, p\rangle$, with $p \in |P|$; the functor $\pi: \eta \downarrow \dot{M} \to P$ takes the said pair to p. Now, for $F \in |\hat{\mathcal{P}}|$ and $\eta \in |\mathcal{P}'|$, $(\hat{M}(F))(\eta)$ is the colimit of the functor $F\circ\pi\colon \eta\downarrow\dot{M} \to SET$; in a somewhat imprecise notation

$$(\hat{M}(F))(\eta) = \varinjlim_{\eta \to \dot{M}(p)} F(p)$$

(p: variable)

With this notation, we have

(1.10) Suppose (a) M. transforms $\mathcal{P}'$-sheaves into $\mathcal{P}$-sheaves, and (b) $\hat{M}$ is left exact. Then there is continuous $\tilde{M}\colon \tilde{\mathcal{P}} \to \tilde{\mathcal{P}}'$ such that $\bar{M} \approx \tilde{M}\circ\varepsilon$. In fact, $\tilde{M}$ can be taken to be $\tilde{M} = a'\circ\hat{M}\circ i$ in the diagram

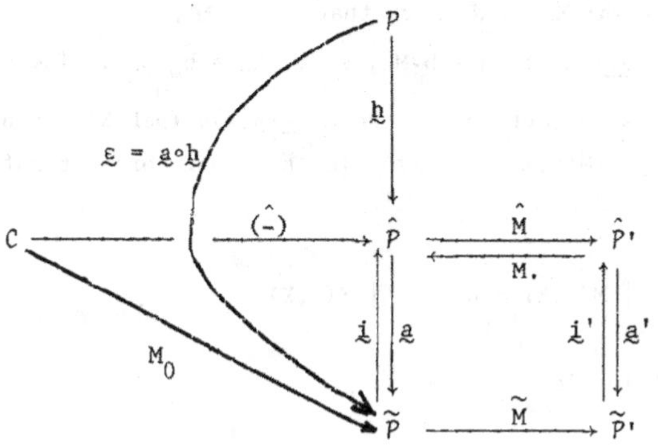

In addition, we have $\bar{M} \approx \tilde{M}\circ\varepsilon \approx a'\circ\hat{M}\circ h$. □

(1.11) The functor

$$\mathrm{Con}(\tilde{\mathcal{P}},\tilde{\mathcal{P}}') \longrightarrow (\mathcal{P},\tilde{\mathcal{P}}')$$

defined by composition with $\varepsilon\colon \mathcal{P} \longrightarrow \tilde{\mathcal{P}}$ is full and faithful [Prop. 4.9.4, Exp. IV, p. 356, SGA4]. □

Now, we return to our previous notation, with fixed $\mathcal{C}$, and $\mathcal{P}$ e.t.c. deduced from it.

(1.12) For any topos $\mathcal{E}$, the functor

$$\bigwedge(\tilde{\mathcal{P}},\mathcal{E}) \xrightarrow{(-)\circ M_0} p(\mathcal{C},\mathcal{E})$$

of Theorem 1.1 is full and faithful.

Proof. By 1.9, $\bigwedge(\tilde{P},E) \longrightarrow (C,E)$ does indeed factor through $p(C,E)$. Let $\tilde{M},\tilde{N} \in |\bigwedge(\tilde{P},E)|$, let $M = \tilde{M}\circ M_0$, $N = \tilde{N}\circ M_0$, and let $h: M \longrightarrow N$ be a natural transformation. Let $\bar{M}: P \longrightarrow E$ be defined as the composite $\bar{M} = \tilde{M}\circ\underset{\sim}{\varepsilon}: P \to \tilde{P} \to E$, and similarly $\bar{N}$. Previously, we denoted by $\bar{M}$ a certain functor $P \longrightarrow E$ derived from a given $M: C \longrightarrow E$; $\bar{M}$ as defined here is the same as that one: for $\bar{M}$ as defined here, and (p,X) as a subobject of $\hat{X}$ in $\hat{P}$ $(p \in P(X))$, we have $\bar{M}(p,X) = \tilde{M}(\underset{\sim}{a}(p,X)) = \tilde{M}(\bigwedge^E\{\underset{\sim}{a}(\hat{A}): A\epsilon p\})$ (see 1.8 (iii)) $= \bigwedge^E\{M(A): A\epsilon p\}$ (since $\tilde{M}$ preserves intersections), which shows the assertion. Since $h: M \longrightarrow N$ is a natural transformation, $h_X|A: M(A) \longrightarrow N(X)$ factors through $N(A) \longrightarrow N(X)$, hence $h_X|\bar{M}(p,X)$ factors through $\bar{N}(p,X) \hookrightarrow N(X)$; this defines a morphism $\bar{M}(p,X) \longrightarrow \bar{N}(p,X)$ that we denote by $\bar{h}_{(p,X)}$. It is clear, in fact, that we have a natural transformation $\bar{h}: \bar{M} \longrightarrow \bar{N}$. By 1.11, there is $\tilde{h}: \tilde{M} \longrightarrow \tilde{N}$ such that $\bar{h} = \tilde{h}\circ\underset{\sim}{\varepsilon}$, i.e. $\bar{h}_{(p,X)} = \tilde{h}_{\underset{\sim}{a}(p,X)}$. We claim that $h = \tilde{h}\circ M_0$, i.e. $h_X = \tilde{h}_{M_0(X)}$. The object $M_0(X) = \underset{\sim}{a}\hat{X}$ is covered by its subobjects of the form $\underset{\sim}{a}(p,X)$ $(p\epsilon P(X))$, hence $\bar{M}(X) = \tilde{M}(M_0(X))$ is covered by $\{\bar{M}(p,X): p\epsilon P(X)\}$ in E. Therefore, it suffices to show that

$$h_X \mid \bar{M}(p,X) = \tilde{h}_{M_0(X)} \mid \bar{M}(p,X) \tag{3}$$

holds for all $p\epsilon P(X)$. The triangle

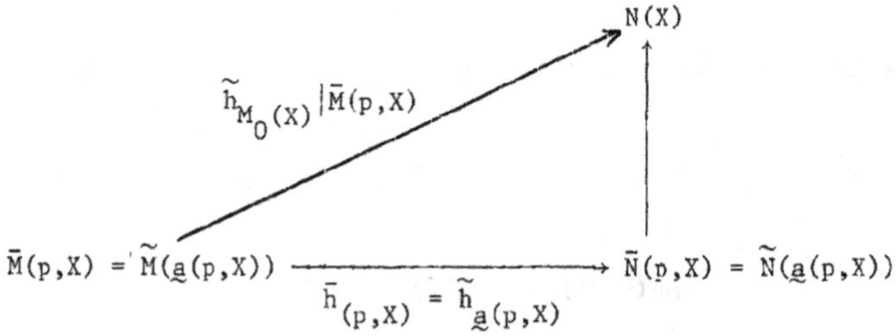

commutes because $\tilde{h}$ is natural; the triangle

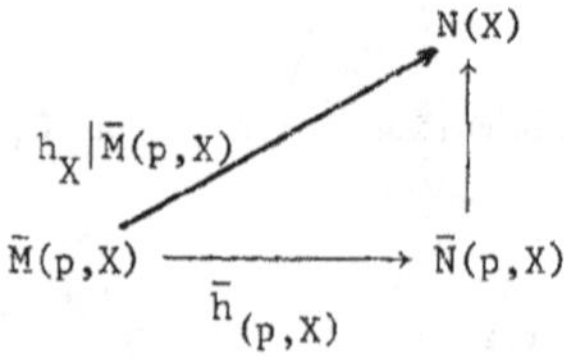

commutes because of the definition of $\bar{h}$ from h; it follows that (3) is true. This proves our claim, and thus the fullness of the functor of 1.12.

Let $\tilde{h}, \tilde{h}': \tilde{M} \longrightarrow \tilde{N}$ be natural transformations such that $h \underset{df}{=} \tilde{h} \circ M_0 = h' \underset{df}{=} \tilde{h}' \circ M_0$; we want to see that $\tilde{h} = \tilde{h}'$. Since both squares in the diagram

$$\begin{array}{ccc} M(X) & \xrightarrow{\quad \tilde{h}_{\hat{\underset{\sim}{a}X}} = \tilde{h}'_{\hat{\underset{\sim}{a}X}} \quad} & N(X) \\ \uparrow & & \uparrow \\ \bar{M}(p,X) & \underset{\tilde{h}'_{\underset{\sim}{a}(p,X)}}{\overset{\tilde{h}_{\underset{\sim}{a}(p,X)}}{\rightrightarrows}} & \bar{N}(p,X) \end{array}$$

commute, for any $p \in P(X)$, we conclude that $\tilde{h} \circ \varepsilon = \tilde{h}' \circ \varepsilon$; by 1.11, it follows that $\tilde{h} = \tilde{h}'$. □

In the final stage of the proof of 1.1, we show that the functor of 1.1 is essentially surjective in case E is a prime-generated topos. Let P' be a prime site (see the beginning of this section), let $E = \tilde{P}'$ be the category of sheaves over P', and let $M: C \longrightarrow E$ be a p-model. For the rest of this section, we fix these items as well. Note that our final aim is to show the existence of $\tilde{M} \in |\bigwedge(\tilde{P}, E)|$ such that $M \simeq \tilde{M} \circ M_0$.

Without loss of generality, we may assume that the canonical functor $\varepsilon_{P'}: P' \longrightarrow \tilde{P}' = E$ is a full inclusion (see the beginning of this section); we make use of this fact to simplify the notation and consider $\eta \in |P'|$ as an object in E as well: $\eta = \varepsilon_{P'}(\eta)$.

Recall the functor $\bar{M}: P \longrightarrow E$ defined from M after the statement of 1.1.

(1.13) $\bar{M}$ satisfies conditions (a) and (b) in 1.10.

Proof. Let $\dot{M}$, $M.$, $\hat{M}$ be derived from $\bar{M}$ as for 1.10.

(ad(a)). Let G be a sheaf $\in |\tilde{P}'|$, and let $F = M.(G)$. To show that F is a sheaf, let $(q,Y) \xrightarrow{\tilde{f}} (p,X)$ be a cover in P, $\xi \in F((q,Y))$, and assume that for all $(r,Z) \underset{\tilde{g}_2}{\overset{\tilde{g}_1}{\rightrightarrows}} (q,Y)$ such that $\tilde{f} \circ \tilde{g}_1 = \tilde{f} \circ \tilde{g}_2$ we have that $F(\tilde{g}_1)(\xi) = F(\tilde{g}_2)(\xi)$. We can assume $f: Y \longrightarrow X$. By the definition of $M.(G)$, $F((q,Y)) = \mathrm{Hom}(\bar{M}(q,Y),G)$, i.e. ξ is an arrow $\xi: \bar{M}(q,Y) \longrightarrow G$ in $\hat{P}'$. $\bar{M}(q,Y) \xrightarrow{\bar{M}(\tilde{f})} \bar{M}(p,X)$ is a cover in E (M is a p-model).

Claim. The family consisting of the single morphism $\bar{M}(q,Y) \xrightarrow{\xi} G$ in $E = \tilde{P}'$ is a compatible family with respect to the covering $\bar{M}(q,Y) \xrightarrow{\bar{M}(\tilde{f})} \bar{M}(p,X)$, i.e. for any $E \rightrightarrows \bar{M}(q,Y)$ in E, if

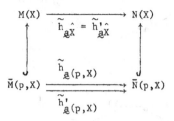

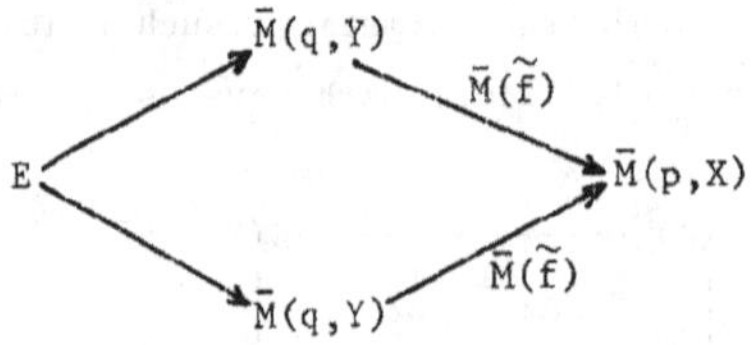

commutes, then

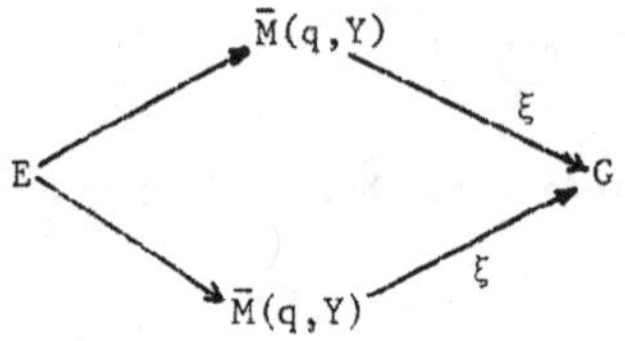

commutes as well.

To prove the <u>Claim</u>, first of all notice that it suffices to show it for $E = \eta \in |P'|$ since these latter generate E. Let $\eta \in |P'|$, $\eta \overset{\alpha_1}{\underset{\alpha_2}{\rightrightarrows}} \bar{M}(q,Y)$ two morphisms in E such that the diagram

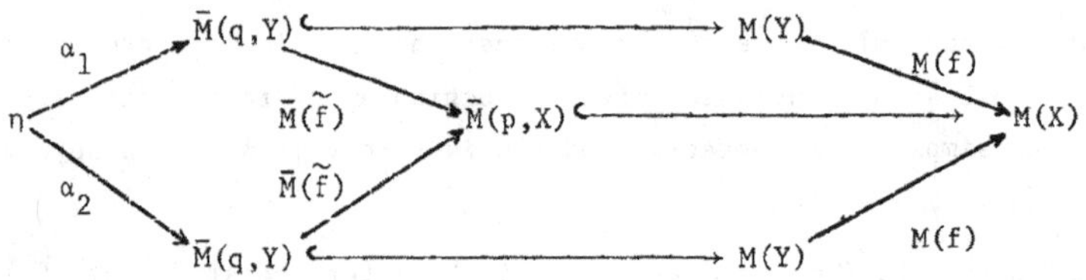

(4)

commutes. Let β be the composite $\eta \xrightarrow{<\alpha_1,\alpha_2>} \bar{M}(q,Y) \times \bar{M}(q,Y) \longrightarrow M(Y\times Y)$. Apply 1.2 to get that the set $r = t(\beta) = \{A \in \mathrm{Sub}_C(Y\times Y)\colon\ \beta$ factors through $M(A) \longrightarrow M(Y\times Y)\}$ is a prime filter over $Y\times Y$. Clearly, the subobjects $B \times 1_Y$, $1_Y \times B$ of $Y\times Y$, for $B\epsilon q$, belong to r. This means that the projections $\pi_1,\pi_2\colon Y\times Y \rightrightarrows Y$ represent morphisms $\tilde{\pi}_1,\tilde{\pi}_2\colon (r,Y\times Y) \rightrightarrows (q,Y)$. Also, β factors through $\bar{M}(r,Y\times Y) \hookrightarrow M(Y\times Y)$, i.e. there is a morphism $\eta \xrightarrow{\gamma} \bar{M}(r,Y\times Y)$ such that

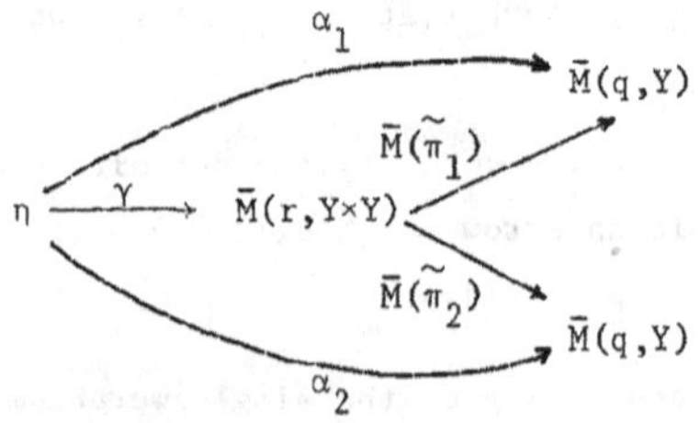

(5)

commutes. Now, consider the kernel pair

$$C \underset{h_2}{\overset{h_1}{\rightrightarrows}} Y \xrightarrow{f} X$$

of f in $\mathcal{C}$, inducing the subobject $C: = (C \overset{h = \langle h_1, h_2 \rangle}{\hookrightarrow} Y\times Y) \in Sub_{\mathcal{C}}(Y\times Y)$. C belongs to r; this follows from the facts that (4) commutes and that M preserves kernel-pairs. It follows that the diagram

$$\begin{array}{ccccc} & & (q,Y) & & \\ & \overset{\tilde{\pi}_1}{\nearrow} & & \overset{\tilde{f}}{\searrow} & \\ (r,Y\times Y) & & & & (p,X) \\ & \underset{\tilde{\pi}_2}{\searrow} & & \underset{\tilde{f}}{\nearrow} & \\ & & (q,Y) & & \end{array}$$

commutes. By our assumptions, this implies that $F(\tilde{\pi}_1)(\xi) = F(\tilde{\pi}_2)(\xi)$. Since $F = Hom(\dot{M}(-),G)$, we obtain that the diagram

$$\begin{array}{ccccc} & & \bar{M}(q,Y) & & \\ & \overset{\bar{M}(\tilde{\pi}_1)}{\nearrow} & & \overset{\xi}{\searrow} & \\ \bar{M}(r,Y\times Y) & & & & G \\ & \underset{\bar{M}(\tilde{\pi}_2)}{\searrow} & & \underset{\xi}{\nearrow} & \\ & & \bar{M}(q,Y) & & \end{array}$$

commutes. It follows, also in view of (5), that

$$\begin{array}{ccccc} & & \bar{M}(q,Y) & & \\ & \overset{\alpha_1}{\nearrow} & & \overset{\xi}{\searrow} & \\ \eta & & & & G \\ & \underset{\alpha_2}{\searrow} & & \underset{\xi}{\nearrow} & \\ & & \bar{M}(q,Y) & & \end{array}$$

commutes; this proves the <u>Claim</u>.

Since G is a sheaf, hence $Hom_{\mathcal{E}}(-,G)$ is a sheaf with respect to the canonical topology on $\mathcal{E}$, it follows that there is a unique arrow $\bar{M}(p,X) \xrightarrow{\zeta} G$ such that

$$\begin{array}{ccc} \bar{M}(q,Y) & & \\ & \overset{\xi}{\searrow} & \\ \tilde{f}\downarrow & & G \\ & \underset{\zeta}{\nearrow} & \\ \bar{M}(p,X) & & \end{array}$$

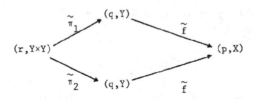

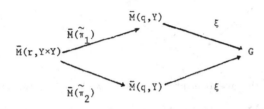

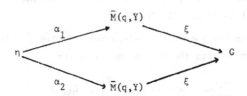

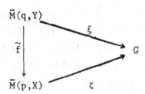

commutes. But remembering that $F = \mathrm{Hom}(\dot{M}(-),G)$, we see that this is precisely what we need for F to be a sheaf.

(ad(b)). Recalling the definition of $\hat{M}$, we see that it suffices to prove that the category $(\eta\downarrow\dot{M})^{op}$ is filtered (see Theorem 1, CWM, p.211).

To show the first condition for filteredness (condition (a) on p. 207 in CWM), we have to prove that for any pair of arrows $\eta \xrightarrow{\alpha_1} \bar{M}(p,X)$, $\eta \xrightarrow{\alpha_2} \bar{M}(q,Y)$ in E there are arrows $(r,Z) \xrightarrow{\tilde{f}_1} (p,X)$, $(r,Z) \xrightarrow{\tilde{f}_2} (q,Y)$ in P and an arrow $\eta \longrightarrow \bar{M}(r,Z)$ in E making

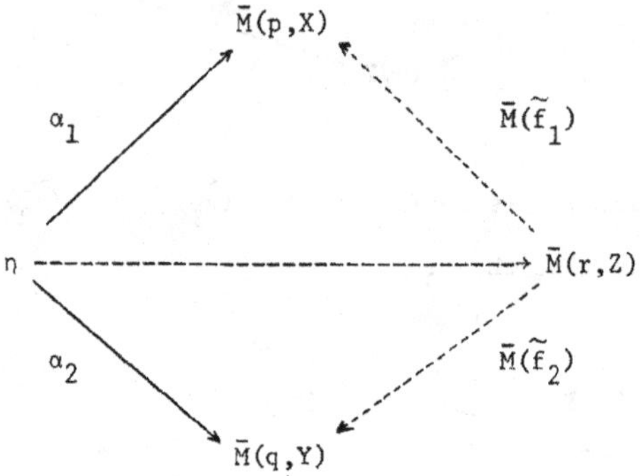

(6)

commute. Now, similarly to the previous proof, we let β be the composite $\eta \xrightarrow{\langle\alpha_1,\alpha_2\rangle} \bar{M}(p,X)\times\bar{M}(q,Y) \hookrightarrow M(X\times Y)$, $r = t(\beta) \in P(X\times Y)$ (see 1.2). We easily see that $(r,X\times Y) \longrightarrow \hat{X}\times\hat{Y} \xrightarrow{\text{proj.}} \dot{X}$ factors through $(p,X) \longrightarrow \hat{X}$, giving rise to $\tilde{f}_1$; similarly for $\tilde{f}_2$; also (by 1.2), β factors through $\bar{M}(r,X\times Y) \hookrightarrow M(X\times Y)$. Now the commutativity of (6) is clear $(Z = X\times Y)$.

Next we verify condition (b) loc. cit. This amounts to being able to complete the diagram of the solid arrows with the dotted ones as shown:

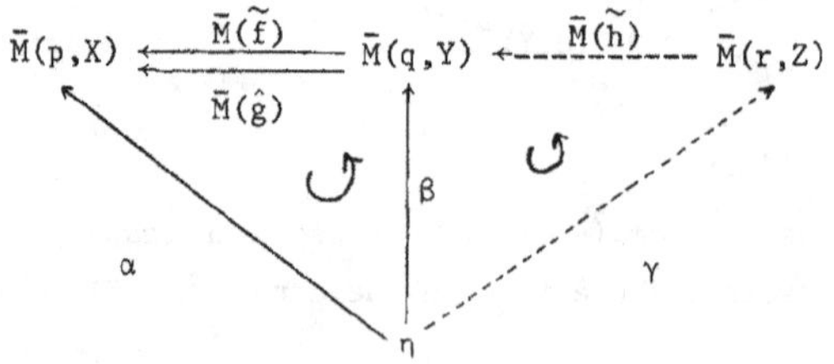

with the additional condition $\tilde{f}\circ\tilde{h} = \tilde{g}\circ\tilde{h}$. We may assume $f,g\colon Y \rightrightarrows X$. Let $B \xrightarrow{h} Y$ be the equalizer of f,g in C, let $r = t(\beta) \in P(Y)$. Then $M(B) \xrightarrow{M(h)} M(Y)$ is the equalizer of $M(f)$ and $M(g)$; it follows by the definition of $t(\beta)$, and $\bar{M}(\tilde{f})\circ\beta = \bar{M}(\tilde{g})\circ\beta$ $(= \alpha)$ that the subobject $B\colon = B \hookrightarrow Y$ belongs to the set r, and in turn, that $\tilde{f}\circ\tilde{h} = \tilde{g}\circ\tilde{h}$; finally, we also get γ as required, by 1.2. □

(1.14) There is $\widetilde{M} \in \bigwedge(\widetilde{P},E)$ such that $M \simeq \widetilde{M}\circ M_0$.

Proof. We use $\bar{M}$ as defined before 1.13. By 1.10, let $\widetilde{M} \in |\mathrm{Con}(\widetilde{P},\widetilde{P}')|$ be such that $\bar{M} \simeq \widetilde{M}\circ\underset{\sim}{\varepsilon}$. We claim that $M \simeq \widetilde{M}\circ M_0$. By 1.10, this is equivalent to saying that $M = \underset{\sim}{a}'\circ\hat{M}\circ\widetilde{(\hat{-})}$. Since $\underset{\sim}{a}'\circ i' \simeq \mathrm{Id}_{\widetilde{P}'}$, it suffices to show that there exists an isomorphism

$$\hat{M}\circ(\hat{-}) \xrightarrow[h]{\sim} i'\circ M.$$

Given $X \in |C|$, we are to define $h_X\colon \hat{M}(\hat{X}) \longrightarrow M(X)$, a natural transformation between functors $(P')^{op} \longrightarrow \mathrm{SET}$. Thus, let $\eta \in |P'|$, and define

$$(h_X)_\eta\colon \hat{M}(\hat{X})(\eta) \longrightarrow M(X)(\eta) = \mathrm{Hom}_E(\eta, M(X)) \qquad (7)$$

as follows. We have

$$\hat{M}(\hat{X})(\eta) = \varinjlim_{\eta \to \bar{M}(p,Y)} \hat{X}((p,Y)).$$

For a specific given $\beta\colon \eta \to \bar{M}(p,Y)$, we can define

$$((h_X)_\eta)_\beta\colon \hat{X}((p,Y)) \to \mathrm{Hom}_E(\eta, M(X))$$

as follows:

$$\underbrace{((p,Y) \xrightarrow{\;\tilde{f}\;} X)}_{\substack{\text{a typical element}\\ \text{of } \hat{X}((p,Y))}} \longmapsto (\eta \to \bar{M}(p,Y) \xrightarrow{\;\bar{M}(\tilde{f})\;} M(X)).$$

It is clear that these maps are compatible with the morphisms of $\eta{\downarrow}\dot{M}$, i.e. whenever we have

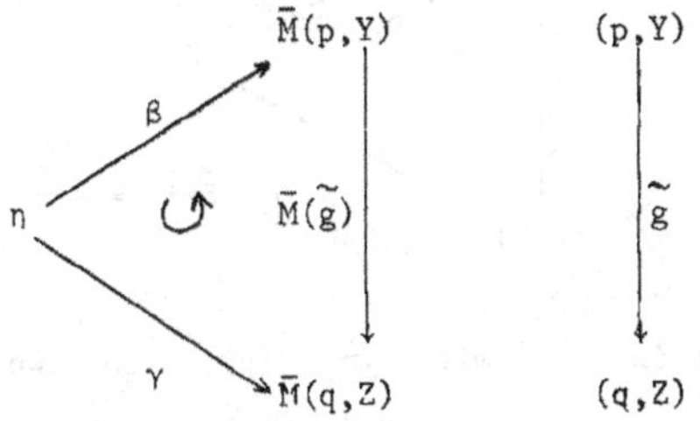

then

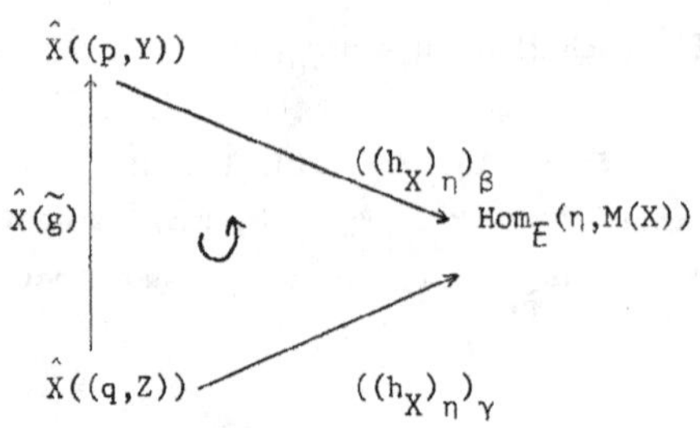

By the universal property of $\varinjlim$, these $((h_X)_\eta)_\beta$ define an arrow as in (7). It is clear that h_X so given is natural, and in fact that h itself is natural.

To show that h is an isomorphism, it suffices to show that each $(h_X)_\eta$ is a bijection.

First we show that $(h_X)_\eta$ is surjective. Let $\beta\colon \eta \to M(X)$ be arbitrary. By 1.2, there is $p \in M(X)$ such that β factors through $\bar{M}(p,X) \longrightarrow M(X)$; now the claimed surjectivity is immediate.

Since $(\eta\dot{\downarrow}M)^{op}$ is filtered (see the proof of 1.13), for proving the injectivity of $(h_X)_\eta$, it suffices to consider two elements of $\hat{X}((p,Y))$ of the form

$(p,Y) \xrightarrow{\tilde{f}} X$, $(p,Y) \xrightarrow{\tilde{g}} X$ such that in

$$\eta \xrightarrow{\beta} \bar{M}(p,Y) \underset{\bar{M}(\tilde{g})}{\overset{\bar{M}(\tilde{f})}{\rightrightarrows}} M(X)$$

we have $\alpha \underset{df}{=} \bar{M}(\tilde{f})\circ\beta = \bar{M}(\tilde{g})\circ\beta$. Let $B \hookrightarrow Y \underset{g}{\overset{f}{\rightrightarrows}} X$ be the equalizer of f and g; then $M(B) \hookrightarrow M(Y) \underset{M(g)}{\overset{M(f)}{\rightrightarrows}} M(X)$ is an equalizer diagram; let us denote by B the subobject $(B \hookrightarrow Y)$ of Y too. Let, as usual, $r = t(\beta)$; note that $B \in r$. By its definition, the set r contains the set p; it follows that $(r,Y) \hookrightarrow \hat{Y}$ factors through $(p,Y) \hookrightarrow \hat{Y}$, and since $B \in r$, we have that in

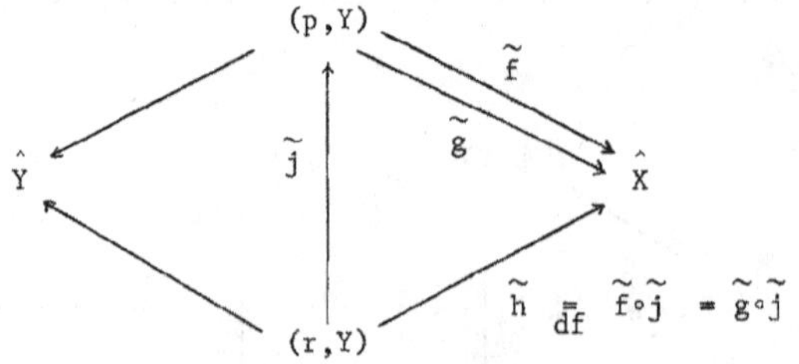

$\tilde{f}\circ\tilde{j} = \tilde{g}\circ\tilde{j}$ holds. This means that in the limit $\varinjlim_{\eta \to \bar{M}(p,X)} \hat{X}(p)$, the two elements $(p,Y) \underset{\tilde{g}}{\overset{\tilde{f}}{\rightrightarrows}} \hat{X}$ represent the same element, namely the one represented by

$(r,Y) \xrightarrow{\tilde{h}} \hat{X}$. This proves the injectivity of the map $(h_X)_\eta$.

We have thus checked that $M \simeq \tilde{M} \circ M_\cap$. It remains to verify that $\tilde{M}$ preserves intersections. To simplify notation, assume $M = \tilde{M} \circ M_\cap$ and $\bar{M} = \tilde{M} \circ \underset{\sim}{\varepsilon}$ (and not just isomorphism).

Let $X \in |C|$, $\eta \xrightarrow{\beta} M(X)$ be any morphism from a (prime) object $\eta \in |P'| \subset |E|$. Let $p = t(\beta) \in P(X)$. We <u>claim</u> that $\underset{\sim}{a}(p,X) \hookrightarrow \underset{\sim}{a}\hat{X}$ is the smallest subobject Σ of $\underset{\sim}{a}\hat{X}$ such that β factors through $\tilde{M}(\Sigma) \hookrightarrow M(X) (= \tilde{M}(\underset{\sim}{a}\hat{X}))$. Since $\tilde{M}(\underset{\sim}{a}(p,X)) = \bar{M}(p,X) = \bigwedge_{A \in p} M(A)$, β does factor through $\tilde{M}(\underset{\sim}{a}(p,X)) \longrightarrow M(X)$. Suppose β factors through $\tilde{M}(\Sigma) \longrightarrow M(X)$. By 1.8 (i), $\Sigma = \bigvee\{\underset{\sim}{a}(p_i,X): i \in I\}$ with some $p_i \in P(X)$; since $\tilde{M}$ preserves sups, $\tilde{M}(\Sigma) = \bigvee\{\tilde{M}(\underset{\sim}{a}(p_i,X)): i \in I\}$; since η is a prime object, there is $i \in I$ such that β factors through $M(p_i) = \tilde{M}(\underset{\sim}{a}(p_i,X)) \hookrightarrow M(X)$; it follows that $p_i \subset p$ (see 1.2), hence $\underset{\sim}{a}(p,X) \le \underset{\sim}{a}(p_i,X) \le \Sigma$, as claimed.

Now, let Σ_i $(i \in I)$ be subobjects of $\underset{\sim}{a}\hat{X}$; we show that $\tilde{M}(\bigwedge_i \Sigma_i) = \bigwedge_i \tilde{M}(\Sigma_i)$. Of course, only the inequality $\bigwedge_i \tilde{M}(\Sigma_i) \le \tilde{M}(\bigwedge_i \Sigma_i)$ requires proof; also, since E is prime-generated, it suffices to show that every $\eta \xrightarrow{\beta} M(X)$ $(\eta \in |P'|)$ that factors through $\bigwedge M(\Sigma_i) \hookrightarrow M(X)$ also factors through $\tilde{M}(\bigwedge_i \Sigma_i)$. Let $p = t(\beta)$; by the previous paragraph, we have $\underset{\sim}{a}(p,X) \le \Sigma_i$ for all $i \in I$, hence $\underset{\sim}{a}(p,X) \le \bigwedge_i \Sigma_i$ and thus $\tilde{M}(\underset{\sim}{a}(p,X)) \le \tilde{M}(\bigwedge_i \Sigma_i)$, as desired.

We have shown the required preservation of intersection of subobjects of objects of the form $\underset{\sim}{a}\hat{X}$ $(X \in |C|)$ of $\tilde{P}$. For objects of the form $\underset{\sim}{a}(p,X)$, $p \in P(X)$, the same conclusion is an immediate consequence since these are domains of subobjects of objects of the previous kind. This latter kind of objects form a family of generators for $\tilde{P}$. Now, the final conclusion follows by an elementary argument showing that a continuous functor between toposes preserves intersections of subobjects of arbitrary objects once we know this of objects that form a family of generators of the domain topos; we omit the easy details. □

By the above, in particular 1.5, 1.6, 1.7(iv), 1.9, 1.12 and 1.14, we have completed the proof of Theorem 1.1.

Concluding this section, we briefly describe an alternative definition of the prime completion. Let $X \in |C|$. A <u>filter on</u> X is a set $\Sigma \subset \mathrm{Sub}_C(X)$ satisfying the first three conditions of the definition of a prime filter. $F(X)$ denotes the set of all filters on X. The <u>category of filters</u> of C, denoted F, is defined similarly to P; a morphism between two filters (Σ,X), (Φ,Y) is an equivalence class (or germ) of morphisms defined exactly as it was done for prime filters. Clearly, P is a full subcategory of F.

For $\Sigma \in F(X)$, $P(\Sigma,X)$ is the set of prime filters $p \in P(X)$ such that $\Sigma \subset p$.

Given a germ $\tilde{f}: (\Sigma,X) \to (\Phi,Y)$, $f: A \to Y$, for every $p \in P(\Sigma,X)$ we have that $f(p)$ $(= \{B \in Sub_C(Y): f^{-1}(B) \in p\})$ belongs to $P(\Phi,Y)$; define the map

$$P(\tilde{f}): P(\Sigma,X) \to P(\Phi,Y)$$

by $(P(\tilde{f}))(p) = f(p)$. In this way, we have actually defined a functor $P: F \to SET$. We define the Grothendieck topology on F as the one generated by all those families

$$\{(\Sigma_i,X_i) \xrightarrow{\tilde{f}_i} (\Phi,Y): i \in I\}$$

that are carried by the functor P into a surjective family:

$$P(\Phi,Y) = \bigcup_{i \in I} Im(P\,\tilde{f}_i) .$$

The functor $i: C \to F$ is defined by mapping $X \in |C|$ into the (set theoretically) minimal filter $(\{1_X\},X)$, and $X \xrightarrow{f} Y$ into the obvious germ $\tilde{f}$. With these definitions, the prime completion

$$M_0: C \to \tilde{P}$$

turns out to be isomorphic to the composite

$$C \xrightarrow{i} F \xrightarrow{\varepsilon_F} \tilde{F}$$

where of course $\tilde{F}$ is the category of sheaves over F as a site. The starting point of proving this fact is to show that F (via i) is the solution of the universal problem of extending C to a category having intersections of arbitrary families of subobjects of any given object.

§ 2. Embeddings and representation

We continue to use all the notation of Section 1 in an unchanged sense. In particular, C is a regular site. X will always denote an object of C.

Definition 2.1. Let K be any subcategory of (C,SET). A functor $F: K \to SET$ is said to have the finite support property (f.s.p.) if the following holds: for any $M \in |K|$ and $a \in F(M)$, there are $X \in |C|$ and $x \in M(X)$ such that for all $N \in |K|$, and all $M \underset{h}{\overset{g}{\rightrightarrows}} N$ in K, $g_X(x) = h_X(x)$ implies $(Fg)(a) = (Fh)(a)$. We call such an x a support of a.

For a discussion of the f.s.p. in a (significant) special case, see [10].

We will fix K to be the full subcategory of $\mathrm{Mod}(C,\mathrm{SET})$ consisting of 'special' models. Let $\kappa_0 = \max(\aleph_0, \mathrm{card}(C))$ where $\mathrm{card}(C) \underset{df}{=} \mathrm{card}(\coprod\{\mathrm{Hom}(X,Y)\colon X,Y \in |C|\})$; $\kappa_{n+1} = 2^{\kappa_n}$ for $n < \omega$, $\lambda = \sup\{\kappa_n : n < \omega\}$. In [CK], the notion of special structure is introduced; in our context a structure is a functor $C \longrightarrow \mathrm{SET}$. We let K be the full subcategory of $\mathrm{Mod}(C,\mathrm{SET})$ consisting of the λ-special models; for the exact meaning of the prefix λ - (which is almost the same as having cardinality λ), see [10]. One doesn't need to know the definition however; everything we use is contained in

<u>Proposition 2.2.</u> (i) Every $M \in K$ is a p-model, $M \in |p(C,\mathrm{SET})|$.
(ii) Whenever $X \in |C|$, $p \in P(X)$, there are $M \in K$ and $x \in M(X)$ such that $p = t_X(x,M)$.
(iii) Whenever $M,N \in K$, $x \in M(X)$, $y \in N(X)$, $p = t_X(x,M)$ and $y \in \bar{N}(p,X) (\subset N(X))$, then there is a homomorphism (natural transformation) $h\colon M \longrightarrow N$ such that $y = h_X(x)$.

<u>Remarks</u>. (i) follows from the fact that every special structure in $\aleph_0$-saturated, an immediate consequence of the definition. (ii) follows from 5.1.8 in CK; (iii) is the same as 1.1(v) in [10]. □

Recall that $\widetilde{P}$ denotes the prime-completion of the regular site C. By 1.1 (since SET is prime-generated!), there is a full and faithful functor

$$\widetilde{(-)}\colon p(C,\mathrm{SET}) \to \bigwedge(\widetilde{P},\mathrm{SET})$$

such that $\mathring{M}_0 \circ \widetilde{(-)} = \mathrm{Id}_{p(C,\mathrm{SET})}$, where $\mathring{M}_0\colon \bigwedge(\widetilde{P},\mathrm{SET}) \to p(C,\mathrm{SET})$ is the functor of 1.1, defined by composition with M_0. In particular, we have that the diagram

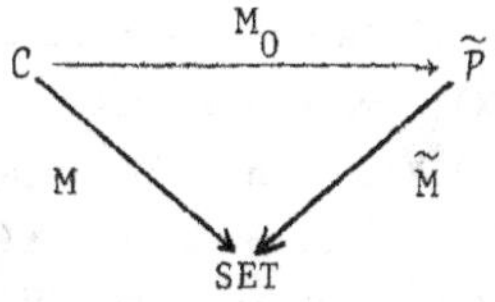

commutes for every $M \in |p(C,\mathrm{SET})|$. Let $\widetilde{K}$ be the image of K under $\widetilde{(-)}$; $\widetilde{K}$ is a full subcategory of $\bigwedge(\widetilde{P},\mathrm{SET})$; let us denote the functor (an isomorphism of categories) $K \longrightarrow \widetilde{K}$ induced by $\widetilde{(-)}$ by the same symbol $\widetilde{(-)}$. $\widetilde{(-)}$ induces an isomorphism $(\widetilde{K},\mathrm{SET}) \xrightarrow{\simeq} (K,\mathrm{SET})$; with $\mathrm{ev}\colon \widetilde{P} \longrightarrow (\widetilde{K},\mathrm{SET})$ the evaluation functor, let $\underset{\sim}{e}$ denote the composite $\widetilde{P} \xrightarrow{\mathrm{ev}} (\widetilde{K},\mathrm{SET}) \xrightarrow{\simeq} (K,\mathrm{SET})$. With $\mathrm{ev}_0\colon C \longrightarrow (K,\mathrm{SET})$ the appropriate evaluation functor, we clearly have that

$$\begin{array}{ccc} \mathcal{C} & \xrightarrow{\ M_0\ } & \widetilde{P} \\ & {\scriptstyle ev_0}\searrow \quad \circlearrowleft \quad \swarrow{\scriptstyle \underset{\sim}{e}} & \\ & (K,SET) & \end{array}$$

commutes.

<u>Theorem 2.3.</u> (<u>Representation theorem for the prime completion</u>).

$\underset{\sim}{e}$ is continuous, full and faithful; a functor $F \in (K,SET)$ is isomorphic to one of the form $\underset{\sim}{e}(\Phi)(\Phi \in |\widetilde{P}|)$ if and only if F has the f.s.p. As a consequence, $\widetilde{P}$ is equivalent to the full subcategory f.s.p.(K,SET) of (K,SET) consisting of the functors with the f.s.p. Here K can be any full subcategory of $Mod(\mathcal{C},SET)$ satisfying the three conditions in 2.2.

<u>Proof.</u> (i) The continuity of $\underset{\sim}{e}$ is clear.

(ii) We prove that $\underset{\sim}{e}$ is conservative; this will imply that it is faithful. Since the objects of the form $\underset{\sim}{a}(p,X)$ $(p \in P(X))$ (call these <u>special</u> <u>objects</u>) generate $\widetilde{P}$, it suffices to show the following: whenever

$$\{G_i \xrightarrow{\alpha_i} G : i \in I\} \tag{1}$$

is a family of morphisms between special objects, which is not a covering in $\widetilde{P}$, the family

$$\{\underset{\sim}{e}(G_i) \xrightarrow{\underset{\sim}{e}(\alpha_i)} \underset{\sim}{e}(G) : i \in I\} \tag{2}$$

is not a covering in (K,SET). Let $G = \underset{\sim}{a}(p,X)$, $G_i = \underset{\sim}{a}(p_i,X_i)$; by 2.2 (ii), let $M \in K$ and $x \in M(X)$ such that $p = t_X(x,M)$. By 1.7 (ii), $\alpha_i = \underset{\sim}{a}(\widetilde{f}_i)$ for some germ $\widetilde{f}_i : (p_i,X_i) \longrightarrow (p,X)$ in P; since (1) is not a covering in $\widetilde{P}$, none of the $\widetilde{f}_i$'s is a cover in P. We have $x \in \bar{M}(p,X)$; we <u>claim</u> that $x \notin Im(\bar{M}(\widetilde{f}_i))$ for all $i \in I$. In fact, we have $Im(\bar{M}(\widetilde{f}_i)) = \bar{M}(f_i(p_i),X)$, since M is a p-model; clearly $p \subset f_i(p_i)$ (set-theoretically); on the other hand $p = t(x)$ means p is the (set-theoretically) largest element of $P(X)$ such that $x \in \bar{M}(p,X)$; so if we had $x \in Im(\bar{M}(\widetilde{f}_i))$, we would have $p = f_i(p_i)$, i.e. $\widetilde{f}_i$ would be a cover in P, contradiction; this shows our <u>claim</u>. What we have shown is that, for a suitable $M \in K$,

$$\{\bar{M}(p_i,X_i) \xrightarrow{\bar{M}(\widetilde{f}_i)} \bar{M}(p,X) : i \in I\}$$

is not a covering in SET; it follows that (2) is not a covering in (K,SET).

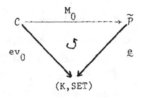

(iii) Next we show that $\underset{\sim}{e}$ is <u>strongly</u> <u>full</u>: $\underset{\sim}{e}$ induces a surjection (hence, an isomorphism) between the subobject-lattices $\mathrm{Sub}_{\widetilde{P}}(\Phi)$ and $\mathrm{Sub}_{(K,SET)}(\underset{\sim}{e}(\Phi))$, for any $\Phi \in |\widetilde{P}|$.

Let first $X \in |C|$ and $\Phi = M_0(X)$; hence $\underset{\sim}{e}(\Phi) = ev_0(X)$; let Z be an arbitrary subobject of $ev_0(X)$. We define a subobject Σ of Φ as follows. Consider an arbitrary model $N \in K$, and any element y of $Z(N)$ (a subset of $N(X)$); define Σ as the sup of the subobjects $\underset{\sim}{a}(t_X(y,N),X)$, for all such N and y:

$$\Sigma = \bigvee\{\underset{\sim}{a}(t_X(y,N),X) : N \in K,\ y \in Z(N)\}.$$

We <u>claim</u> that $Z = \underset{\sim}{e}(\Sigma)$.

Note first that for any $M \in K$, $(\underset{\sim}{e}(\Sigma))(M) = (ev(\Sigma))(\widetilde{M}) = \widetilde{M}(\Sigma) = \bigvee\{\widetilde{M}\underset{\sim}{a}(t(y),X) : \ldots\}$ $= \bigvee\{\bar{M}(t(y),X) : \ldots\}$. Let $x \in Z(M)$; of course, $x \in \bar{M}(t(x),X)$; it follows that $x \in \underset{\sim}{e}(\Sigma)(M)$; hence $Z(M) \subset \underset{\sim}{e}(\Sigma)(M)$ and thus $Z \le \underset{\sim}{e}(\Sigma)$ (as subobjects of $ev_0(X)$).

Now, let $M \in K$ and $x \in \underset{\sim}{e}(\Sigma)(M) \subset M(X)$. By definition of Σ, there is $N \in K$ and $y \in Z(N)$ such that $x \in \bar{M}(t_X(y,N),X)$. By 2.2 (iii), there is a homomorphism $h: N \longrightarrow M$ such that $h_X(y) = x$. It follows that $x = Z(h)(y) \in Z(M)$ (Z is a subfunctor of $ev_0(X)$), proving $\underset{\sim}{e}(\Sigma) \le Z$, and thus the <u>claim</u>.

The claim is the strong fullness condition for Φ of the form $M_0(X)$. The same now follows for any object $\Phi = \underset{\sim}{a}(p,X)$ since $\underset{\sim}{a}(p,X)$ is the domain of a subobject of $M_0(X)$. Hence, we have the strong fullness condition for objects G that form a family of generators for $\widetilde{P}$. Strong fullness in general now follows by the following easy argument: for an arbitrary $\Phi \in |\widetilde{P}|$, $Z \in \mathrm{Sub}(\underset{\sim}{e}(\Phi))$, let $\{G_i \xrightarrow{\alpha_i} \Phi : i \in I\}$ be a covering in $\widetilde{P}$ with G_i "special"; form the pullbacks

$$\begin{array}{ccc}
\underset{\sim}{e}(G_i) & \xrightarrow{\underset{\sim}{e}(\alpha_i)} & \underset{\sim}{e}(\Phi) \\
\uparrow & & \uparrow \\
Z_i & \longrightarrow & Z\ ;
\end{array}$$

we have that

$$Z = \bigvee\{\mathrm{Im}(\underset{\sim}{e}(\alpha_i)|Z_i) : i \in I\}; \tag{3}$$

there is $\Sigma_i \in \mathrm{Sub}(G_i)$ such that $Z_i = \underset{\sim}{e}(\Sigma_i)$; define $\Sigma = \bigvee\{\mathrm{Im}(\alpha_i|\Sigma_i) : i \in I\}$; since $\underset{\sim}{e}$ is continuous, we obtain by (3) that $\underset{\sim}{e}(\Sigma) = Z$ as required.

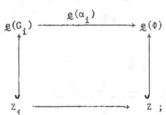

(iv) The following two facts are easy consequences of the strong fullness of $\underset{\sim}{e}$, together with its being continuous and conservative. One is that $\underset{\sim}{e}$ is full in the ordinary sense; the other is that the essential image of $\underset{\sim}{e}$ is closed under quotients: if $\underset{\sim}{e}(\Phi) \xrightarrow{f'} F$ is an epimorphism in (K,SET), then F is isomorphic to $\underset{\sim}{e}(\Psi)$ for some $\Psi \in \widetilde{P}$. Both facts are essentially proved in the proof of 3.2 in [10].

(v) We show that for any $\Phi \in |\widetilde{P}|$, $\underset{\sim}{e}(\Phi)$ has the f.s.p. There is a family

$$\{\underset{\sim}{a}(p_i,X_i) \xrightarrow{\alpha_i} \Phi : i \in I\}$$

covering Φ in $\widetilde{P}$. Let $a \in \underset{\sim}{e}(\Phi)(M) = \widetilde{M}(\Phi)$; $\widetilde{M}(\Phi)$ is covered by the family of morphisms $\bar{M}(p_i,X_i) \xrightarrow{\widetilde{M}(\alpha_i)} \widetilde{M}(\Phi)$; hence there are $i \in I$ and $x \in \bar{M}(p_i,X_i)$ such that

$$\widetilde{M}(\alpha_i)(x) = a. \tag{4}$$

We claim that x is a support of a. Let $\alpha = \alpha_i$, $(p,X) = (p_i,X_i)$. Suppose $M \underset{h}{\overset{g}{\rightrightarrows}} N$ are such that $g_X(x) = h_X(x)$. Clearly, with $\widetilde{M} \underset{\widetilde{h}}{\overset{\widetilde{g}}{\rightrightarrows}} \widetilde{N}$ the corresponding homomorphisms in $\widetilde{K}$, we have

$$\widetilde{g}_{\underset{\sim}{a}(p,X)}(x) = \widetilde{h}_{\underset{\sim}{a}(p,X)}(x). \tag{5}$$

Also, $\underset{\sim}{e}(\Phi)(g) = \widetilde{g}_\Phi$; similarly for h. $\widetilde{g}$ and $\widetilde{h}$ being natural, the two appropriate squares in

$$\begin{array}{ccc}
\bar{M}(p,X) & \underset{\widetilde{h}_{\underset{\sim}{a}(p,X)}}{\overset{\widetilde{g}_{\underset{\sim}{a}(p,X)}}{\rightrightarrows}} & \bar{N}(p,X) \\
\Big\downarrow \widetilde{M}(\alpha) & & \Big\downarrow \widetilde{N}(\alpha) \\
\widetilde{M}(\Phi) & \underset{\widetilde{h}_\Phi}{\overset{\widetilde{g}_\Phi}{\rightrightarrows}} & \widetilde{N}(\Phi)
\end{array}$$

commute. By (4) and (5), it follows that $\widetilde{g}_\Phi(a) = \widetilde{h}_\Phi(a)$, hence $(\underset{\sim}{e}(\Phi)(g))(a) = (\underset{\sim}{e}(\Phi)(h))(a)$, as required.

(vi) We show that every $F \in |(K,SET)|$ with the f.s.p. is a quotient of an object of the form $\underset{\sim}{e}(\Phi)$, $\Phi \in \widetilde{P}$; hence, by (iv), F is in the essential image of $\underset{\sim}{e}$.

Let $M \in K$, $a \in F(M)$; let $x \in M(X)$ be a support of a; let $p = t_X(x,M)$. We claim that there is a morphism

$$f: \underset{\sim}{e}(\underset{\sim}{a}(p,X)) \to F$$

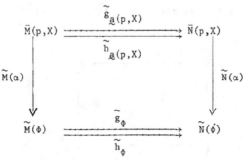

in (K,SET), f depending on a, such that $f_M(x) = a$. To define $f = (f_N)_{N\in K}$, $f_N\colon \bar{N}(p,X) \longrightarrow F(N)$, let $N\in K$ and $y \in \bar{N}(p,X)$. By 2.2 (iii), there is $g\colon M \longrightarrow N$ such that $g_X(x) = y$. Define $f_N(y)$ to be $(F(g))(a) \in F(N)$. To see that $f_N(y)$ is well-defined, just note that by x being a support of a, if $h\colon M \longrightarrow N$ is another homomorphism such that $h_X(x) = y$, we will have $F(h)(a) = F(g)(a)$. This defines the functions f_N, for every $N\in K$. To check the naturality of $f = (f_N)_{N\in K}$, one wants to see that the diagram

$$\begin{array}{ccc} \bar{N}(p,X) & \xrightarrow{\ f_N\ } & F(N) \\ {\scriptstyle \tilde{h}_{\underset{\sim}{a}(p,X)}}\downarrow & & \downarrow{\scriptstyle F(h)} \\ \bar{N}'(p,X) & \xrightarrow[\ f_{N'}\ ]{} & F(N') \end{array}$$

commutes for any $h\colon N \longrightarrow N'$. Let $y \in \bar{N}(p,X)$, let $y' = \tilde{h}_{\underset{\sim}{a}(p,X)}(y) \in \bar{N}'(p,X)$, let $g\colon M \longrightarrow N$ be such that $g_X(x) = y$ and $f_N(y) = F(g)(a)$; then for $g' = h\circ g$ we have $g'_X(x) = y'$ and hence $f_{N'}(y') = F(g')(a) = (F(h)\circ F(g))(a) = F(h)(f_N(y))$, proving the required commutativity. It is clear that $f_M(x) = a$; this proves the claim.

Let η_a denote $\underset{\sim}{a}(p,X)$ for $p = t_X(x,M)$, with $x = x_a$ a selected support of $a \in F(M)$; let $f_a\colon \underset{\sim}{g}(\eta_a) \to F$ be the morphism f constructed in the claim. The family $\langle f_a\colon a \in F(M),\ M\in K\rangle$ induces a map $\underset{\sim}{g}(\Sigma) \xrightarrow{\ \beta\ } F$, with $\Sigma = \coprod\{\eta_a\colon a \in F(M),\ M\in K\}$. Since $(f_a)_M(x_a) = a$ for every $a \in F(M)$, $M\in K$, clearly β is an epimorphism. By part (iv) above, it follows that F is isomorphic to $\underset{\sim}{g}(\Phi)$ for some object Φ of $\tilde{P}$.

This proves the theorem. □

Next we study the canonical embedding of $\tilde{C}$, the category of sheaves over the site C, in $\tilde{P}$, the prime completion of C. Our results will be general, but also somewhat technical. In the next section we draw less technical consequences, but for certain particular sites C only.

Let $\varepsilon_C\colon C \longrightarrow \tilde{C}$ be the canonical functor (Yoneda followed by associated sheaf); by our assumptions on C, ε_C is full and faithful. By the universal property of ε_C, there is an essentially unique continuous functor $\tilde{M}_0\colon \tilde{C} \longrightarrow \tilde{P}$ such that

$$\begin{array}{ccc} C & \xrightarrow{\ M_0\ } & \tilde{P} \\ & {\scriptstyle \varepsilon_C}\searrow \quad \nearrow{\scriptstyle \tilde{M}_0} & \\ & \tilde{C} & \end{array}$$

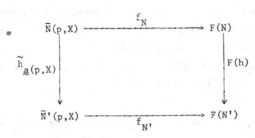

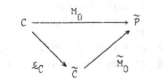

commutes.

<u>Proposition 2.4.</u> $\widetilde{M}_0: \widetilde{C} \longrightarrow \widetilde{P}$ is continuous and conservative.

<u>Proof.</u> $\widetilde{M}_0$ is continuous by definition. To see the conservativeness of $\widetilde{M}_0$, it suffices to do the following. Let $\{X_i \longrightarrow X: i\in I\}$ be a family of monomorphisms in C which is not in $\mathrm{Cov}_C(X)$; we want to show that $\{\hat{\mathfrak{a}X_i} \longrightarrow \hat{\mathfrak{a}X}: i\in I\} \notin \mathrm{Cov}_{\widetilde{P}}(\hat{\mathfrak{a}X})$. Let us also write X_i for the subobject of X determined by the monomorphism $X_i \longrightarrow X$. Since $\widetilde{C}$ has enough points, there is $M \in |\mathrm{Mod}(C,\mathrm{SET})|$ and $x \in M(X)$ such that $x \notin M(X_i)$ for all $i\in I$. Let $p = t_X(x,M)$; then $X_i \notin p$ and hence (by 1.8 (ii)), $\mathfrak{a}(p,X) \nleq \hat{\mathfrak{a}X_i}$; since $\mathfrak{a}(p,X)$ is a prime, $\mathfrak{a}(p,X) \nleq \bigvee\{\hat{\mathfrak{a}X_i}: i\in I\}$, hence a fortiori $\bigvee\{\hat{\mathfrak{a}X_i}: i\in I\} \neq 1_{\hat{\mathfrak{a}X}}$, as required. □

To simplify notation, below we consider ε_C an inclusion (i.e. we write X for $\varepsilon_C(X)$, e.t.c); moreover, we write $\dot{X}$ for $M_0(X)$, $\dot{E}$ for $\widetilde{M}_0(E)$, e.t.c. Let $\mathrm{Sub}^{(f)}(\dot{X})$ (f for 'filter') denote the set of subobjects of $\dot{X}$ of the form $\bigwedge\{\dot{A}_i: i\in I\}$, with $A_i \in \mathrm{Sub}_C(X)$ $(i\in I)$, $\mathrm{Sub}^{(p)}(\dot{X})$ the set of subobjects of $\dot{X}$ of the form $\mathfrak{a}(p,X)$ with $p\in P(X)$. Clearly, $\mathrm{Sub}^{(p)}(\dot{X}) \subset \mathrm{Sub}^{(f)}(\dot{X})$.

<u>Lemma 2.5.</u> (i) Let $G \in \mathrm{Sub}^{(f)}(\dot{X})$, $E \in \mathrm{Sub}_{\widetilde{C}}(X)$, and suppose that $G \leq \dot{E}$. Then there is $A \in \mathrm{Sub}_C(X)$ such that $G \leq \dot{A}$ and $A \leq E$.

(ii) Let $G \in \mathrm{Sub}^{(f)}(\dot{X})$, $E \in |\widetilde{C}|$, and $X \overset{f_1}{\underset{f_2}{\rightrightarrows}} E$ two morphisms in $\widetilde{C}$. Suppose that $\dot{f}_1|G = \dot{f}_2|G$. Then there is $A \in \mathrm{Sub}_C(X)$ such that $G \leq \dot{A}$ and $f_1|A = f_2|A$.

<u>Proof.</u> (ad(i)). Suppose the conclusion fails. By 6.1.3 in MR, we have subobjects $B_j \in \mathrm{Sub}_C(X)$ $(j\in J)$ such that $E = \bigvee\{B_j: j\in J\}$. Consider the following set of sentences, in full first-order logic over the language the graph of C, augmented with an individual constant $\underline{x}$ of sort X:

$$T' \underset{df}{=} T_C \cup \{\underset{\sim}{A}(\underline{x}): A \in \mathrm{Sub}_C(X),\ G \leq \dot{A}\}$$
$$\cup \{\neg \underset{\sim}{B}_j(\underline{x}): j\in J\}.$$

We <u>claim</u> that T' is consistent. By compactness, it suffices to show that $T'_A = T_C \cup \{\underset{\sim}{A}(\underline{x})\} \cup \{\neg \underset{\sim}{B}_j(\underline{x}): j\in J\}$ is satisfiable, for any $A \in \mathrm{Sub}_C(X)$ such that $G \leq \dot{A}$. By our indirect hypothesis, $A \nleq \bigvee_{j\in J} B_j$; by $\widetilde{C}$ having enough points, there are $M \in |\mathrm{Mod}(C,\mathrm{SET})|$ and $x \in M(X)$ such that $x \in M(A) - \bigcup_{j\in J} M(B_j)$; this means precisely the satisfiability of T'_A. By the existence theorem on $\aleph_0$-saturated models,

T' has an $\aleph_0$-saturated model (M,x); in particular, M is a p-model of C. M gives rise to $\widetilde{M} \in \bigwedge(\widetilde{P},SET)$, by 1.1. Since $G = \bigwedge\{\dot{A}: A \in Sub_C(X), G \le \dot{A}\}$, we have $\widetilde{M}(G) = \bigcap\{M(A): A \in Sub_C(X), G \le \dot{A}\}$. Since $(M,x) \models T'$, it follows that $x \in \widetilde{M}(G) - \widetilde{M}(\dot{E})$, contradicting $G \le \dot{E}$.

(ad(ii)). This is an easy consequence of part (i). Let $E' \longrightarrow X$ be the equalizer of f_1 and f_2 in $\widetilde{C}$, let us also write E' for the subobject of X determined by the monomorphism $E' \longrightarrow X$. Since $\widetilde{M}_0 = (\dot{-})$ is continuous, $\dot{E}' \longrightarrow \dot{X}$ is the equalizer of $\dot{f}_1$ and $\dot{f}_2$. Since $\dot{f}_1|G = \dot{f}_2|G$, we have $G \le \dot{E}'$ (in $Sub_{\widetilde{P}}(\dot{X})$). By part (i), there is $A \in Sub_C(X)$ such that $G \le \dot{A}$ and $A \le E'$; in particular, $f_1|A = f_2|A$. □

<u>Theorem 2.6.</u> (i) For any $G \in Sub^{(p)}(\dot{X})$, $E \in |\widetilde{C}|$ and $g: G \longrightarrow \dot{E}$ in $\widetilde{P}$, there are $A \in Sub_C(X)$ and $A \xrightarrow{f} E$ in $\widetilde{C}$ such that $G \le A$ and $g = \dot{f}|G$.

(ii) Let $I = \{1,\dots,n\}$ be a finite index-set. Given $G_i \in Sub^{(p)}(\dot{X})$, $g_i: G_i \to \dot{E}$ in $\widetilde{P}$ $(i \in I)$ such that $g_i|(G_i \wedge G_j) = g_j|(G_i \wedge G_j)$ $(i,j \in I)$, then there are $A_i \in Sub_C(X)$ and $f_i: A_i \to E$ in $\widetilde{C}$ $(i \in I)$ such that $G_i \le A_i$, $g_i = \dot{f}_i|G_i$ and $f_i|(A_i \wedge A_j) = f_j|(A_i \wedge A_j)$.

<u>Proof.</u> (ad(i)). Let $\{Y_i \xrightarrow{h_i} E: i \in I\}$ be a covering of E in $\widetilde{C}$ with objects Y_i of C $(\subset \widetilde{C})$.

Let $G = \underline{a}(p,X)$. We consider the following diagram in $\widetilde{P}$, for any $i \in I$:

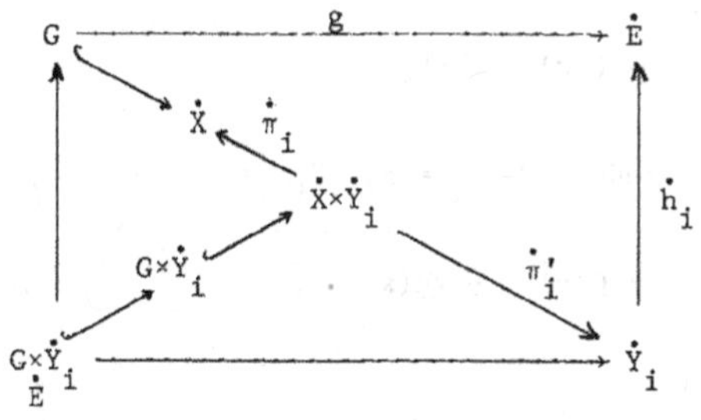

In this, the outer square is a pullback; $G \hookrightarrow \dot{X}$ is the structure morphism of $G \in Sub(\dot{X})$, $G \times_{\dot{E}} \dot{Y}_i \longrightarrow G \times \dot{Y}_i$ is the canonical monomorphism, π_i and π'_i are projections. The subobject of $\dot{X} \times \dot{Y}_i$ determined by the composite monomorphism $G \times_{\dot{E}} \dot{Y}_i \longrightarrow \dot{X} \times \dot{Y}_i$ is called F_i. Denote the graph of g as a subobject of $\dot{X} \times \dot{E}$ (via $G \hookrightarrow \dot{X}$) by R. Then F_i is the same as the canonical interpretation in $\widetilde{P}$ of the formula $\underline{R}(x,h_i(y))$; $F_i = [\underline{R}(x,h_i(y))] \in Sub(\dot{X} \times \dot{Y}_i)$. Let us record, in logical language, the facts expressing that "R is functional with domain G and codomain E": with x,z variables of sorts $\dot{X}$ and $\dot{E}$, respectively, we have ($\models$ meaning truth in $\widetilde{P}$)

$$\models \underset{\sim}{R}(x,z) \Rightarrow \underset{\sim}{G}(x), \tag{6}$$

$$\models \underset{\sim}{R}(x,z) \wedge \underset{\sim}{R}(x,z') \Rightarrow z \approx z', \tag{7}$$

$$\models \underset{\sim}{G}(x) \Rightarrow \exists z \underset{\sim}{R}(x,z). \tag{8}$$

The fact that the $\dot{h}_i$ form a covering of E is expressed as

$$\models \exists y \bigvee_{i \in I} (z \approx \dot{h}_i(y)). \tag{9}$$

By 6.1.3 in MR, we have $F_i = \bigvee_k H_{ik}$ for some $H_{ik} \in \mathrm{Sub}^{(p)}(\dot{X}\times\dot{Y}_i)$ (since the objects of the form $\underset{\sim}{a}(p,X)$ generate $\widetilde{P}$, and since the image of $\underset{\sim}{a}(p,X)$ under any morphism $\underset{\sim}{a}(p,X) \longrightarrow \underset{\sim}{a}\hat{Y}$ in $\widetilde{P}$ is of the form $\underset{\sim}{a}(q,Y)$). This equality can be written as

$$\models R(x,h_i(y)) \Leftrightarrow \bigvee_k H_{ik}(x,y). \tag{10}$$

Now, by elementary logic, we conclude from (8), (9) and (10) that

$$\models \underset{\sim}{G}(x) \Rightarrow \bigvee_{i,k} \exists y H_{ik}(x,y) .$$

Since G is a prime, there are indices i and k such that

$$\models \underset{\sim}{G}(x) \Rightarrow \exists y \underset{\sim}{H}_{ik}(x,y) .$$

Fix these i and k, and redenote $F_i = F$, $H_{ik} = H$, $h_i = h$, $Y_i = Y$. So, we have

$$\models \underset{\sim}{G}(x) \Rightarrow \exists y \underset{\sim}{H}(x,y) , \tag{11}$$

and from (10), of course

$$\models \underset{\sim}{H}(x,y) \Rightarrow \underset{\sim}{R}(x,h(y)). \tag{12}$$

As a consequence of (7) and (12), we have

$$\models \underset{\sim}{H}(x,y) \wedge \underset{\sim}{H}(x,y') \Rightarrow \dot{h}(y) \approx \dot{h}(y').$$

With $E' = [x,y,y' : h(y) \approx h(y')] \in \mathrm{Sub}_{\widetilde{C}}(X\times Y\times Y)$ and $G' = [\underset{\sim}{H}(x,y) \wedge \underset{\sim}{H}(x,y')] \in$ $\in \mathrm{Sub}_{\widetilde{P}}(\dot{X}\times\dot{Y}\times\dot{Y})$, the last fact is equivalent to saying that

$$G' \leq \dot{E}' \tag{13}$$

(since $(\dot{-})$ is continuous). Notice too that $G' \in \mathrm{Sub}^{(f)}(\dot{X}\times\dot{Y}\times\dot{Y})$, since $H \in \mathrm{Sub}^{(p)}(\dot{X}\times\dot{Y})$, hence $H \in \mathrm{Sub}^{(f)}(\dot{X}\times\dot{Y})$, thus $[x,y,y'\colon \underset{\sim}{H}(x,y)] \in \mathrm{Sub}^{(f)}(\dot{X}\times\dot{Y}\times\dot{Y})$ (obtained by pullback), similarly $[x,y,y'\colon \underset{\sim}{H}(x,y')] \in \mathrm{Sub}^{(f)}(\dot{X}\times\dot{Y}\times\dot{Y})$, and G' is the intersection of these latter two subobjects. We apply 2.5 (i) to (13); taking into account the definition of G', we obtain $C \in \mathrm{Sub}_C(X\times Y)$ such that $H \le \dot{C}$, i.e.

$$\models \underset{\sim}{H}(x,y) \Rightarrow \dot{\underset{\sim}{C}}(x,y) \tag{14}$$

and $[\dot{\underset{\sim}{C}}(x,y) \wedge \dot{\underset{\sim}{C}}(x,y')] \le \dot{E}'$, i.e.

$$\models \dot{\underset{\sim}{C}}(x,y) \wedge \dot{\underset{\sim}{C}}(x,y') \Rightarrow \dot{h}(y) \approx \dot{h}(y'). \tag{15}$$

We <u>claim</u> that we have

$$\models R(x,z) \Longleftrightarrow \underset{\sim}{G}(x) \wedge \exists y(\dot{\underset{\sim}{C}}(x,y) \wedge z \approx \dot{h}(y)); \tag{15'}$$

in fact this statement is a logical consequence of the ones we have established before. In particular, to establish the left-to-right implication, we argue as follows. Suppose $R(x,z)$. By (6), we have $\underset{\sim}{G}(x)$. By (11), let y be such that $\underset{\sim}{H}(x,y)$. By (14), we have $\dot{\underset{\sim}{C}}(x,y)$, and by (12), we have $\underset{\sim}{R}(x,h(y))$. By the last fact, $R(x,z)$ and (7), we have $z = h(y)$. We have established: $\underset{\sim}{G}(x) \wedge \exists y(\dot{\underset{\sim}{C}}(x,y) \wedge z = \dot{h}(y))$. For the converse direction: Suppose $\underset{\sim}{G}(x)$, $\dot{\underset{\sim}{C}}(x,y)$, and $z = \dot{h}(y)$. By (11), let y' be such that $H(x,y')$; hence by (12), also $\underset{\sim}{R}(x,h(y'))$, and by (14), $\dot{\underset{\sim}{C}}(x,y')$. By (15), it follows that $h(y) = h(y')$, hence $z = h(y')$ and so by $\underset{\sim}{R}(x,h(y'))$ we conclude $R(x,z)$ as required.

Of course, this proof uses the fact that for finitary coherent logic, any ordinary logical inference 'using elements' results in valid conclusions in any topos; compare MR.

Let E'' be the subobject

$$E'' = [\exists z \exists y(\underset{\sim}{C}(x,y) \wedge z \approx h(y))] \in \mathrm{Sub}_{\tilde{C}}(X).$$

(15) and (8), and the fact that $(\dot{-})$ continuous, imply that

$$G \le \dot{E}''.$$

Applying 2.5 (i) again, there is $A \in \mathrm{Sub}_C(X)$ such that $G \le \dot{A}$ and $A \le E''$, i.e.

$$\models \underset{\sim}{G}(x) \Rightarrow \dot{\underset{\sim}{A}}(x) \tag{16}$$

$$\models \dot{\underset{\sim}{A}}(x) \Rightarrow \exists z \exists y(\dot{\underset{\sim}{C}}(x,y) \wedge z \approx \dot{h}(y)) \tag{17}$$

Consider

$$D = [\underset{\sim}{A}(x) \wedge \exists y(\underset{\sim}{C}(x,y) \wedge z \approx h(y))] \in \mathrm{Sub}(X\times E).$$

We <u>claim</u> that D is functional with domain A and codomain E, i.e.

$$\widetilde{C} \models \underset{\sim}{D}(x,z) \Rightarrow \underset{\sim}{A}(x), \tag{18}$$

$$\widetilde{C} \models \underset{\sim}{D}(x,z) \wedge \underset{\sim}{D}(x,z') \Rightarrow z \approx z', \tag{19}$$

$$\widetilde{C} \models \underset{\sim}{A}(x) \Rightarrow \exists z\, \underset{\sim}{D}(x,z). \tag{20}$$

First of all, since $\widetilde{M}_0 = (\dot{-})$ is continuous and conservative, it suffices to verify the same things in $\widetilde{P}$, with dots put over A and D. (19) is clear by definition; (20) follows from (17), and (19) is a consequence of (15). Let $f: A \longrightarrow E$ be a morphism in $\widetilde{C}$ whose graph (as a subobject of $X\times E$) is D. Comparing the definition of D with (15'), and remembering that R was the graph of g, we conclude by (16) that $g = \dot{f}|G$ as required.

(ad(ii)). This is an easy consequence of part (i) and 2.5. By part (i), there are $A_i \in \mathrm{Sub}_C(X)$ and $f_i: A_i \longrightarrow E$ in $\widetilde{C}$ $(i\in I)$ such that $G_i \leq A_i$ and $g_i = \dot{f}_i|G_i$ $(i\in I)$. We now restrict the f_i to meet the other requirements. Let i,j be two fixed indices $\in I$. Since $\dot{f}_i|(G_i\wedge G_j) = \dot{f}_j|(G_i\wedge G_j)$, by 2.5 (ii) (applied to $A_i\wedge A_j$ instead of X, and $G = G_i\wedge G_j$), we get $A \in \mathrm{Sub}_C(X)$ such that $G_i \wedge G_j \leq \dot{A}$, $A \leq A_i \wedge A_j$, and $f_i|A = f_j|A$. By an application of 2.5 (i), we find $A_i', A_j' \in \mathrm{Sub}_C(X)$ such that $G_i \leq A_i'$, $G_j \leq A_j'$, $A_i' \leq A_i$, $A_j' \leq A_j$ and $A_i' \wedge A_j' \leq A$. We let $f_i' = f_i|A_i'$, $f_j' = f_j|A_j'$; we have achieved that $\dot{f}_i'|G_i = g_i$, $\dot{f}_j'|G_j = g_j$, $\dot{f}_i'|A_i' \wedge A_j' = \dot{f}_j'|A_i' \wedge A_j'$. Replacing f_i, f_j by f_i', f_j', respectively, for the fixed indices i,j, and keeping the other f_k, we now repeat the same procedure to the new system $\langle f_k: k\in I\rangle$ successively for all pairs of indices. □

We conclude this section by showing that we can talk about the prime completion of a coherent topos; i.e., if C_1, C_2 are two regular sites such that $\widetilde{C}_1 \simeq \widetilde{C}_2$, then also $\widetilde{P}_{C_1} \simeq \widetilde{P}_{C_2}$ ($\widetilde{P}_C$ meaning the prime completion of $\widetilde{C}$ $\widetilde{P}$ above).

Let C be a regular site, R the full subcategory of coherent objects of $\widetilde{C}$ (see e.g. MR). Then R, with the precanonical topology, is a regular site, and the inclusion $R \longrightarrow \widetilde{C}$ is isomorphic to $R \xrightarrow[\varepsilon_R]{} \widetilde{R}$; i.e. we can consider $\widetilde{R} = \widetilde{C}$, with the inclusion $R \longrightarrow \widetilde{C}$ being the canonical ε-functor. Also, $C \xrightarrow[\varepsilon_C]{} \widetilde{C}$ factors through $R \longrightarrow \widetilde{C}$; if we consider ε_C an inclusion (as we did above) then simply C is a (full) subcategory of R; we adopt this point of view below.

The inclusion $C \longrightarrow R$ induces a 'reduct' functor $\mathrm{Mod}(R,\mathrm{SET}) \xrightarrow{f} \mathrm{Mod}(C,\mathrm{SET})$

which is an equivalence of categories. This equivalence induces an equivalence of the special models of $\mathcal{R}$. and those of $\mathcal{C}$; more precisely, $\lambda_{\mathcal{C}} = \lambda_{\mathcal{R}} = \lambda$ (see the beginning of this section) and if $K_{\mathcal{C}}$, $K_{\mathcal{R}}$ denote the categories of the λ-special models of $\mathcal{C}$ and $\mathcal{R}$, respectively, then ρ restricted to $K_{\mathcal{R}}$ maps onto $K_{\mathcal{C}}$. Although this claim is very easy to establish, its verification requires looking at the details of the definition of special structures, so we omit it.

The functor $\rho\colon K_{\mathcal{R}} \xrightarrow{\sim} K_{\mathcal{C}}$ induces an equivalence $(K_{\mathcal{C}},SET) \xrightarrow[\bar{\rho}]{\sim} (K_{\mathcal{R}},SET)$. Denoting the full subcategory of (K,SET) consisting of the functors with the f.s.p. by f.s.p.(K,SET), we obtain an equivalence f.s.p.$(K_{\mathcal{C}},SET) \xrightarrow[\bar{\rho}]{\sim}$ f.s.p.$(K_{\mathcal{R}},SET)$; this is again easy to check.

Now, consider the following diagram

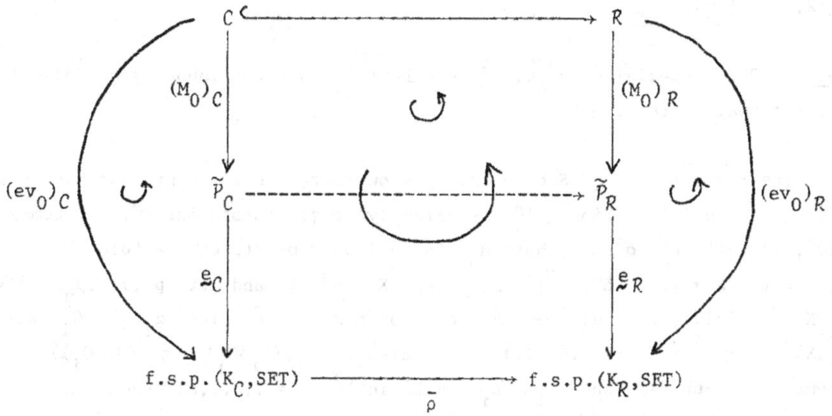

Here we used our earlier notation, once in relation to $\mathcal{C}$, once to $\mathcal{R}$. The functor $\widetilde{P}_{\mathcal{C}} \dashrightarrow \widetilde{P}_{\mathcal{R}}$ is obtained by applying the universal property of $\widetilde{P}_{\mathcal{C}}$ (Theorem 1.1). By Theorem 2.3, $\underset{\sim}{e}_{\mathcal{C}}$ and $\underset{\sim}{e}_{\mathcal{R}}$ (as used here) are equivalences. It follows that $\widetilde{P}_{\mathcal{C}} \dashrightarrow \widetilde{P}_{\mathcal{R}}$ is an equivalence as well.

Now, our initial statement concerning $\mathcal{C}_1$ and $\mathcal{C}_2$ is clear.

§ 3. <u>Uses of the prime completion.</u>

Our first use of the prime completion is to produce full continuous embeddings of certain coherent toposes into functor categories. Theorem 3.2 below is a generalization of the main result, Corollary 2.7, in [10], and it is an immediate consequence of the results of Section 2. A simple but artificial example will show that the generalization is real. However, since we do not have interesting such examples, the added generality in 3.2 seems to have only a limited interest. On the other hand, it is not inconceivable that the general results of Section 2 can lead to other results of the type of 3.2.

Definition 3.1. We call a regular site C a special site if every $X \in |C|$ is covered by finitely many prime filters, i.e. there are $n < \omega$ and $p_1,\dots,p_n \in P(X)$ such that whenever $A_i \in p_i$ $(i = 1,\dots,n)$, then $\bigvee_{i=1}^{n} A_i = 1_X$.

Remarks. The condition is equivalent to saying that in $\widetilde{P}$, $\hat{g}X$ is covered by the subobjects $g(p_i,X_i)$.

If $A \in \mathrm{Sub}_C(X)$, and (the domain object of) A is prime, then the principal filter on $\mathrm{Sub}_C(X)$ generated by A is a prime filter. Therefore, if the regular site C is prime-generated, then it is special. If E is a prime-generated coherent topos, then, by Lemma 2.3 in [10], $C = \mathrm{Coh}(E)$ = the full subcategory of E consisting of the coherent objects with the precanonical topology is a prime generated site, and of course, C is regular and $\widetilde{C} \simeq E$. Hence, the following result generalizes Corollary 2.7 in [10].

Theorem 3.2. For a special site C, $\widetilde{C}$ can be fully and continuously embedded into (K,SET) for a small category K.

Proof. Using the notation of Section 2, we show that, under the present hypotheses, $\widetilde{M}_0 = (\dot{-}): \widetilde{C} \longrightarrow \widetilde{P}$ is full. Since $|C|$ is a family of generators for $\widetilde{C}$, by Lemma 2.4 in [10], it suffices to show that $\widetilde{M}_0$ is full on hom-sets of the form $\mathrm{Hom}_{\widetilde{C}}(X,E)$, with $X \in |C|$, $E \in |\widetilde{C}|$. Fix such X and E and let $p_1,\dots,p_n \in P(X)$ covering X (see 3.1). Let $g: \dot{X} \longrightarrow \dot{E}$ be a morphism in $\widetilde{P}$, let $g_i = g|G_i$ with $G_i = g(p_i,X) \in \mathrm{Sub}^{(p)}(\dot{X})$. By 2.6 (ii) (since clearly $g_i|(G_i \wedge G_j) = g_j|(G_i \wedge G_j)$) we have some $A_i \in \mathrm{Sub}_C(X)$ and $f_i: A_i \longrightarrow E$ in $\widetilde{C}$ $(i = 1,\dots,n)$ such that $A_i \in p_i$, $g_i = \dot{f}_i|G_i$ and $f_i|(A_i \wedge A_j) = f_j|(A_i \wedge A_j)$. Since the p_i cover X, $\{A_i \hookrightarrow X: i = 1,\dots,n\} \in \mathrm{Cov}_{\widetilde{C}}(X)$. Since the representable presheaf $\mathrm{Hom}_{\widetilde{C}}(-,X)$ is a sheaf, there is a unique $f: X \longrightarrow E$ such that $f|A_i = f_i$ $(i = 1,\dots,n)$. We have $\dot{f}|G_i = \dot{f}_i|G_i = g_i = g|G_i$; since the G_i cover $\dot{X}$ in $\widetilde{P}$, it follows that $\dot{f} = g$ as required.

This proves that $\widetilde{M}_0$ is full; it is faithful and continuous by 2.4. Composing $\widetilde{M}_0$ with the functor of Theorem 2.3, we obtain the required embedding. □

Example. We describe a coherent theory T in the one-sorted language having the unary predicate symbols A_i^0 and A_i^1 $(i \in \omega)$. The axioms of the theory are:

$$\forall x(A_{i+1}^{\alpha}(x) \to A_i^{\alpha}(x)) \quad (i \in \omega,\ \alpha = 0,1),$$

$$\forall x(A_i^0(x) \vee A_j^1(x)) \quad (i,j \in \omega),$$

$$\forall x A_0^{\alpha}(x) \quad (\alpha = 0,1),$$

$$\exists x(A_i^0(x) \wedge A_j^1(x)) \quad (i,j \in \omega).$$

Let C be the logical category (in the terminology of MR) derived from T, $C = R_T$; and let C be made a regular site with the precanonical topology. The classifying topos of T is $\tilde{C}$. We also have the canonical interpretation $[\varphi]$ of formulas φ of T as certain subobjects of U^n, for some n, where U is the 'universe' object $= [x{=}x]$. Let $A_i^\alpha = [A_i^\alpha(x)] \in \mathrm{Sub}_C(U)$. With $\vec{i} = \langle i_1,\ldots,i_n\rangle$, $\vec{j} = \langle j_1,\ldots,j_n\rangle$ let $C_{\vec{i},\vec{j}} = [\bigwedge_{k=1}^{n} A_{i_k}^0(x_k) \wedge \bigwedge_{k=1}^{n} A_{j_k}(x_k)] \in \mathrm{Sub}(U^n)$; call a subobject of U^n of the form $C_{\vec{i},\vec{j}}$ a <u>special</u> subobject. With $m \le n$, a <u>diagonal morphism</u> $U^m \longrightarrow U^n$ is one of the form $[\langle x_1,\ldots,x_m\rangle \longmapsto \langle x_{i_1},\ldots,x_{i_n}\rangle]$, with $i_1,\ldots,i_n$ certain indices such that $\{i_1,\ldots,i_n\} = \{1,\ldots,m\}$; a diagonal morphism is a monomorphism. An easy computation ("elimination of quantifiers") shows that every subobject of U^n in C is the image of a special subobject of some U^m $(m \le n)$ under some diagonal morphism. Let p^α be the filter on $\mathrm{Sub}_C(U)$ generated by $\{A_i^\alpha\colon i\in\omega\}$ $(\alpha = 0,1)$. We have that p^0, p^1 are prime filters on U, as easily checked. More generally, let $\varepsilon\colon \{1,\ldots,n\} \to \{0,1\}$; let p^ε be the filter on U^n generated by $\{[\bigwedge_{k=1}^{n} A_i^{\varepsilon(k)}(x_k)]\colon i_1,\ldots,i_n \in \omega\}$. Then for every special subobject C of U^n, and any ε, $C(\wedge)p^\varepsilon \underset{df}{=} \{C\wedge A\colon A\in p^\varepsilon\}$, which is a filter on $\mathrm{Sub}_C(C)$ under the obvious identification of subobjects of C with those of U^n, is a prime filter on C. To verify this, one uses the easily seen facts that $A_i^0 \wedge A_j^1 \nleq A_{i+1}^0$, $A_i^0 \wedge A_j^1 \nleq A_{j+1}^1$. Putting all these facts together, one easily concludes that C is a special site. However, C is not prime-generated, and hence $\tilde{C}$ is not prime-generated either; in fact, e.g. U does not have any subobject in C which is a prime. □

It would be interesting to know if the following is true: if a coherent topos $\tilde{C}$ can be fully and continuously embedded into some category of the form (K,SET), then the canonical embedding $\tilde{M}_0\colon \tilde{C} \longrightarrow \tilde{P}$ is full.

Our second use of the prime completion is a result relating the category of models of a theory to the prime completion of the classifying topos of the theory, also called the <u>topos</u> <u>of</u> <u>types</u> <u>of</u> the theory. The main role in this result is played by a technical theorem of Daniel Lascar; our result does nothing more than bring out in a certain sense the 'content' of Lascar's theorem.

In model theory, from early on it has been an important aim to recover detailed syntactical information from 'global' semantical behaviour, in many different situations; it is enough to remind the reader of the classic example of Beth's definability theorem to indicate what we have in mind. Since the category of models of a theory can be reasonably viewed as a codification of an essential body of semantical properties of the theory, it is reasonable to ask to what extent can syntactical aspects of a theory be recovered from the category of models. Now, as we are prepared to argue, the topos of types is a reasonable codification of the 'discrete' (non topological) syntactical structure of types of the theory just as the classifying topos is a reasonable codification of the full syntactic content of the theory. Our recasting of Lascar's theorem below therefore can be paraphrased into saying that for at least certain countable complete theories (including ones with Skolem functions, and many others, see below), the category of models contains the essential information on the discrete structure of types.

Let T be a finitary first order theory. As we explained in § 5 in [10], it makes a difference what fragment of full finitary first order logic we consider with T; see also the Introduction. Once a theory T is conceived in this way, one has attached to it its classifying topos $\mathcal{E}(T)$. This notion is dealt with for theories in the 'coherent fragment' (coherent theories) in MR in detail, and in § 5 in [10] the simple way of generalizing this concept to arbitrary theories is explained. The 'logical category' derived from T, $\mathcal{R}_T$, sits inside $\mathcal{E}(T)$, and in fact, with $\mathcal{R}_T$ considered a site with the precanonical topology, $\mathcal{R}_T$ is a regular site and $\widetilde{\mathcal{R}}_T \simeq \mathcal{E}(T)$.

The <u>topos of types</u> of T, denoted $\widetilde{\mathcal{P}}(T)$, is defined as the prime completion of the coherent topos $\mathcal{E}(T)$; this is the same as $\widetilde{\mathcal{P}}_{Coh(\mathcal{E}(T))}$ or $\widetilde{\mathcal{P}}_{\mathcal{R}_T}$ (see the end of the last section).

From now on, assume that T is a complete theory in full finitary logic; allowing many sorted theories, this is equivalent to saying that in $\mathcal{C} = Coh(\mathcal{E}(T))$, every subobject $A \in Sub_{\mathcal{C}}(X)$ has a Boolean complement B such that $A \vee B = 1_X$ and $A \wedge B = 0_X$, and the terminal object of $\mathcal{C}$ is an <u>atom</u>, i.e. it has exactly two distinct subobjects. We claim that in this case $\widetilde{\mathcal{P}}(T)$ is an atomic topos, i.e. it has a family of generators consisting of atoms (compare [3] and [10]); namely, it is easy to check that in this case $p,q \in P(X)$, $X \in |\mathcal{C}|$, $p \subset q$ imply $p = q$; therefore (by 1.8), $\underset{\sim}{a}(q,X)$ is an atom in $\widetilde{\mathcal{P}}(T)$.

There is a single λ-special model N of T up to isomorphism ($\lambda = \lambda_{\mathcal{C}}$; see CK, 5.1.17); hence now in 2.3 K can be taken to be the monoid End(N) of all elementary embeddings of N into itself.

For an arbitrary category $\mathcal{D}$, a functor $\mathcal{D} \to SET$ is called <u>upcontinuous</u> if it preserves small filtered (directed) colimits existing in $\mathcal{D}$; Upcon($\mathcal{D}$,SET) is the full subcategory of ($\mathcal{D}$,SET) with objects the upcontinuous functors. It was André Joyal's basic observation that upcontinuous functors might be important, since

$$ev: \mathcal{E}(T) \to (\text{Mod } T, SET)$$

factors through Upcon(Mod T,SET). Whereas Upcon($\mathcal{D}$,SET) is defined for an abstract category $\mathcal{D}$, and hence, if Mod T $\simeq$ Mod T', then Upcon(Mod T, SET) $\simeq$ Upcon(Mod T',SET) the same cannot be said a priori of f.s.p.(Mod T,SET).

Daniel Lascar introduces the following notions. For a model M of T, and a finite subset X of M, let $Aut_X(M)$ denote the group of X-automorphisms γ of M; $\gamma(x) = x$ for $x \in X$. $\gamma, \delta \in Aut_X(M)$ are called <u>weakly conjugate</u> if there is an elementary extension $N \models T$ of M and extensions γ', δ' of γ, δ, respectively, such that γ', δ' are conjugate in the group $Aut_X(N)$. A subgroup of $Aut_X(M)$ is called <u>very normal</u> if $\gamma \in H$, $\delta \in Aut_X(M)$ and γ, δ being weakly conjugate imply that $\delta \in H$. The theory T is called G-<u>trivial</u> if for some (any) special model N of T, and for any finite $X \subset N$, the only very normal subgroup of $Aut_X(N)$ is $Aut_X(N)$ itself. Note that this is certainly true if every $\gamma \in Aut_X(N)$ is weakly conjugate to Id_N, which is the case in all known G-trivial cases, and which is something quite accessible to verification.

One large class of G-trivial theories is that of all complete theories with Skolem functions (see CK, p.143). Another is those stable theories in which all types over arbitrary sets are stationary [12]. The theory of dense linear orders without endpoints is another G-trivial theory. On the other hand, the theory of algebraically closed fields with given characteristic is not G-trivial.

<u>Lascar's Theorem</u> [7]. Suppose T is a countable complete G-trivial theory. Then every upcontinuous functor $\mathrm{Mod}\, T \longrightarrow SET$ has the f.s.p. (see 2.1).

<u>Theorem 3.3.</u> If T and T' are G-trivial theories, and $\mathrm{Mod}\, T \simeq \mathrm{Mod}\, T'$, then $\widetilde{P}(T') \simeq \widetilde{P}(T)$.

<u>Proof.</u> Some general terminology first. Given a (Grothendieck) topos $\mathcal{E}$, and a family G of objects of $\mathcal{E}$, there is a smallest subcategory $\mathcal{E}'$ of $\mathcal{E}$ such that (i) $G \subset |\mathcal{E}'|$, (ii) $\mathcal{E}'$ is a topos, (iii) the inclusion $\mathcal{E}' \longrightarrow \mathcal{E}$ is continuous and strongly full, (iv) $\mathcal{E}'$ is closed under isomorphisms in $\mathcal{E}$: if $E' \in |\mathcal{E}'|$, $E' \simeq E$ in $\mathcal{E}$, then $E \in |\mathcal{E}'|$. Indeed, note that conditions (ii) and (iii) can be replaced by the following equivalent closure conditions: $\mathcal{E}'$ is closed under finite products (in $\mathcal{E}$), subobjects, disjoint sums, and quotients. Let us call such $\mathcal{E}'$ the <u>subtopos</u> of $\mathcal{E}$ <u>generated</u> by G, and denote it by $\langle G \rangle$. If G happens to be a family of generators for $\mathcal{E}$ in the usual sense, then $\langle G \rangle = \mathcal{E}$, as it is easily seen.

Let now T be a G-trivial countable complete theory. Let us write $\underset{\sim}{e}[\widetilde{P}]$ for the essential image of $\underset{\sim}{e}$: $\widetilde{P}(T) \to (K, SET)$, with $K = \mathrm{End}(N)$ as explained above, and $\underset{\sim}{e}$ the functor of 2.3; in particular, $\underset{\sim}{e}[\widetilde{P}]$ is closed under isomorphisms in $(\mathrm{End}(N), SET)$. Let G be the family of those functors $F: \mathrm{End}(N) \to SET$ such that

(*) there is an upcontinuous $G: \mathrm{Mod}\, T \to SET$ with $F = G|\mathrm{End}(N)$.

We claim that $\underset{\sim}{g}[\tilde{P}] = \langle G\rangle$; in other words, *for a G-trivial theory* T, *the topos of types is recovered, up to equivalence, from the category of models, as the subtopos of* (End(N),SET) *generated by those functors that are restrictions of upcontinuous functors* Mod T → SET.

We have to show that $|\underset{\sim}{g}[\tilde{P}]| = |\langle G\rangle|$. If $p \in P(X)$ $(X \in |C| = |Coh(E(T))|)$, then (p,X) gives rise to a functor $[p,X]$: Mod T → SET with the f.s.p. such that

$$\underset{\sim}{g}(\underset{\sim}{a}(p,X)) = [p,X] \mid End(N). \tag{1}$$

Indeed, define for any model $M \in$ Mod T,

$$[p,X](M) = \bar{M}(p,X) = \bigcap\{M(A): A\in p\}$$

and for $M \xrightarrow{h} N$,

$$[p,X](h) = h_X|\bar{M}(p,X): \bar{M}(p,X) \to \bar{N}(p,X) .$$

$[p,X]$ has the f.s.p. simply because $x \in [p,X](M)$ has itself as a support; it is also clear that (1) holds.

What this says is that $\underset{\sim}{g}(\underset{\sim}{a}(p,X)) \in G$. Since the $\underset{\sim}{a}(p,X)$ generate $\tilde{P}(T)$, it follows that $|\underset{\sim}{g}[\tilde{P}]| \subset |\langle G\rangle|$.

Conversely, assume that $F \in G$, i.e. we have (*). By Lascar's theorem, G has the f.s.p. A fortiori, F has the f.s.p. By 2.3, $F \in |\underset{\sim}{g}[\tilde{P}]|$. We have shown $G \subset |\underset{\sim}{g}[\tilde{P}]|$; hence $|\langle G\rangle| \subset |\underset{\sim}{g}[\tilde{P}]|$, proving our claim.

Now, the assertion of the theorem easily follows, essentially because the λ-special model (now $\lambda = \beth_\omega$) can be singled out purely categorically from Mod T; in particular, if T, T' are countable complete theories, Mod T $\xrightarrow[F]{\simeq}$ Mod T' is an equivalence functor, N is a (the) λ-special model of T, then F(N) is a (the) λ-special model of T'. To see this, first note that a model M of T has cardinality $\kappa(>\aleph_0)$ iff there is an elementary chain of models M_α $(\alpha < \kappa)$, $M_{\alpha+1}$ a proper extension of M_α for all $\alpha < \kappa$, whose union (direct limit) is M; i.e. we have a category definition of cardinality. Next, M is κ-saturated just in case the following holds: whenever $N \prec N'$, both models of T of cardinality $< \kappa$, and $f: N \to M$ is an elementary embedding, then f can be extended to an elementary embedding $f': N' \to M$; finally, a λ-special model is nothing but the union of an elementary chain $\langle M_n : n < \omega\rangle$ of models, such that M_n is κ_n-saturated and card $M_n \le \kappa_n$; here $\kappa_0 = \aleph_0$, $\kappa_{n+1} = 2^{\kappa_n}$, and $\lambda = \lim_{n\to\omega} \kappa_n$. These remarks clearly establish what we want. Thus, our claim above proves the theorem. □

<u>Remarks</u>. There is some awkwardness in the above proof; in particular, it would be nicer if we could say that $\widetilde{P}(T)$ is equivalent to f.s.p.(End(N),SET). Whereas we know this to be true in some special cases, we do not know it for an arbitrary G-trivial theory. We note that the idea of considering restrictions of upcontinuous functors Mod T $\to$ SET in the manner of the above proof is due to Lascar [7].

It would be interesting if we could strengthen 3.3 to saying that for any G-trivial theory T, $\mathrm{Mod}(T') \simeq \mathrm{Mod}(T)$ implies $\widetilde{P}(T') \simeq \widetilde{P}(T)$ for <u>any</u> theory T'. We can prove this in a special case, namely when T is $\le \aleph_0$-categorical. A countable complete T is $\le \aleph_0$-categorical iff $\mathcal{E}(T)$ is atomic (Ryll-Nardzewski's theorem, see CK); in this case $\widetilde{P}(T) \simeq \mathcal{E}(T)$ as it is easily seen. We have

<u>Theorem 3.4.</u> If T is an $\le \aleph_0$-categorical G-trivial theory, then T is distinguished by its category of models in the sense that for any countable theory T', Mod T' $\simeq$ Mod T implies that $\mathcal{E}(T') \simeq \mathcal{E}(T)$.

<u>Proof:</u> We are going to apply the 'comparison theorem', Theorem 9.2.9 in MR. This theorem says the following. Let $\mathcal{E}_1, \mathcal{E}_2$ be coherent toposes, $I\colon \mathcal{E}_1 \longrightarrow \mathcal{E}_2$ a continuous functor mapping coherent objects of $\mathcal{E}_1$ into coherent objects in $\mathcal{E}_2$. Consider the induced functor $\hat{I}\colon (\mathrm{Con}(\mathcal{E}_2,\mathrm{SET}) \to \mathrm{Con}(\mathcal{E}_1,\mathrm{SET})$ (defined by composition). Then, if $\hat{I}$ is an equivalence of categories, so is I. If $\mathcal{E}_i = \mathcal{E}(T_i)$, then of course $\mathrm{Con}(\mathcal{E}_i,\mathrm{SET}) \simeq \mathrm{Mod}\ T_i$.

To prepare the stage for the application of this theorem, let N,N' be λ-special models of T,T', respectively; let $F\colon \mathrm{Mod}\ T \xrightarrow{\simeq} \mathrm{Mod}\ T'$ be an equivalence functor; by what we said above, we may assume that $N' = F(N)$; let $C = \mathrm{End}(N) \subset \mathrm{Mod}\ T$, $C' = \mathrm{End}(N') \subset \mathrm{Mod}\ T'$; so we have $F|C\colon C \xrightarrow{\sim} C'$; we also identify T,T' with the appropriate regular sites R_T, $R_{T'}$, respectively.

We consider the following commutative diagram

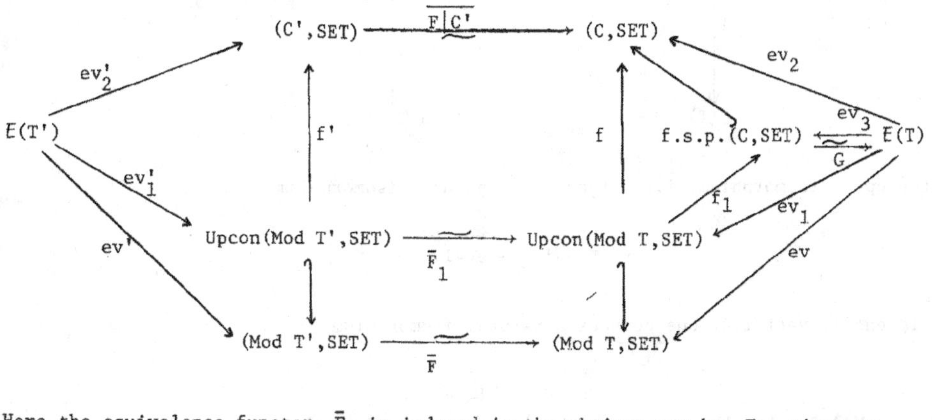

Here the equivalence functor $\bar{F}$ is induced in the obvious way by F; since up-continuity is a 'categorical notion', one has $\bar{F}_1$ induced. By 2.3 (or, already by 3.2 in [10]), we have that ev_2 factors through f.s.p.(C,SET) so that the resulting ev_3 is an equivalence; its quasi-inverse is denoted by G. ρ and ρ' are restriction functors. By the proof of 3.3, ρ factors through f.s.p.(C,SET), resulting in the functor ρ_1.

We define $I: \mathcal{E}(T') \longrightarrow \mathcal{E}(T)$ as the composite

$$I = G\circ\rho_1\circ\bar{F}_1\circ ev_1' .$$

As a composite of continuous functors between toposes, I is continuous.

We claim that I maps coherent objects of $\mathcal{E}(T')$ into coherent object in $\mathcal{E}(T)$. Since in a coherent atomic topos, the coherent objects are precisely the finite sums of atoms (3.1(i) in [10]), it suffices to show that I maps an atom into an atom.* ev_2' maps an atom in $\mathcal{E}(T)$ into an atom in (C',SET), by the strong fullness of ev_2' (2.3). Now, the assertion follows by the commutativity up to isomorphism of the diagram

$$\begin{array}{ccc}
\mathcal{E}(T') & \xrightarrow{\;ev_2'\;} & (C',SET) \\
\downarrow & & \wr\downarrow \overline{F|C} \\
\mathcal{E}(T) & \xrightarrow[\;ev_2\;]{} & (C,SET)
\end{array} .$$

Also, by the construction, the diagram

$$\begin{array}{ccc}
\mathcal{E}(T') & \xrightarrow{\;ev'\;} & (\text{Mod } T',SET) \\
I\downarrow & & \downarrow\bar{F} \\
\mathcal{E}(T) & \xrightarrow[\;ev\;]{} & (\text{Mod } T,SET)
\end{array}$$

commutes up to isomorphism, i.e. there is a natural isomorphism

$$\nu: \bar{F}\circ ev' \xrightarrow[\sim]{\cdot} ev\circ I.$$

As it is easily verified, one deduces a natural isomorphism

$$\mu: F \xrightarrow[\sim]{\cdot} \hat{I}$$

*Mod T' $\simeq$ Mod T clearly implies that T' is $\aleph_0$-categorical.

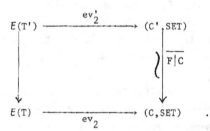

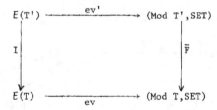

(where $\hat{I}$: (Mod T',SET) → (Mod T,SET) is defined by composition) defined by

$$\mu = (\mu_M)_{M \in |\text{Mod } T|},$$

$$\mu_M = ((\mu_M)_X)_{X \in |T'|},$$

$$(\mu_M)_X = (\nu_X)_M .$$

Since F is an equivalence, so is $\hat{I}$. By the comparison theorem, I is an equivalence. □

REFERENCES

SGA4[1] M. Artin, A. Grothendieck and J.L. Verdier, Théorie des Topos et Cohomologie Etale des Schémas, Springer Lecture Notes Math., no's 269 and 270, 1972.

[2] M. Barr, Exact categories, in: Springer Lecture Notes Math., No. 226, pp. 1-120, 1971.

[3] M. Barr and R. Diaconescu, Atomic toposes, J. Pure and Applied Algebra 17 (1980), 1-24.

CK[4] C.C. Chang and H.J. Keisler, Model Theory, North Holland, 1973.

[5] A. Joyal and G.E. Reyes, Forcing and generic models in categorical logic, to appear.

[6] D. Lascar and B. Poizat, An introduction to forking, J. Symbolic Logic, 44 (1979), 330-350.

[7] D. Lascar, On the category of models of a complete theory, to appear.

CWM[8] S. MacLane, Categories for the working mathematician, Springer Verlag, 1971.

MR[9] M. Makkai and G.E. Reyes, First order categorical logic, Springer Lecture Notes Math., no. 611, 1977.

[10] M. Makkai, Full continuous embeddings of toposes, to appear.

[11] M. Morley, Applications of topology to $L_{\omega_1\omega}$, Proc. Tarski Symposium, Proc. Sympos. AMS, v. 25, pp.233-240, 1974.

[12] S. Shelah, Classification theory and the number of non-isomorphic models, North Holland, 1978.

SOME DECISION PROBLEMS FOR SUBTHEORIES OF TWO-DIMENSIONAL PARTIAL ORDERINGS

by

A. B. Manaster and J. B. Remmel
Department of Mathematics
University of California, San Diego
La Jolla, CA 92093

In this paper we present a natural subclass of dense two-dimensional partial orderings with a decidable theory and provide a complete axiomatization of that theory. We then see that several ways to restrict the theory by removing axioms lead to undecidable theories. We close the paper with an open question about the decidability of a similar restriction of the theory.

Two-dimensional partial orderings (denoted 2dPOs) were defined by Dushnik and Miller [2]. The theory of 2dPOs was shown undecidable in Manaster and Rosenstein [8]. The notion of dense 2dPOs arose in our study of the model companion of the theory of 2dPOs [4]. The specific 2dPOs denoted $D_2^{h,v}$ discussed below provide examples of $\aleph_0$-categorical 2dPOs which are not recursively categorical and which have recursive presentations with no non-trivial recursive automorphisms (see [4] and [5]). In the next few paragraphs we present terse definitions of the notions needed in this paper. More discursive discussions are in [4], [5], and [6].

$\mathbb{Q}$ denotes the rational numbers. $<_{\mathbb{Q}}$ denotes the usual linear ordering of $\mathbb{Q}$. ${}^2\mathbb{Q} = \{(x,y): x \in \mathbb{Q} \;\&\; y \in \mathbb{Q}\}$ is the rational plane. $<_{{}^2\mathbb{Q}}$ is the product order on ${}^2\mathbb{Q}$. Since we only consider countable structures, a *2dPO* is a partial ordering isomorphic to a subset of ${}^2\mathbb{Q}$ with the induced ordering. A *dense 2dPO* is one which is isomorphic to a topologically dense subset of ${}^2\mathbb{Q}$. A *line segment* of a (topologically) dense subset S of ${}^2\mathbb{Q}$ is a non-trivial interval of S which is linearly ordered by $<_{{}^2\mathbb{Q}}$. It is easy to see that any such line segment is either horizontal or vertical. A *line* of S is a maximal line segment of S. Let P be a dense 2dPO. An embedding of P into ${}^2\mathbb{Q}$ is appropriate if it maps P onto a dense subset of ${}^2\mathbb{Q}$. Since the image of a subset of P is a line or line segment of $^2\mathbb{Q}$ under one appropriate embedding if and only if the image is a line or, respectively line segment, under all appropriate embeddings, we may define the notions of line and line segment for 2dPOs via appropriate embeddings.

$<_1$ denotes the first coordinate pre-ordering on $^2\mathbb{Q}$ and subsets of $^2\mathbb{Q}$. Thus $(a,b) <_1 (c,d)$ iff $a <_{\mathbb{Q}} c$ iff (a,b) is to the left of (c,d). $x =_1 y$ is defined by $\neg(x <_1 y \vee y <_1 x)$. Similarly $<_2$ denotes the second coordinate pre-ordering on $^2\mathbb{Q}$, and $x =_2 y$ is defined by $\neg(x <_2 y \vee y <_2 x)$. We see that a horizontal line of a dense subset S of $^2\mathbb{Q}$ is $\{x \in S: x =_2 y\}$ for some $y \in {}^2\mathbb{Q}$. Since reflections of appropriate embeddings are again appropriate embeddings, we see that the notions of $<_1$, $=_1$, $<_2$, $=_2$, horizontal, and vertical, cannot simply be described via appropriate embeddings. However once a pair of incomparable elements is fixed, say $i|j$, in a dense 2dPO, and we restrict attention to embeddings e such that $e(i) <_1 e(j) <_2 e(i)$, then all of these notions are fixed. In fact we saw in [4] that all of these notions are definable in the theory of dense 2dPOs given $i <_1 j <_2 i$. Throughout the remainder of this paper, when we refer to these notions for a dense 2dPO we assume that an incomparable pair $i <_1 j <_2 i$ has been fixed.

Let L be a horizontal line segment of a dense 2dPO P. L is called *unbounded* (in P) if for every point p of P there are points ℓ and r on L to the left and to the right of p. L is called *full* (in P) if it is as dense as possible in P; that is if whenever $\ell_1 \in L$, $\ell_2 \in L$, and $\ell_1 <_1 s_1 <_1 s_2 <_1 \ell_2$ there must exist $\ell_3 \in L$ such that $s_1 <_1 \ell_3 <_1 s_2$. Analogous definitions apply to vertical line segments of dense 2dPOs.

$D_2 = D_2^{0,0}$ is any dense 2dPO which has no line segments. In [4] we saw that any two presentations of D_2 are isomorphic so that the notation is reasonable. In general we define $D_2^{h,v}$ for $h \geq 0$ and $v \geq 0$ to be any dense 2dPO satisfying all of the following conditions. $D_2^{h,v}$ has exactly h+v distinct lines. Each line of $D_2^{h,v}$ is full and unbounded. No two lines of $D_2^{h,v}$ intersect. $D_2^{h,v}$ has exactly h horizontal lines and v vertical lines.

It is not difficult to see that any two presentations of $D_2^{h,v}$ are isomorphic. Let F be $\{D_2^{h,v}: 0 \leq h \text{ and } 0 \leq v\}$. F is a critical class in the study of recursively categorical dense 2dPOs (see [5]). Let K be the theory of F. In Section 1 we will see that K is decidable, find a finite axiomatization of K, and show that for each $h \leq 0$ and $v \leq 0$ the theory of $D_2^{h,v}$ is decidable. In Section 2 we will see that all of our axioms for K beyond the basic axioms for dense 2dPOs are essential for the decidability of K.

Section 1: DECIDABILITY RESULTS.

A finite axiomatization of the theory of dense 2dPOs is in [4]. The main observations about dense 2dPOs required to formulate the axioms are the definability of $<_1$ and $<_2$ from an incomparable pair and the partitioning of a dense 2dPO into 25 regions, nine of which must be non-empty, by any non-colinear pair of points. The 25 regions are indicated in Figure 1; regions 1-9 must be non-empty.

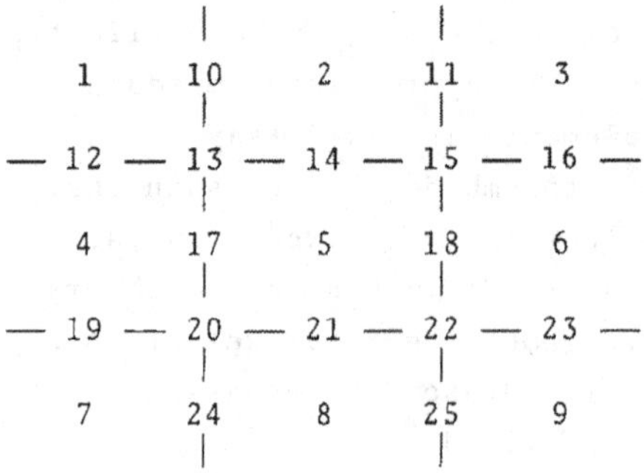

FIGURE 1

The importance of the notion of discrete in the theory of finite linear orderings led us to find an analogue for K. The assertion that the set of horizontal lines is discretely ordered in P by $<_2$ is stronger than the assertion that this set of lines is discretely ordered by $<_2$. Being discretely ordered in P means that for any $x \in P$ if x lies below any horizontal lines, then there is a horizontal line L above x with no horizontal line below L and above x, and if x lies above any horizontal lines, then there is a horizontal line L below x with no horizontal lines between x and L. A similar definition gives the meaning of the vertical lines being discretely ordered in P by $<_1$.

There do exist dense 2dPOs P in which the set of horizontal lines is discretely ordered in P by $<_2$, the set of vertical lines is discretely ordered in P by $<_1$, and yet both sets of lines are infinite. One such 2dPO, called $D^{\omega+\omega^*,\omega+\omega^*}$ or U, will be described here and used in Section 2. The $<_2$ order type of the horizontal lines of U is $\omega+\omega^*$; similarly, the $<_1$ order type of the vertical lines of U is $\omega+\omega^*$. The hypotheses of discreteness in U implies that there are no points between the ω and ω^* sequences of horizontal lines and that there are no points between the ω^* and ω sequences of vertical lines. To construct U, start with a topologically dense subset of ${}^2\mathbb{Q}$ which does not intersect any of the lines $x = 1/n$ for $n \in \{\pm 1, \pm 2, \pm 3, \ldots\}$, does not intersect the x-axis or the y-axis, does not intersect a $<_1$ dense set of vertical lines, and does not intersect a $<_2$ dense set of horizontal lines. To finish a construction of U, add to this set full and unbounded sets of points on each of the lines $x = 1/n$ and $y = 1/n$ for $n \in \{\pm 1, \pm 2, \pm 3, \ldots\}$ in such a way that no other lines are formed and no two lines intersect.

Theorem 1. The following axioms, when formalized, are a complete set of axioms for K.

(1) P is a dense 2dPO.

(2) No two lines of P intersect.

(3) Each line of P is full.

(4) Each line of P is unbounded.

(5) If P has any horizontal lines, the set of horizontal lines is discretely ordered in P by $<_2$ with both a first and last element.

(6) If P has any vertical lines, the set of vertical lines is discretely ordered in P by $<_1$ with both a first and last element.

Proof. Let K' be the theory generated by axioms (1)-(6). It is clear that each $D_2^{h,v}$ satisfies axioms (1)-(6) so that $K' \subseteq K$. To prove that $K \subseteq K'$ we shall first introduce a list of basic predicates for K' and show that an elimination of quantifiers is possible in K' over these predicates. We will then be able to observe that the sentences true in every model of axioms (1)-(6) are precisely those sentences true in all the $D_2^{h,v}$. Moreover, as a result of our analysis, the decidability of K, as well as other interesting properties of the $D_2^{h,v}$s, will easily follow.

Our list of basic predicates is as follows:

$H(x)$: $(\exists y)[x =_2 y \ \& \ x \neq y]$

H defines the set of points lying on horizontal lines.

$V(x)$: $(\exists y)[x =_1 y \ \& \ x \neq y]$

V defines the set of points lying on vertical lines.

$$d_1(x,y) \geq n: \ (\exists z_1)\ldots(\exists z_n)[x <_1 z_1 <_1 \cdots <_1 z_n <_1 y \ \& \ \mathop{\&}_{i=1}^{n} V(z_i)]$$

For any non-negative integer n, $d_1(x,y) \geq n$ defines the pairs for which x lies left of y and at least n vertical lines lie between x and y.

$$d_1(x,y) = n: \ d_1(x,y) \geq n \ \& \ \neg d_1(x,y) \geq n+1$$

$d_1(x,y) = n$ defines the pairs (x,y) for which x lies to the left of y and there are exactly n vertical lines between x and y.

$$d_1(x,+\infty) \geq n\colon\ (\exists z_1)\ldots(\exists z_n)[x <_1 z_1 <_1 \cdots <_1 z_n \ \&\ \mathop{\&}_{i=1}^{n} V(z_i)]$$

$d_1(x,+\infty) \geq n$ defines the set of points which lie to the left of at least n vertical lines.

Obvious modifications of the preceding give definitions of $d_1(x,+\infty) = n$, $d_1(-\infty,y) \geq n$, $d_1(-\infty,y) = n$, and analogues with d_2 in place of d_1.

Lemma. Every formula is equivalent in K' to a propositional combination of the predicates just listed and $=_1$, $=_2$, $<_1$, $<_2$.

Proof. The usual manipulations in elimination of quantifier proofs together with consideration of the pre-ordering properties of $<_1$ and $<_2$ and the relations between d_1 and $<_1$ and between d_2 and $<_2$ show that the general form reduces to the cases suggested by the next formula. In that formula $(\neg)$ indicates that $\neg$ may occur or may not occur in that position, $(=,\geq)$ indicates that one of $=$ or $\geq$ occurs, $(\&\ldots)$ indicates that the parenthetical clause may occur or not, and any or all of ℓ, r, b, a may or may not occur where first displayed. If the first ℓ or b does not occur, later occurrences are understood to be occurrences of $-\infty$. Similarly if the first r or a does not occur, later occurrences represent $+\infty$.

$$(*)\quad (\exists x)[(\neg)V(x)\ \&\ (\neg)H(x)$$
$$\&\ \ell <_1 x <_1 r\ \&\ d_1(\ell,x)(=,\geq)n_1\ \&\ d_1(x,r)(=,\geq)m_1(\&\ x =_1 v\ \&\ x \neq v)$$
$$\&\ b <_2 x <_2 a\ \&\ d_2(b,x)(=,\geq)n_2\ \&\ d_2(x,a)(=,\geq)m_2(\&\ x =_2 h\ \&\ x \neq h)].$$

There are certain obvious cases where $(*)$ is equivalent in K' to the contradictory sentence $d_1(-\infty,\infty) = 0\ \&\ d_1(-\infty,\infty) \geq 1$ because $(*)$ can never be satisfied by a model of K'. For example, $(*)$ may contain $V(x)\ \&\ H(x)$ or may contain $\neg V(x)\ \&\ x =_1 v\ \&\ x \neq v$. Moreover, K' clearly implies that for all x one of the following 3 possibilities holds: (a) $\neg V(x)\ \&\ \neg H(x)$, (b) $\neg V(x)\ \&\ H(x)$, or (c) $V(x)\ \&\ \neg H(x)$, so that we may assume $(*)$ starts with one of these three combinations. We shall indicate the general pattern of the various possibilities by analyzing only those cases in which the first two conjuncts are $\neg V(x)\ \&\ H(x)$. This given, first we consider those cases in which $(\&\ x =_2 h\ \&\ x \neq h)$ does not occur. In such cases $(*)$ is equivalent over K' to

(**) $\ell <_1 r$ & $d_1(\ell,r)(=,\geq)_1\ n_1+m_1$ & $b <_2 a$ & $d_2(b,a)(=,\geq)_2\ n_2+m_2+1$

Here $(=,\geq)_k$ represents $=$ if both formulas involving d_k in (*) have $=$ and represents $\geq$ if at least one of these formulas has $\geq$. The equivalence of (*) and (**) in this case will be considered in detail in the next paragraph. If (& $x =_2 h$ & $x \neq h$) occurs, (*) is equivalent to

(***) $\ell <_1 r$ & $d_1(\ell,r)(=,\geq)_1\ n_1+m_1$ & $b <_2 h <_2 a$ & $H(h)$
& $d_2(b,h)(=,\geq)_3\ n_2$ & $d_2(h,a)(=,\geq)_4\ m_2$.

$(=,\geq)_1$ has the same denotation it did in (**). $(=,\geq)_3$ denotes $=$ if $d_2(b,x) = n_2$ is a conjunct of (*) and denotes $\geq$ if $d_2(b,x) > n_2$ is a conjunct of (*). The denotation of $(=,\geq)_4$ is defined similarly.

In considering the equivalence of (*) to (**) for 2dPOs which are models of K' we shall further restrict our attention to the special case (*s) of (*) displayed below. This special case was chosen to have enough variety to indicate the general pattern.

(*s) $(\exists x)[\neg V(x)$ & $H(x)$
& $x <_1 r$ & $d_1(-\infty,x) = n_1$ & $d_1(x,r) \geq m_1$
& $b <_2 x <_2 a$ & $d_2(b,x) \geq n_2$ & $d_2(x,a) \geq m_2]$.

According to the previous paragraph we want to see that this formula is equivalent over K' to

(**s) $d_1(-\infty,r) \geq n_1+m_1$ & $b <_2 a$ & $d_2(b,a) \geq n_2+m_2+1$.

Since (*s) is easily seen to imply (**s) in any dense 2dPO, we consider the converse implication for any model U of K'. Since the collection of horizontal lines is discretely ordered in U and there are at least n_2+m_2+1 horizontal lines between b and a, let L be the horizontal line which has exactly n_2 lines below it which are above b. Any point x on L will satisfy the last three conjuncts of the matrix of (*s) since there must be at least m_2 lines above L which are below a. Since $d_1(-\infty,r) \geq n_1+m_1$ and the set of vertical lines is discretely ordered with a leftmost element, there is a line M which has exactly n_1 lines to its left. Let x be any point on L to the left of M but to the right of all lines to the left of M.

x exists since L is full if $n_1 \geq 0$ and since L is unbounded if $n_1 = 0$. $d_1(-\infty,x) = n_1$ and $d_1(x,r) \geq m_1$ since $d_1(-\infty,r) \geq n_1+m_1$.

It now follows by our elimination of quantifiers that if φ is any sentence in the language of 2dPOs then φ is equivalent over K' to a sentence of the form

$$\bigwedge_{j=1}^{n} \left(\bigvee_{k=0}^{m} \varphi_{k,j}\right)$$

where each $\varphi_{k,j}$ is either $d_i(-\infty,\infty) \geq p$ or $d_i(-\infty,\infty) = p$ for some $i \in \{1,2\}$ and some $p \geq 0$. It is now easy to see that the only sentences true in all models of K' are sentences which assert $d_1(-\infty,\infty) \geq 0$ and/or $d_2(-\infty,\infty) \geq 0$. Since these are the same sentences which are true in all the $D_2^{h,v}$'s we have $K \subseteq K'$. Thus (1)-(6) axiomatize K.

We proved in [4] that the theory of dense 2dPOs is finitely axiomatizable so that it follows from our elimination of quantifiers and our remarks above that we can effectively decide whether $K \models \varphi$ for any sentence φ. Thus we have the following.

Corollary 1. K is a decidable finitely axiomatizable theory.

Corollary 2. For each h and v, the theory of $D_2^{h,v}$ is decidable. A complete set of axioms for this theory consists of axioms (1)-(4) together with the assertion that there exist exactly h horizontal and v vertical lines.

We finish this section by fulfilling a commitment we made in [5]. We shall use the above elimination of quantifiers to show that $Th(D_2^{h,v})$ has decidable atoms; that is, we can effectively decide whether a given formula $\varphi(x_1,\ldots,x_n)$ is an atom in the Lindenbaum algebra of $Th(D_2^{h,v})$. Given that the algebraic closure of a subset A of a model M is the union of all finite sets in M defined by formulas with parameters from A, it will be clear that the algebraic closure of any $A \subseteq D_2^{h,v}$ is just A itself. Metakides and Remmel [9] and Remmel [11] were able to generalize many of the classical theorems on the lattice of r.e. sets to apply to the lattice of r.e. substructures of any atomic decidable model M of a decidable theory T with decidable atoms. Thus the $D_2^{h,v}$'s provide an interesting class of structures to which their techniques apply.

We shall use notational conventions suggested by those we used in (*). It follows from the elimination of quantifiers procedure that any formula whose only free variables are $x_1,\dots,x_k$ is equivalent over K to a disjunction of formulas of the following types (t). Let σ and τ be any permutations of $\{1,\dots,k\}$.

$$\begin{aligned}
\text{(t)}\quad & x_{\sigma 1}(<_1,=_1)x_{\sigma 2}(<_1,=_1)x_{\sigma 3}(<_1,=_1)\dots(<_1,=_1)x_{\sigma k} \;\&\; \underset{i=1}{\overset{k}{\&}} (\neg)V(x_{\sigma i}) \\
& \& \; d_1(-\infty,x_{\sigma 1})(=,\geq)n_0 \;\&\; \underset{i=1}{\overset{k-1}{\&}} d_1(x_{\sigma i},x_{\sigma(i+1)})(=,\geq)n_i \\
& \qquad\qquad \& \; d_1(x_{\sigma k},+\infty)(=,\geq)n_{k+1} \\
& \& \; x_{\tau 1}(<_2,=_2)x_{\tau 2}(<_2,=_2)x_{\tau 3}(<_2,=_2)\dots(<_2,=_2)x_{\tau k} \;\&\; \underset{i=1}{\overset{k}{\&}} (\neg)H(x_{\tau i}) \\
& \& \; d_2(-\infty,x_{\tau 1})(=,\geq)m_0 \;\&\; \underset{i=1}{\overset{k-1}{\&}} d_2(x_{\tau i},x_{\tau(i+1)})(=,\geq)m_i \;. \\
& \qquad\qquad \& \; d_2(x_{\tau k},+\infty)(=,\geq)n_{k+1}
\end{aligned}$$

Given any formula we can effectively find an equivalent disjunction of (t) formulas in the theory of any $D_2^{h,v}$. Since the formula is an atom just in case that disjunction has only one disjunct and every occurrence of $(=,\geq)$ represents an occurrence of $=$, the theory of $D_2^{h,v}$ does have decidable atoms. It also easily follows by the form of the atoms that the algebraic closure of every subset of $D_2^{h,v}$ is itself.

Section 2. UNDECIDABILITY RESULTS.

We saw in [4] that the theory of dense 2dPOs is finitely axiomatizable but undecidable. In this section we shall see that the theories which result when any one of the axioms (2)-(6) is removed from K are undecidable. We shall assume that the reader is familiar with the Rabin-Scott method [10] for establishing undecidability. Thus in each case we shall omit all of the details except to verify the crucial step of the Rabin-Scott method, which is to interpret, by fixed formulas, any countable model of a theory known to be undecidable in a model of the given theory. Finally, we end this section with several open questions concerning the decidability of closely related theories.

Theorem 2. The theory obtained from K by omitting (2) is undecidable.

Proof. We use the undecidability of graph theory, by which we mean the theory of an irreflexive symmetric binary relation [1,9]. Let $U = D^{\omega^*+\omega,\omega^*+\omega}$ be the extension of D_2 formed by adding to it an $\omega^*+\omega$ sequence of horizontal lines and an $\omega^*+\omega$ sequence of vertical lines, as described in Section 1. It is easy to see that although $U \notin F$, U is a model of K. Let the sequence of horizontal lines be denoted $H_1, H_2, H_3, \ldots, H_{-3}, H_{-2}, H_{-1}$ in order. Similarly, let $V_1, V_2, V_3, \ldots, V_{-3}, V_{-2}, V_{-1}$ denote the sequence of vertical lincs.

Let G be any graph on $\mathbb{N}^+ = \{1,2,3,\ldots\}$. Extend U to I_G by adding the following points to U. For each $n \in \mathbb{N}^+$, add the intersection of H_n and V_n. For each pair $(m,n) \in G$ with $m < n$ add the intersection of H_m and V_n. Notice that I_G is a model of the theory obtained from K by omitting (2). In the interpretation, the domain of G in I_G is the set of intersections of lines with no intersection above on the same line. G is interpreted in I_G by the relation of pairs of points in the domain such that the horizontal line through one intersects the vertical line through the other. Formally, $\delta(x)$ and $\gamma(x,y)$ below give the required interpretations.

$$\delta(x): H(x) \mathbin{\&} V(x) \mathbin{\&} (\forall y)[x <_2 y =_1 x \rightarrow \neg H(y)]$$

$$\gamma(x,y): \delta(x) \mathbin{\&} \delta(y) \mathbin{\&} x \neq y \mathbin{\&} (\exists z)[x =_1 z =_2 y \vee x =_2 z =_1 y].$$

Theorem 3. The theory obtained from K by omitting (3) is undecidable.

Proof. We again use the undecidability of graph theory. We now use suitably chosen substructures of U where U is again $D^{\omega^*+\omega,\omega^*+\omega}$. For each $n \in \mathbb{N}^+$ we remove a gap on H_n between V_n and V_{n+1}. Next for each $(m,n) \in G$ with $m < n$, where G is a given graph on $\mathbb{N}^+$, we remove a gap from V_n between H_m and H_{m+1}. Call the resulting 2dPO F_G. Notice that F_G is a model of the theory obtained from K by omitting (3). Since we are allowing the remaining lines to be dense in themselves, even through they may not be full in the 2dPO, equality in G will also be interpreted by a formula in F_G. The domain of the interpretation of G in F_G is the set of those points in F_G which are on a horizontal line segment of F_G between two adjacent vertical lines of F_G such that that horizontal line segment has a gap in F_G. Equality on this domain is the intrepretation of $=_2$ on the domain. Finally two points in the interpretation of G are adjacent if there is one vertical gap in F_G which is on the vertical line just to the right of one of the points and is horizontally just above the other. Formally,

$\delta(x)$: $H(x) \mathbin{\&} (\exists y)(\exists z)[V(y) \mathbin{\&} V(z) \mathbin{\&} \neg(\exists w)[y <_1 w <_1 z \mathbin{\&} V(w)] \mathbin{\&} y <_1 x < z$

$\mathbin{\&} (\exists u)(\exists v)[y <_1 u <_1 v <_1 z \mathbin{\&} \neg(\exists w)[u <_1 w <_1 v \mathbin{\&} w =_2 x]]]$

$\varepsilon(x,y)$: $\delta(x) \mathbin{\&} \delta(y) \mathbin{\&} x =_2 y$

$\gamma(x,y)$: $\delta(x) \mathbin{\&} \delta(y) \mathbin{\&} \neg\varepsilon(x,y)$

$\mathbin{\&} \big((\exists h)(\exists w)[H(h) \mathbin{\&} V(v) \mathbin{\&} x <_2 v <_2 h \mathbin{\&} \neg(\exists w)[x <_2 w <_2 h \mathbin{\&} H(w)]$

$\mathbin{\&} y <_1 v \mathbin{\&} \neg(\exists w)[y <_1 w <_1 v \mathbin{\&} V(w)]$

$\mathbin{\&} (\exists b)(\exists a)[x <_2 b <_2 a <_2 h \mathbin{\&} \neg(\exists w)[b <_2 w <_2 a \mathbin{\&} w =_1 v]]$

$\vee (\exists h)(\exists v)[H(h) \mathbin{\&} V(v) \mathbin{\&} y <_2 v <_2 h \mathbin{\&} \neg(\exists w)[y <_2 w <_2 h \mathbin{\&} H(v)]$

$\mathbin{\&} x <_1 v \mathbin{\&} \neg(\exists w)[x <_1 w <_1 v \mathbin{\&} V(w)]$

$\mathbin{\&} (\exists b)(\exists a)[y <_2 b <_2 a <_2 h \mathbin{\&} \neg(\exists w)[b <_2 w <_2 a \mathbin{\&} w =_1 v]]\big)$

Theorem 4. The theory obtained from K by omitting (4) is undecidable.

Proof. We do not see how to use the undecidability of graph theory to prove this theorem. Instead, we use the undecidability of the theory of two equivalence relations [12]. Let C^1 and C^2 be two equivalence relations on $\mathbb{N}^+$. Let $\{C_1^1, C_2^1, C_3^1, \ldots\}$ be the set of equivalence classes of C^1, and let $\{C_1^2, C_2^2, C_3^2, \ldots\}$ be the set of equivalence classes of C^2. Restrict U to U_{C^1,C^2} by bounding the horizontal lines $H_1, H_2, H_3, \ldots$ as follows. If $n \in C_m$, arrange that H_n has some points to the left of V_{m+1} but no points to the left of V_m. Similarly, if $n \in C_m^2$ arrange that H_n has some points to the right of $V_{-(m+1)}$ but no points to the right of V_{-m}. U_{C_1,C_2} is a model of the theory obtained from K by omitting (4).

To interpret (C^1,C^2) in U_{C^1,C^2} we use the following definitions. The domain is the set of points on horizontal lines which are not unbounded. As in the proof of Theorem 2, equality is $=_2$ on this domain. Two points of the domain are in the same $C^1(C^2)$ equivalence class if their lines have left (right) boundaries between the same adjacent vertical lines. Formally, we can use the following formulas:

$\delta(x)$: $H(x)$ & $(\exists y)\neg(\exists z)[y <_1 z =_2 x]$

$\varepsilon(x,y)$: $H(x)$ & $H(y)$ & $x =_2 y$

$\gamma_1(x,y)$: $\delta(x)$ & $\delta(y)$ & $(\exists a)(\exists v)[V(u)$ & $V(v)$ & $\neg(\exists w)[u <_1 w < v$ & $V(w)]$

& $(\exists w)[x =_2 w < v]$ & $\neg(\exists w)[x =_2 w < u]$

& $(\exists w)[y =_2 w < v]$ & $\neg(\exists w)[y =_2 w < u]]$

$\gamma_2(x,y)$: $\delta(x)$ & $\delta(y)$ & $(\exists u)(\exists v)[V(u)$ & $V(v)$ & $\neg(\exists w)[u <_1 w <_1 v$ & $V(w)]$

& $(\exists w)[u <_1 w =_2 x]$ & $\neg(\exists w)[v <_1 w =_2 x]$

& $(\exists w)[u <_1 w =_2 y]$ & $\neg(\exists w)[v <_1 w =_2 x]]$.

Theorem 5. The theory obtained from K by omitting either (5) or (6) is undecidable.

Proof. In the presence of both (5) and (6), the basic denseness property of dense 2dPOs ensures that in every model $M \models K$, any interval $[x,y] = \{z \in M | x < z < y\}$ which contains a pair of incomparable points contains a point $u \in M$ off every line, i.e., $M \models \neg V(u)$ & $\neg H(u)$ & $x < u < y$. It thus is easy to see that $\{x \in M | M \models \neg V(x)$ & $\neg H(x)\}$ is a substructure of M isomorphic to D_2. However, if we drop, say (5), from K then it is easy to construct models M of axioms (1)-(4) and (6) where the points on the horizontal lines are enough to satisfy the denseness properties required by axiom (1) and the points off every line determine a substructure of M isomorphic to any given 2dPO. Since, as stated at the start of this section, the theory of 2dPOs is undecidable, it follows that dropping (5), or similarly (6), from K results in an undecidable theory. Formally we let

$$\delta(x) = \neg V(x) \ \& \neg H(x) \quad \text{and}$$

$$x < y = x < y.$$

In view of the proof of Theorem 5 to have any chance of having a decidable theory given axioms (1)-(4) but not both (5) and (6), we must add some axiom like the following which ensures that the points off every line always determine a substructure isomorphic to D_2.

(7) The points off every line are "dense", i.e., every interval $[x,y]$ which contains a pair of incomparable points contains points off every line.

$$\forall x \forall y[x<y \ \& \ \exists z_1 \ \exists z_2(x<z_1, z_2<y \ \& \ z_1|z_2) \rightarrow$$

$$\exists u(x<u<y \ \& \neg V(u) \ \& \neg H(u)]$$

We do not know whether the theories axiomatized by (1)-(4),(7); (1)-(4),(5),(7); (1)-(4),(6),(7); or even (1)-(4),(7) plus an axiom which says there are no vertical lines, are decidable. All of these theories are rich enough to interpret the theory of linear orderings within them. However, in view of Leonard and Lauchli's decision procedure for the theory of linear orderings [3], it is still possible that they may be decidable.

REFERENCES

1. Church, A. and Quine, W. V., Some theorems on definability and decidability, J. of Symbolic Logic 17 (1952), 179-187.

2. Dushnik, B. and Miller, E. W., Partially ordered sets, Amer. J. of Math. 63 (1941), 600-610.

3. Läuchli, H. and Leonard, J., On the elementary theory of linear order, Fundamenta Mathematicae 59 (1966), 109-116.

4. Manaster, A. B. and Remmel, J. B., Partial orderings of fixed finite dimension: model companions and density, J. of Symbolic Logic (to appear).

5. ________, Some recursion theoretic aspects of dense two-dimensional partial orderings, to appear in "Aspects of Effective Algebra: Proceedings of a mini-conference, Monash University, 1979", Upside-Down A Book Co., Melbourne, Australia.

6. ________, Recursively categorical decidable dense planar orderings, (in preparation).

7. Manaster, A. B. and Rosenstein, J. G., Two-dimensional partial orderings: recursive model theory, J. of Symbolic Logic 45 (1980), 121-132.

8. ________, Two-dimensional partial orderings: undecidability, J. of Symbolic Logic 45 (1980), 133-143.

9. Metakides, G. and Remmel, J. B., Recursion theory on orderings, I: a model theoretic setting, J. of Symbolic Logic 44 (1979), 383-402.

10. Rabin, M. O., A simple method for undecidability proofs and some applications, Logic, Methodology, and Philosophy of Science (Proceedings 1964 International Congress), North-Holland, Amsterdam (1965), 58-68.

11. Remmel, J. B., Recursion theory on orderings II, J. of Symbolic Logic 45 (1980), 317-333.

12. Rogers, H. J. Jr., Logical reduction and decision problems, Annals of Mathematics 64 (1956), 264-284.

COUNTER-EXAMPLES VIA MODEL COMPLETIONS

Terrence Millar
Department of Mathematics
University of Wisconsin
Madison, WI 53706/USA

Most of the decidable theories produced in the literature to have certain model and recursion theoretic properties allow elimination of quantifiers. In fact, typically the desired recursion and associated model theoretic requirements can be coded via universal sentences, with the desired theory being the model completion of the set of those universal sentences. It is therefore useful to know when a universal theory has a complete, decidable, model completion. The first section of this paper provides a necessary and sufficient condition for this to be true, while the second section is an application of the first section to an example concerning recursively saturated models.

I.

In order to motivate the theorem which will follow, fix a complete, decidable theory T that allows elimination of quantifiers. Assume without loss of generality that the language of T has the distinct variables $\{x_i, y_i \mid i < \omega\}$, and let $\{\phi_i \mid i < \omega\}$ be an effective enumeration of all quantifier free formulas of $L(T)$. Because T allows elimination of quantifiers, it follows that there exists an $f : \omega \to \omega$ such that for all $i < \omega$

(1) $$T \vdash \forall x[\exists \bar{y}\ \phi_i(\bar{x}, \bar{y}) \longleftrightarrow \phi_{f(i)}(\bar{x})] .$$

Because T is decidable, such an f exists that is recursive. Notice that a trivial consequence of (1) is that for all $i, j < \omega$

$$T \vdash \forall \bar{z}[\exists \bar{y}(\phi_i(\bar{x}, \bar{y}) \wedge \phi_j(\bar{x}, \bar{z})) \longleftrightarrow (\phi_j(\bar{x}, \bar{z}) \wedge \phi_{f(i)}(\bar{x}))]$$

as long as $\text{range } \bar{y} \cap \text{range } \bar{z} = \emptyset$.

Now fix a language with variables $\{x_i, y_i \mid i < \omega\}$ and with an effective enumeration $\{\phi_i \mid i < \omega\}$ of all quantifier free formulas.

Theorem 1. A universal consistent theory T' has a complete, decidable, model completion iff there exists a recursive $f : \omega \to \omega$ such that for all $i, j < \omega$

(0) $T' \vdash [\forall \bar{x}\ \forall \bar{y}\ \neg\ \phi_i(\bar{x}, \bar{y})]$ iff $f(i) = 0$;

(1) $T' \vdash [\phi_i(\bar{x}, \bar{y}) \to \phi_{f(i)}(\bar{x})]$;

(2) If $\phi_i = \phi_i(\bar{y})$ and $f(i) \neq 0$, then $\phi_{f(i)} = (x_0 = x_0)$; and

(3) $T' \cup \{\phi_i(\bar{x}, \bar{y}), \phi_j(\bar{x}, \bar{z})\}$ is consistent iff $T' \cup \{\phi_{f(i)}(\bar{x}), \phi_j(\bar{x}, \bar{z})\}$ is consistent, provided $\text{range } \bar{y} \cap \text{range } \bar{z} = \emptyset$.

Proof. Assume first that T' has a complete decidable, model completion T. Then by a theorem of Robinson [4], T allows elimination of quantifiers. So fix a recursive f such that for all $i < \omega$ satisfying $T \vdash \exists \bar{x} \exists \bar{y} \phi_i$

$$T \vdash [\exists \bar{y} \phi_i(\bar{x}, \bar{y}) \longleftrightarrow \phi_{f(i)}(\bar{x})],$$

where if in addition $\phi_i = \phi_i(\bar{y})$ then we require that $\phi_{f(i)} = (x_o = x_o)$, and $f(i) = 0$ if ϕ_i is not consistent with T. We will assume without loss that ϕ_0 is not consistent with T. We now claim that in fact this f satisfies the theorem. (0) and (2) are satisfied by our choice of f. Suppose that (1) failed. Then

$$T' \cup \{\exists \bar{x}[\exists \bar{y} \phi_i(\bar{x}, \bar{y}) \wedge \neg \phi_{f(i)}(\bar{x})]\}$$

would be consistent. Since T is the model completion of T' and is complete, it would then follow that

$$T \vdash \exists \bar{x}[\exists \bar{y} \phi_i(\bar{x}, \bar{y}) \wedge \neg \phi_{f(i)}(\bar{x})].$$

But this contradicts the choice of f. Finally, to prove (3), assume first that $T' \cup \{\phi_i(\bar{x}, \bar{y}), \phi_j(\bar{x}, \bar{z})\}$ is consistent. Then by (1), $T' \cup \{\phi_{f(i)}(\bar{x}), \phi_j(\bar{x}, \bar{z})\}$ is also consistent. On the other hand, if $T' \cup \{\phi_{f(i)}(\bar{x}), \phi_j(\bar{x}, \bar{z})\}$ is consistent, then as before

$$T \vdash \exists \bar{x} \exists \bar{z} [\phi_{f(i)}(\bar{x}) \wedge \phi_j(\bar{x}, \bar{z})].$$

But then by the choice of f,

$$T \vdash \exists \bar{x} \exists \bar{y} \exists \bar{z}[\phi_i(\bar{x}, \bar{y}) \wedge \phi_j(\bar{x}, \bar{z})].$$

In particular, $T \cup \{\phi_i(\bar{x}, \bar{y}), \phi_j(\bar{x}, \bar{z})\}$ is consistent. Therefore, since T is the model completion of T', $T' \cup \{\phi_i(\bar{x}, \bar{y}), \phi_j(\bar{x}, \bar{z})\}$ is also consistent. This demonstrates that (0) - (3) hold.

Next assume that there is such a recursive f satisfying (0) - (3), we must show that T' has a complete, decidable, model completion T. The axioms for T are:

I. $\forall \bar{x} \forall \bar{y}[\neg \phi_i(\bar{x}, \bar{y})]$ for $i < \omega$ such that $f(i) = 0$; and

II. $\forall \bar{x} \exists \bar{y}[\phi_{f(i)}(\bar{x}) \longrightarrow \phi_i(\bar{x}, \bar{y})]$, for $i < \omega$ such that $f(i) \neq 0$.

First we will show that T is consistent. Let

$$C \equiv_{df} \{\forall \bar{x}_i \exists \bar{y}_i [\psi_{f(i)}(\bar{x}_i) \longrightarrow \psi_i(\bar{x}_i, \bar{y}_i)] \mid 1 \le i \le n\}$$

be a finite subset of the axioms in II. It is enough, for arbitrary such C, to find a model for $T' \cup C$. We will construct the diagram of a model of T' for which the sentences in C hold. Since T' is universal it is sufficient to insure that the diagram is consistent with T' in order to have the structure a model of T'. Let $\{a_i \mid i < \omega\}$ be distinct new constants, and $\{\theta_i \mid i < \omega\}$ an indexing of all sentences in the language of $T' \cup \{a_i \mid i < \omega\}$. Let $g_i : \omega \to \omega^{m_i}$ be an onto function such that each element of ω^{m_i} has infinitely many pre-images, where m_i is the length of $\bar{x}$ in $\theta_i(\bar{x}, \bar{y})$, $1 \le i \le n$. In the usual Henkin style, the universe of the model will be equivalent classes of the a_i's — we will assume that the reader is familiar with this procedure (we will also omit the steps for providing "witnesses" for existential sentences in the diagram). We specify the diagram Δ by induction, where $\Delta \equiv_{df} \bigcup_{t<\omega} \Delta_t$.

Step 0: $\Delta_0 \equiv_{df} \emptyset$.

Step $r = (n+1)t$: Assume inductively that $\bigcup_{j<r} \Delta_j \cup T'$ is consistent and that only a finite number of a_i's occur in formulas of $\bigcup_{j<r} \Delta_j$. By this assumption, either $\Delta_r \equiv_{df} \{\theta_t\}$ or $\Delta_r \equiv_{df} \{\neg\theta_t\}$ preserves the induction assumptions, choose it to do so.

Step $r = (n+1)t + i$ (where $1 \leq i \leq n$): If $\psi_{f(i)}(\overline{a}_{g_i(r)}) \notin \bigcup_{j<r} \Delta_j$, then let $\Delta_r \equiv_{df} (a_0 = a_0)$, where $\overline{a}_{g_i(r)} = (a_{k_1}, \ldots, a_{k_{m_r}})$, and $g_i(r) = \langle k_1, \ldots, k_{m_r}\rangle$. If $\psi_{f(i)}(\overline{a}_{g_i(r)}) \in \bigcup_{j<r} \Delta_j$, then let $\overline{a}'$ be a p_i-tuple of a_k's none of which occur in the sentences of $\bigcup_{j\leq r} \Delta_j$, where $p_i = lh(\overline{y})$, '$\overline{y}$' as in $\psi_i(\overline{x}, \overline{y})$. Define $\Delta_r \equiv_{df} \{\psi_i(\overline{a}_{g_i(r)}, \overline{a}')\}$. By (1) $\{\psi_{f(i)}(\overline{a}_{g_i(r)}), \psi_i(\overline{a}_{g_i(r)}, \overline{a}')\} \cup T'$ is consistent; therefore by the induction hypotheses and (3) it is easy to see that $T' \cup \bigcup_{j\leq r} \Delta_j$ must be consistent.

This ends the induction. By the construction T' is consistent, and for any i, $1 \leq i \leq n$, and for every tuple $\overline{a}$ from $\{a_k \mid k < \omega\}$ satisfying $\bigcup_{j<\omega} \Delta_j \vdash \psi_{f(i)}(\overline{a})$, there is an $\overline{a}'$ such that $\bigcup_{j<\omega} \Delta_j \vdash \psi_i(\overline{a}, \overline{a}')$. So the consistency of $C \cup T'$ is established. Notice that the proof of consistency can easily be modified to show that every model $\mathfrak{B}$ of T' is isomorphically embeddable in a model of T. For it is enough to show for every C, as before, that $T' \cup C \cup \Delta_{\mathfrak{B}}$ has a model. The proof of this is essentially the same as that given above, with "$T' \cup \Delta_{\mathfrak{B}}$" substituted for "$T'$". The point of this observation is that the theorem will now follow from the axioms in I if we can in addition show that T is submodel complete and complete.

An equivalent condition to submodel completeness is that for all models of T, submodels $C \subset \mathfrak{A}, \mathfrak{B}$, and existential sentences ψ with parameters from C,

$$\mathfrak{B} \models \psi \quad \text{if} \quad \mathfrak{A} \models \psi .$$

So arbitrarily fix such $\mathfrak{A}$, $\mathfrak{B}$, C and $\psi = \exists\overline{y}\, \theta(\overline{c}, \overline{y})$ ($\theta(\overline{x}, \overline{y})$ quantifier free), $c \subset |C|^r$, satisfying $\langle \mathfrak{A}, \overline{c}\rangle \models \psi$. Also fix an $i < \omega$ such that $\theta(\overline{x}, \overline{y})$ is $\phi_i(\overline{x}, \overline{y})$. Clearly $\phi_i(\overline{x}, \overline{y})$ is consistent with T, so by (1) $\langle \mathfrak{B}, \overline{c}\rangle \models \phi_{f(i)}(\overline{c})$ also. But then because $\mathfrak{B}$ is a model of T $\mathfrak{B} \models \forall\overline{x}\exists\overline{y}[\phi_{f(i)}(\overline{x}) \longrightarrow \phi_i(\overline{x}, \overline{y})]$ (an axiom from II). Thus $\langle \mathfrak{B}, \overline{c}\rangle \models \exists\overline{y}\, \phi_i(\overline{c}, \overline{y})$, as desired.

Finally we must demonstrate that T is complete. Let ψ be any sentence in $L(T)$ consistent with T. Since we have shown that T admits elimination of quantifiers (submodel completeness), $(\psi \wedge y_1 = y_1)$ is equivalent under T

to a quantifier free formula $\phi(y_1)$. Then by the axioms in II, this implies that

$$T \vdash \forall x_0 \exists \bar{y}_1 [(x_0 = x_0) \longrightarrow \phi(\bar{y}_1)]$$

and so $T \models \psi$. This finishes the proof of the theorem.

Theorem 1 furnishes a "uniform" approach for producing particular theories, and it also eliminates the need of directly providing the tedious "existential closure" axioms for those theories.

II.

Conventions: $2^{<\omega}$ will denote the set of all sequences $\underline{f}: n \to 2$, $n < \omega$. For elements $f, g \in 2^{<\omega} \cup 2^{\omega}$, we write $\underline{f \lessdot g}$ if f is an initial segment of g (where elements of 2^{ω} are also treated as sequences, of length ω). Also $\underline{f < g}$ if there is an $h \in 2^{<\omega}$ such that $\underline{h}^\frown \langle 0 \rangle \lessdot f$ and $\underline{h}^\frown \langle 1 \rangle \lessdot g$. ($\underline{h}^\frown \underline{g}$ is the sequence f such that $\underline{f}(i) = \underline{h}(i)$, $i < \ell h(\underline{h})$, and $\underline{f}(j + \ell h(\underline{h})) = \underline{g}(j)$ for $j < \ell h(\underline{g})$). $f{\restriction}m$ denotes the $\underline{g} \lessdot f$ of length m, $m < \ell h(f)$. If $F \subset 2^{<\omega}$ and $g \in 2^{\omega}$ satisfies $\underline{f} \in F$ for all $\underline{f} \lneq g$ then g is a <u>branch</u> of F. If θ is a formula, then $\theta^0 \equiv \theta$ and $\theta^1 \equiv \neg\theta$.

In [1] Barwise and Schlipf introduced the notion of recursively saturated models. Let L be a countable language which is effectively presented, and let $\mathfrak{A}$ range over structures for the language L.

Definition. $\mathfrak{A}$ is recursively saturated if, for every recursive set $\Phi(x, y_1, \ldots, y_k)$ of formulas from L and $c_1, \ldots, c_k \in |\mathfrak{A}|$, if $Th(\langle \mathfrak{A}, c_1, \ldots, c_k \rangle)$ is consistent with $\Phi(x, \underline{c}_1, \ldots, \underline{c}_k)$, then $\Phi(x, \underline{c}_1, \ldots, \underline{c}_k)$ is realized in $\langle \mathfrak{A}, c_1, \ldots, c_k \rangle$.

Countable, recursively saturated models provide a new expository tool in model theory [1]. Because of certain similarities between recursively saturated models and saturated models, Barwise asked whether a complete theory which is not ω-categorical must have a model which is not recursively saturated[1]. This section will deal with a related partial result.

It is well known that every recursively saturated model is ω-homogeneous. Another interesting property of recursively saturated models is that if Γ, Σ are complete types of the theory of such an $\mathfrak{M}$, $\Gamma \leq_T \Sigma$, and Σ is realized in $\mathfrak{M}$, then Γ is also realized in $\mathfrak{M}$. The third fact motivating the construction in the main result of this section is:

Theorem 2. If T has a complete extension T' in a language $L' = L(T) \cup \{c_i \mid 1 \leq i \leq n\}$ such that T' does not have an atomic model, then T has a model which is not recursively saturated.

---

[1] In [2] a negative answer was announced to the question thus stated. However, there were several errors in the construction, and the result here represents that portion which has been "repaired" to date.

Proof. Fix such a T' extending T. Then for some $m < \omega$ and consistent (w.r.t. T') formula $\theta(c_1, \ldots, c_n, x_1, \ldots, x_m)$ of $L(T')$, there is no principal type of T' containing $\theta(c_1, \ldots, c_n, x_1, \ldots, x_m)$. Fix an effective enumeration $\{\psi_i(y_1, \ldots, y_n, x_1, \ldots, x_m) \mid i < \omega\}$ of all formulas of $L(T)$ in the displayed variables. Let $\Phi \equiv_{df} \{\phi_i \mid i < \omega\}$, where the ϕ_i are defined by:

$$\phi_0 \equiv_{df} \theta(y_1, \ldots, y_n, x_1, \ldots, x_m)$$

$$\phi_{n+1} \equiv_{df} \{\psi_{n+1}(y_1, \ldots, y_n, x_1, \ldots, x_m) \longleftrightarrow \exists z_1 \ldots \exists z_m[\psi_{n+1}(\overline{y}, \overline{z}) \wedge \bigwedge_{i \leq n} \phi_i(\overline{y}, \overline{z})]\}.$$

Obviously Φ is a recursive set of formulas, and it is easy to check that $\Phi(\overline{c}_1, x_1, \ldots, x_m)$ generates an m-type of T'. However, since no principal type of T' contains $\theta(c_1, \ldots, c_n, x_1, \ldots, x_m)$, $\Phi(c_1, \ldots, c_n, x_1, \ldots, x_m)$ must be a non-principal m-type of T'. So there is a model $\langle \mathfrak{A}, a_1, \ldots, a_n \rangle$ of T' omitting $\Phi(c_1, \ldots, c_n, x_1, \ldots, x_m)$, $\mathfrak{A} \models T$. Therefore $\mathfrak{A}$ cannot be recursively saturated, since $\Phi(y_1, \ldots, y_n, x_1, \ldots, x_m)$ is recursive. Thus the theorem follows.

We now wish to refine the notion of 'recursively saturated':

Definition. $\mathfrak{A}$ is n-recursively saturated if

(1) The theory of $\mathfrak{A}$ has a non-principal 1-type; and

(2) For every recursive set $\Phi(x_1, \ldots, x_m, y_1, \ldots, y_n)$ of formulas from L and every $c_1, \ldots, c_n \in |\mathfrak{A}|$, if $\Phi(x_1, \ldots, x_m, \underline{c}_1, \ldots, \underline{c}_n)$ is consistent with $Th(\langle \mathfrak{A}, c_1, \ldots, c_n \rangle)$, then $\Phi(x_1, \ldots, x_m, \underline{c}_1, \ldots, \underline{c}_n)$ is realized in $\langle \mathfrak{A}, c_1, \ldots, c_n \rangle$.

If every model of a complete theory T is n-recursively saturated, then we will say that T is n-recursively saturated. Note that if $\mathfrak{A}$ is n-recursively saturated for every $n < \omega$, then $\mathfrak{A}$ is recursively saturated. The restriction to theories with non-principal 1-types is made to avoid trivialities, since otherwise a decidable theory with no non-principal n-types and no recursive non-principal types would automatically be n-recursively saturated. Also, by expanding the language one can always transform a theory with some non-principal type into a 'similar' theory with a non-principal 1-type.

0-recursively saturated theories are easy to find. Simply take any complete decidable theory with a non-principal 1-type, no recursive non-principal types, and only a countable number of types altogether. For consider any recursive set of formulas $\Phi(x_1, \ldots, x_m)$ consistent with such a T. Then since T has only countably many m-types, $\Phi(x_1, \ldots, x_m)$ must be contained in a recursive m-type $\Sigma(x_1, \ldots, x_m)$ of T. But such a recursive type must be principal, by choice of T, and therefore realized in every model of T. Since $\Phi \subset \Sigma$, T is 0-recursively saturated. The main result of this section is:

Theorem 3. There exists a 1-recursively saturated theory.

Proof. In [3] we produced, given an r.e. set v_1, a recursive tree $F \subset 2^{<\omega}$, a recursive $Ter \subset F$, and a recursive $H : \omega \to \omega$ such that

Lemma 1. (i) The F, Ter, and H are recursive;

(ii) If $\underline{f} \in F$, then $\underline{f}' \in F$ for all $\underline{f}' \lhd \underline{f}$; $\underline{f} \in (F - Ter)$ iff $\underline{f}^\frown \langle 0 \rangle \in F$;

(iii) There is exactly one branch h of F such that for all $\underline{f} \underset{\neq}{\lhd} h$ there is an $\underline{g} \in F$ satisfying $\underline{f} \lhd \underline{g} \ntriangleleft h$: call this h the limit branch of F;

(iv) If h is the limit branch of F, then $h \equiv_T v_1$;

(v) For every branch $g \in 2^\omega$ of F other than h, $h < g$ and $g(p) = 0$ for all but finitely many $p < \omega$;

(vi) If $\underline{g} < \underline{f} \in Ter$ and $\underline{g} \in F$, then $lh(\underline{g}) < lh(\underline{f})$;

(vii) There is exactly one $\underline{f} \in Ter$ of length $H(r)$ for all $r < \omega$;

(viii) If $\{\underline{f}_i \mid i < \omega\}$ is an enumeration of Ter by length, then $H(r) = lh(\underline{f}_r)$, $r < \omega$.

We will also assume in what follows that $\langle 1 \rangle \notin F$; this is easy to arrange.

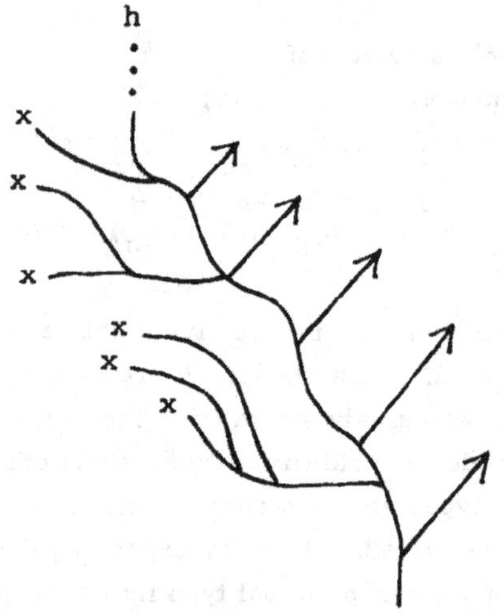

$F \subset 2^{<\omega}$

So fix r.e. sets v_1, v_2 which are Turing incomparable, and also fix the corresponding F_i, H_i, Ter_i $i = 1, 2$ respectively, given by Lemma 1. Thus F_i has exactly one limit branch, $h_i \equiv_T v_i$. All terminating paths $\underline{g}$ through F_i will belong to Ter_i and also satisfy $\underline{g} < h_i$. We will need one more

Definition. $R_0 \equiv_{df} \{\langle\ \rangle\}$;

$$R_{n+1} \equiv_{df} \{\underline{f}^\frown\langle 0\rangle \mid \underline{f}^\frown\langle k\rangle \in F_1,\ k = 0, 1;\ \ lh(\underline{f}) = H(r)$$

for some $r < lh(\underline{f})$; and there exists $\underline{g} \lhd \underline{f}[g \in R_m]\}$; the elements of R_i can be viewed as "best" approximations to h_i at a given level.

Lemma 1'. (i) $\{R_i \mid i < \omega\}$ is uniformly recursive; and
(ii) $\forall m < \omega\ \exists \underline{h} \in R_m[\underline{h} \lessdot h_1]$.

Proof. (i) is obvious, since F_1 is recursive. (ii) is true because h_1 is a limit branch of F_1 by Lemma 1.

The language for the desired theory will contain the unary predicate symbols $\{P_n \mid n < \omega\}$ and the binary predicate symbols $\{S_n \mid n < \omega\}$. We adopt the following abbreviations:

$$\mathrm{conj}(\underline{f}, x) \equiv_{df} \bigwedge_{i < lh(\underline{f})} P_i(x)^{\underline{f}(i)}$$

$$\mathrm{conj}(\underline{f}, x, y) \equiv_{df} \bigwedge_{i < lh(\underline{f})} S_i(x, y)^{\underline{f}(i)}$$

for all $\underline{f} \in 2^{<\omega}$.

We now specify the axioms of a universal consistent theory T' which satisfies (0) - (3) in theorem 1:

I. $(S_i(x,y) \longrightarrow S_i(y,x)) \wedge \neg S_0(x,x)$ $\quad i < \omega$;

II. $\mathrm{conj}_1(\underline{f}, x) \longrightarrow P_j(x)$ $\quad \underline{f} \notin (F_1 - Ter_1)$, $\quad j \geq lh(\underline{f})$;

III. $\mathrm{conj}(\underline{f}, x, y) \longrightarrow S_j(x, y)$ $\quad \underline{f} \notin (F_2 - Ter_2)$, $\quad j \geq lh(\underline{f})$;

IV. $\neg\mathrm{conj}(\underline{f}, x)$ $\quad \underline{f} \notin F_1$ such that there is no $\underline{h} \lessdot \underline{f}$, $\underline{h} \in Ter_1$;

V. $\mathrm{conj}(\underline{f}, x, y) \longleftrightarrow (x = y)$ $\quad f \notin F_2$ such that there is no $\underline{h} \lessdot \underline{f}$, $\underline{h} \in Ter_2$;

VI. $\mathrm{conj}(\underline{f}, x, y) \longrightarrow (\mathrm{conj}(\underline{h}, x) \longleftrightarrow \mathrm{conj}(\underline{h}, y))$ for all $\underline{f} \in (F_2 - Ter_2)$ and $\underline{h} \in F_1$ satisfying:

$$\exists r \leq lh(\underline{f})[lh(\underline{f}) = H_2(r) \text{ and no } \underline{h}' \trianglelefteq \underline{h} \text{ satisfies } \underline{h}' \in R_{r+2}];$$

VII. $\mathrm{conj}(\underline{f}, x, y) \longrightarrow \neg\mathrm{conj}(\underline{h}, x) \vee \neg\mathrm{conj}(\underline{h}, y)$

$$\underline{f} \in Ter_2, \quad \underline{h} \in R_{s+1}, \quad \text{where } H_2(s) = lh(\underline{f});$$

VIII. $\mathrm{conj}(\underline{f}, x, y) \wedge \mathrm{conj}(\underline{g}, y, z) \wedge x \neq y \longrightarrow \mathrm{conj}_2(\underline{f}, x, z)$ $\quad \underline{f}, \underline{g} \in 2^{<\omega}$, $\underline{f} < \underline{g}$.

Lemma 2. T' satisfies (0) - (3) in theorem 1.

The proof of this lemma will be deferred until the end. By theorem 1, let T be the decidable, complete, model completion of T'. We will show that T is 1-recursively saturated by a series of lemmas.

Lemma 3. If $\Sigma_{ij}(x_i, x_j)$ is a 2-type of T for each $1 \leq i < j \leq n$, then there is at most one n-type $\Gamma(x_1, \ldots, x_n)$ of T containing all of the $\Sigma_{ij}(x_i, x_j)$'s.

Proof. This follows because T allows elimination of quantifiers (eloq.).

Lemma 4. Each 2-type $\Sigma(x,y)$ of T is uniquely determined by the $f_1, f_2, g \in 2^\omega$ such that

$$\mathrm{conj}(\underline{f}_1, x), \quad \mathrm{conj}(\underline{f}_2, y), \quad \mathrm{conj}(\underline{g}, x, y) \in \Sigma(x,y)$$

for all $\underline{f}_1 \lessdot f_1$, $\underline{f}_2 \lessdot f_2$, and $\underline{g} \lessdot g$ respectively.

Proof. Again, T allows eloq.

Lemma 5. If $\mathrm{conj}(\underline{f}, x) \in \Gamma(x)$ for all $\underline{f} \lessdot f \in 2^\omega$, where $\Gamma(x)$ is a 1-type of T, then either

(i) $\forall \underline{f} \lessdot f[\underline{f} \in F_1]$; or

(ii) $\exists \underline{f} \lessdot f[\underline{f} \in \mathrm{Ter}_1$ and $\forall n \geq \mathrm{lh}(\underline{f}) | f(n) = 0)]$.

Proof. By the axioms in II, IV and lemma 1.

Lemma 6. If $(x \neq y)$, $\mathrm{conj}(\underline{f}, x, y) \in \Sigma(x,y)$ for all $\underline{f} \lessdot f \in 2^\omega$ then either

(i) $\forall \underline{f} \lessdot f[\underline{f} \in F_2]$; or

(ii) $\exists \underline{f} \lessdot f[\underline{f} \in \mathrm{Ter}_2$ and $\forall n \geq \mathrm{lh}(\underline{f})$ $(f(n) = 0)]$.

Proof. By the axioms in III, V and Lemma 1.

Lemma 7. T has only countably many types.

Proof. This follows from Lemmas 3 - 6 and the fact that each F_i $i = 1, 2$ has only one limit branch (by Lemma 1).

Lemma 8. T has a non-principal 1-type.

Proof. By the axioms for T' and the fact that T is the model completion of T', $T \cup \{\mathrm{conj}(\underline{f}, x)\}$ is consistent for every $\underline{f} \in F_1$. Since F_1 has a limit branch, the result follows.

Lemma 9. If $\Gamma(x)$ is a 1-type of T, then it is either recursive or it is of the same Turing degree as v_1.

Proof. This follows from the fact that T is decidable, the axioms in IV, V and lemma 1 (iii), (v).

Now suppose we have arbitrary $\mathfrak{G} \models T$, $b \in |\mathfrak{G}|$, and $\Phi(y, x_1, \ldots, x_n)$, a recursive set of formulas, such that $\mathrm{Th}(\langle \mathfrak{G}, b \rangle)$ is consistent with $\Phi(\underline{b}, x_1, \ldots, x_n)$. Then we must show that $\Phi(\underline{b}, x_1, \ldots, x_n)$ is realized in $\langle \mathfrak{G}, b \rangle$, in order to show that $\mathfrak{G}$ is 1-recursively saturated. By Lemma 7 T has only countably many types, and so the same is true of the theory $\Gamma(\underline{b})$, where Γ is the 1-type of T realized by b in $\mathfrak{G}$. Therefore there is an n-type

$\Sigma(\underline{b}, x_1, \ldots, x_n)$ of $\Gamma(\underline{b})$ containing $\Phi(\underline{b}, x_1, \ldots, x_n)$ such that $\Sigma \leq_T \Gamma$. Fix such a Σ. It is sufficient then to prove that there exists $\theta(y, x_1, \ldots, x_n) \in L(T)$ such that

$$\Gamma(\underline{b}) \vdash \exists x_1 \ldots \exists x_n\ \theta(\underline{b}, x_1, \ldots, x_n) \text{ and}$$

$$\Gamma(\underline{b}) \vdash [\theta(\underline{b}, x_1, \ldots, x_n) \longrightarrow \sigma(\underline{b}, x_1, \ldots, x_n)], \text{ for every } \sigma \in \Sigma.$$

Thus it will be enough to define (by Lemma 3):

$$\theta(y, x_1, \ldots, x_n) \equiv_{df} [\bigwedge_{1 \leq i \leq n} \theta_i(y, x_i) \wedge \bigwedge_{1 \leq i < j \leq n} \psi_{ij}(y, x_i, x_j)],$$

provided

(1) for each i, $1 \leq i \leq n$ $\theta_i(y, x_i) \in \Sigma(y, x_1, \ldots, x_n)$ and $\Gamma(\underline{b}) \vdash [\theta_i(y, x_i) \longrightarrow \sigma(y, x_i)]$ for every $\sigma(y, x_i) \in \Sigma(y, x_1, \ldots, x_n)$; and

(2) for each i, j $1 \leq i < j \leq n$, $\psi_{ij}(y, x_i, x_j) \in \Sigma(y, x_1, \ldots, x_n)$ and $\Gamma(b) \vdash [\psi_{ij}(b, x_i, x_j) \longrightarrow \sigma(x_i, x_j)]$ for every $\sigma(x_i, x_j) \in \Sigma(y, x_1, \ldots, x_n)$.

There are now two cases, depending on whether or not (*) $\text{conj}(\underline{h}, y) \in \Gamma(y)$ for all $\underline{h} \lessdot h_1$. The cases are similar and so we will do just one — assume (*) holds. First fix an i, $1 \leq i \leq n$ and we will find the corresponding θ_i. If $(y = x_i) \in \Sigma(y, x_1, \ldots, x_n)$ then we take θ_i to be $(y = x_i)$. So assume otherwise. Fix $f, g \in 2^\omega$ satisfying

$$\text{conj}(\underline{f}, x_i),\ \text{conj}(\underline{g}, y, x_i) \in \Sigma(y, x_1, \ldots, x_n)$$

for all $\underline{f} \lessdot f$ and $\underline{g} \lessdot g$ respectively. Because (a) $\Sigma \leq_T \Gamma$, (b) h_1 and h_2 are Turing incomparable, and (c) Γ is Turing equivalent to h_1 (lemma 9), it follows that $g \neq h_2$. Thus by lemma 1 and 6 there is a $\underline{g} \lessdot g$ such that $\underline{g} \in Ter_2$ or g is the only branch of F_2 satisfying $\underline{g} \lessdot g$. Fix such a $\underline{g}$. Then by either the axioms in II or VI

$$T \vdash [\text{conj}(\underline{g}, y, x_i) \longrightarrow \text{conj}(\underline{g}', y, x_i)] \text{ for all } \underline{g}' \lessdot g.$$

If we are in the latter case, i.e. g is a branch of F_2, then by the axioms in VI

$$\Gamma(\underline{b}) \vdash [\text{conj}(\underline{g}, \underline{b}, x_i) \longrightarrow \text{conj}(\underline{h}, x_i)] \text{ for all } \underline{h} \lessdot h_1.$$

In that case, by lemma 4 it suffices to define

$$\theta_i(y, x_i) \equiv_{df} \text{conj}(\underline{g}, y, x_i).$$

So assume that g is not a branch of F_2, and fix $r < \omega$ such that $lh(\underline{g}) = H_2(r+1)$ (we are in the case $\underline{g} \in Ter_2$). By Lemma 1' there is an $\underline{h} \lessdot h_1$, $\underline{h} \in R_{r+2}$. Therefore by the axioms in VII, $\text{conj}(\underline{h}, x_i) \notin \Sigma(y, x_1, \ldots, x_n)$ and so $f \neq h_1$. By lemmas 1 and 5 fix an $\underline{f} \lessdot f$ such that either $\underline{f} \in Ter_1$ or f is the only branch of F_1 such that $\underline{f} \lessdot f$. Then by the axioms in either II or IV

$$T \vdash [\text{conj}(\underline{f}, x_i) \longrightarrow \text{conj}(\underline{f}', x_i)] \text{ for all } \underline{f}' \lessdot f.$$

We now define

$$\theta_i \equiv_{df} [\mathrm{conj}(\underline{g}, y, x_i) \wedge \mathrm{conj}(\underline{f}, x_i)] .$$

Finally we must define the ψ_{ij}'s. Fix i, j $1 \le i < j \le n$. Again we assume $(x_i \ne x_j) \in \Sigma(y, x_1, \ldots, x_n)$, since otherwise we simply define ψ_{ij} to be $(x_i = x_j)$. In fact it is enough to find $\psi'(x_i, x_j) \in \Sigma(y, x_1, \ldots, x_n)$ satisfying $T \vdash [\psi'(x_i, x_j) \rightarrow \mathrm{conj}(\underline{g}, x_i, x_j)]$ for all $\underline{g} \in 2^{<\omega}$ satisfying $\mathrm{conj}(\underline{g}, x_i, x_j) \in \Sigma(y, x_1, \ldots, x_n)$, since then by Lemma 4 we can define $\psi_{ij}(y, x_i, x_j) \equiv_{df} [\psi'(x_i, x_j) \wedge \theta_i(y, x_i) \wedge \theta_j(y, x_j)]$. So fix $g \in 2^{\omega}$ such that $\mathrm{conj}_2(\underline{g}, x_i, x_j) \in \Sigma(y, x_1, \ldots, x_n)$ for all $\underline{g} < g$. Just as in the case of the g for the θ_i's, it follows from the fact that $\Sigma \le_T \Gamma$ that g is recursive. Then we proceed in the same way we did for the θ_i case —— the details are left to the reader. This completes the proof that T is 1-recursively saturated.

Proof of Lemma 3.

Let B be a set of quantifier-free sentences in a language obtained from $L(T)$ by the addition of constant symbols. We define: B is <u>locally consistent</u> with T' if B is consistent and if the closure of B under finite conjunctions and disjunctions does not contain a sentence that is equivalent to an instance of a negation of an axiom of T' (remember that T' is universal). Similarly a set of quantifier-free formulas C is locally consistent with T' if the set of sentences obtained by replacing variables by new constants (uniformly) is locally consistent with T'. A set of formulas B is <u>terraced</u> if it is a maximum consistent subset, for fixed variables and/or constants $u_1, \ldots, u_n$ and $r < \omega$, of

$$\{(u_i = u_j)^k, P_p(u_i)^k, S_q(u_i, u_j)^k \mid k = 0, 1; 1 \le i, j \le n; p < H_1(r); q < H_2(r)\};$$

locally consistent with T'. $\overline{u}$ are the <u>parameters</u> of B and r is the <u>index</u>. Define $f(i) = 0$ if $\phi_i(\overline{x}; \overline{y})$ is not consistent with T'. If $\phi_i(\overline{x}, \overline{y})$ is consistent with T' but has no occurrences of any x_i's, then define $f(i) = j$, where j is the smallest index such that ψ_j is $(x_0 = x_0)$. If neither of the above two circumstances hold for i, then let r be equal to the largest subscript of any predicate symbol occurring in $\phi_i(\overline{x}; \overline{y})$ or equal to 1, whichever is greater. Finally let C_ϕ be the set of terraced sets C with index r and parameters $\overline{x} \cup \overline{y}$ such that $C \cup \{\phi_i(\overline{x}; \overline{y})\}$ is locally consistent with T'. $f(i)$ is then defined to be the least index of a formula $\bigvee_{j<m} \wedge E_j$, where for every $C \in C_\phi$ there is a $j < m$ such that $E_j \subset C$ and E_j is a terraced set with index r and parameters $\overline{x}$.

We must now verify (0) - (3) of Theorem 1. That f is recursive follows easily from the

<u>Claim</u>: If D is a terraced set, then it is consistent with T'. This will follow directly from the proof that (3) of the theorem is satisfied. By the definition of f and the claim, we are therefore reduced to showing that (3) is satisfied.

So assume it is not the case that $T' \vdash \neg(\phi_{f(i)}(\overline{x}) \wedge \phi_k(\overline{x},\overline{z}))$. We will also assume without loss of generality that $T' \vdash [\phi_k(\overline{x};\overline{z}) \wedge \phi_{f(i)}(\overline{x}) \rightarrow (\alpha \neq \beta)]$ for α, β equal to any $x_i \in \overline{x}$ with different subscripts, any $z_i \in \overline{z}$ with different subscripts, or α an $x_i \in \overline{x}$ and β a $z_i \in \overline{z}$. We make a similar assumption concerning $\phi_i(\overline{x};\overline{y})$ and $\phi_{f(i)}(\overline{x})$. So, since T' is universal and the ϕ_r's are quantifier-free, we can assume C is a model of $T' \cup \{\phi_k(\overline{a};\overline{c}) \wedge \phi_{f(i)}(\overline{a})\}$ and $|C| = \overline{a} \cup \overline{c}$. Let $\mathfrak{A}$ be the submodel with universe $\overline{a}$.

<u>Lemma 3'.</u> $\Delta_{\mathfrak{A}} \cup \{\phi_i(\overline{a};\overline{b})\}$ is consistent with T'.

<u>Proof.</u> We will specify a set of sentences B that will uniquely determine the diagram of a model for $\Delta_{\mathfrak{A}} \cup \{\phi_i(\overline{a};\overline{b})\} \cup T'$. By an obvious induction we can assume that $\overline{b} = \langle b \rangle$. It follows from the definition of the function f that there is a terraced set C with index r (as in the definition of f) and parameters $\overline{a} \cup \overline{b}$ such that $\Delta_{\mathfrak{A}} \cup C$ is locally consistent with T' and

$$C \vdash \phi_i(\overline{a};\overline{b}) .$$

As a first step, put $C \cup \Delta_{\mathfrak{A}} \subset B$. Next we will determine the membership of the various $S_i(a_j, b)^k$ in the set B. Because of the axioms in I, if $S_i(a_j, b)^k$ is in B we automatically insist that $S_i(b, a_j)^k$ is also. Assume inductively on v that

$$S_i(a_j, b)^k \in B$$

has been determined for $j < v$, $k = 0, 1$ and $i < \omega$, such that B is locally consistent with T' (where $\overline{a} = \langle a_1, \ldots, a_n \rangle$).

Fix $\underline{f} \in 2^{H_2(r)}$ such that $S_i(a_v, b)^{\underline{f}(i)} \in C$ for $i < H_2(r)$. There are now several cases:

I. $\underline{f} \notin F_2\text{-}Ter_2$;

II. $\exists v' < v$ and $\underline{g}, \underline{h} \in 2^{<\omega}$ satisfying

(i) $lh(\underline{g}) = lh(\underline{h})$;

(ii) $\underline{h} \neq \underline{g}$; and

(iii) $B^* \vdash conj(\underline{g}, b, a_{v'}) \wedge conj(\underline{h}, a_v, a_{v'})$ (B^* is B up to this step); and

III. Otherwise.

We determine the desired membership with respect to I - III according to:

I. $S_i(a_v, b) \in B$ for all $i \geq H_2(r)$;

II. Fix such a v', $\underline{g}$ and $\underline{h}$. Then specify

$S_i(a_v, b)^k \in B$ iff $S_i(a_{v'}, b)^k \in B^*$ if $\underline{g} < \underline{h}$, and

$S_i(a_v, b)^k \in B$ iff $S_i(a_v, a_{v'})^k \in B^*$ otherwise, for $k = 0, 1$, $i < \omega$,

III. Define

$$h = \max_{<}\{g \in 2^{\omega} \mid \exists v' < v[\underline{f} < g \text{ and } S_i(a_{v'}, b)^{g(i)} \in B\ ,\ i < \omega]\}$$

or, if no such h exists, then let h be the $<$-least element of 2^{ω} such that $\underline{f} \lhd h$ (if h is not a branch of F_2, then there is an $\underline{h} \lhd h$ such that $\underline{h} \in \text{Ter}$). Then require $S_i(b, a_v)^{h(i)} \in B$ for $i < \omega$.

We now check that the induction hypothesis has been preserved. If the defining condition is I or III, the argument is easy and it is left to the reader. Likewise for the preservation of the axioms in I - VII under defining condition II. If a problem occurs with respect to condition II and the axioms in VIII, then there must be v', u and $\underline{f}_i$, $\underline{g}_i$ $i = 1, 2$ of equal length such that

$$B^* \vdash \text{conj}(\underline{f}_1, b, a_{v'}) \wedge \text{conj}(\underline{f}_2, a_v, a_{v'}) \wedge \text{conj}(\underline{g}_1, b, a_u) \wedge \text{conj}(\underline{g}_2, a_v, a_u)$$

and $\underline{f}_1 \neq \underline{f}_2$ or $\underline{g}_1 \neq \underline{g}_2$ with

$$\min_{<}(\underline{f}_1, \underline{f}_2) \neq \min_{<}(\underline{g}_1, \underline{g}_2).$$

Since all of the cases are argued similarly, we will do just one, by assuming that $\underline{f}_1 < \underline{f}_2$, $\underline{g}_1 < \underline{g}_2$, and $\underline{f}_1 < \underline{g}_1$.

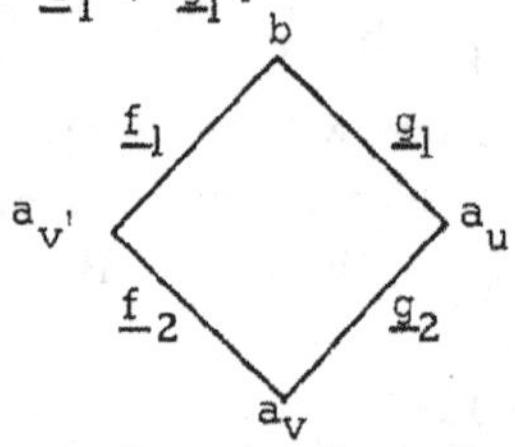

Therefore, since $\underline{f}_1 < \underline{g}_1$, we have by the induction hypothesis and the axioms in VIII that

$$B^* \vdash \text{conj}(\underline{f}_1, a_{v'}, a_u).$$

But since $\underline{f}_1 < \underline{g}_1 < \underline{g}_2$ we have a contradiction with the induction hypothesis via the axioms in VIII and $a_{v'}$, a_u, a_v (since $\underline{f}_1 < \underline{f}_2$). This completes the induction.

Let B^* be the set B as determined after the above induction; we now determine the membership of the $P_i(b)^k$, again by cases:

I. $\exists g$ a branch of F_2 and an i such that for all $\underline{f} \lhd g$

$$B^* \vdash \text{conj}(\underline{f}, a_i, b);$$

II. Not I and there is an i, t and $\underline{f} \in 2^{H_2(t)}$ such that

$$B^* \vdash \text{conj}(\underline{f}, a_i, b) \text{ and } \underline{f} \in \text{Ter}_2$$

III. Otherwise.

Again the desired determination is made according to these alternatives:

I. Fix such an i and require that $P_j(b)^k \in B$ iff $P_j(a_i)^k \in B^*$, $k = 0, 1$ and $j < \omega$;

II. Fix such an i so that the corresponding $\underline{f}$ is of maximum length and let

$h \in 2^{\omega}$ satisfy $\mathrm{conj}(\underline{h}, a_i) \in B^*$ for all $\underline{h} < h$. If $\exists \underline{h} < h[\underline{h} \in R_{t+1}]$, then $P_j(b)^{\underline{h}(j)}$, $\neg P_i(b)$, $P_k(b) \in B^*$, for $j < \mathrm{lh}(\underline{h})$ and $i = \mathrm{lh}(h) < k < \omega$; otherwise $P_j(b)^{h(j)} \in B^*$ for all $j < \omega$.

III. $P_j(b) \in B$ for $r < j < \omega$.

We now check that B so determined is consistent with T'. Only the axioms in VI and VII with respect to cases I and II present any non-trivial difficulties, the rest of the details again are left to the reader. If an axiom in VI fails, then there must be $\underline{f}_1, \underline{f}_2 \in F_2\text{-}\mathrm{Ter}_2$ such that

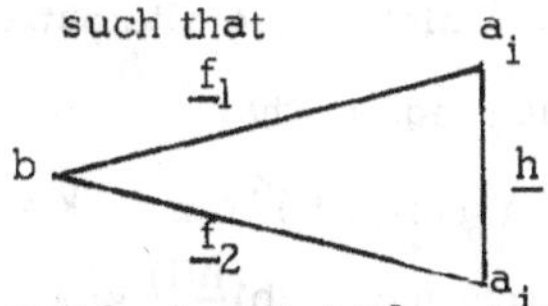

$$B^* \vdash \mathrm{conj}(\underline{f}_1, b, a_i) \wedge \mathrm{conj}(\underline{f}_2, b, a_j) \wedge \neg[\mathrm{conj}(\underline{g}, a_i) \longleftrightarrow \mathrm{conj}(\underline{g}, a_j)]$$

and $\ell h(\underline{f}_1) = \ell h(\underline{f}_2) = H_2(n)$, and no $\underline{g}' \leq \underline{g}$ satisfies $\underline{g}' \in R_{n+2}$. We will show then that $\mathfrak{a}$ is not a model of T', which will give us the desired contradiction. Let $\underline{h} \in 2^{H_2(n)}$ satisfy $B^* \vdash \mathrm{conj}(\underline{h}, a_i, a_j)$. Then by the axioms in VIII and the consistency of B^* it follows that $\min_{<}(\underline{f}_1, \underline{f}_2) \leq \underline{h}$. By Lemma 1 (vi) it is thus easy to see that $\underline{h} \in F_2\text{-}\mathrm{Ter}_2$. With this and the axioms in VI we have

$$T' \vdash \mathrm{conj}(\underline{h}, x, y) \rightarrow [\mathrm{conj}(\underline{g}, x) \longleftrightarrow \mathrm{conj}(\underline{g}, y)],$$

and so $\mathfrak{a}$ would fail to be a model for T'.

Finally, if an axiom in VII were to fail, then there would be i and j, $\underline{f} \in F_2\text{-}\mathrm{Ter}_2$, $\underline{g} \in \mathrm{Ter}_2$, $\underline{h} \in R_{n+1}$, such that $\ell h(\underline{f}) = \ell h(\underline{g}) = H_2(n)$ and

$$B^* \vdash \mathrm{conj}(\underline{f}, b, a_i) \wedge \mathrm{conj}(\underline{g}, b, a_j) \wedge \mathrm{conj}(\underline{h}, a_i) \wedge \mathrm{conj}(\underline{h}, a_j).$$

In this case it is enough to show that $\mathfrak{a}$ fails to satisfy an axiom in VII or VIII. By the assumptions and Lemma 1 (vi) $\underline{g} < \underline{f}$. Thus by the axioms in VIII

$$B^* \vdash \mathrm{conj}(\underline{g}, a_i, a_j),$$

which gives the desired contradiction to the axioms in VII, since $\underline{g} \in \mathrm{Ter}_2$. This completes the proof of Lemma 3'. Let $\mathfrak{B}$ be the model of $\Delta_{\mathfrak{a}} \cup \{\varphi_i(\bar{a}, \bar{b})\} \cup T'$ with diagram determined by B.

It is now enough to establish;

<u>Lemma 3''.</u> $T' \cup \Delta_{\mathfrak{B}} \cup \Delta_{\mathfrak{C}}$ is consistent.

<u>Proof.</u> By the obvious induction we may assume that $\bar{b} = \langle b \rangle$ and $\bar{c} = \langle c \rangle$. If $\bar{a} = \langle\ \rangle$ then the proof is trivial——so assume that $\bar{a} = \langle a_1, \ldots, a_n \rangle$, $n \geq 1$. It is enough to specify a set of sentences D that uniquely determines a model for $\Delta_{\mathfrak{B}} \cup \Delta_{\mathfrak{C}} \cup T'$. Immediately require that

$$\Delta_{\mathfrak{B}} \cup \Delta_{\mathfrak{C}} \subset D.$$

$S_i(b,c)^k$ membership for $k = 0,1$ and $i < \omega$ is determined according to the following alternatives:

I. $\exists i$ and $\underline{f}, \underline{g} \in 2^{<\omega}$ such that $\underline{f} < \underline{g}$ (or $\underline{g} < \underline{f}$) and

$$\Delta_{\mathcal{B}} \cup \Delta_{\mathcal{C}} \vdash \mathrm{conj}(\underline{f}, b, a_i) \wedge \mathrm{conj}(\underline{g}, c, a_j);$$

II. $\exists n$ such that $\Delta_{\mathcal{B}} \cup \Delta_{\mathcal{C}} \vdash \bigwedge_{i<n} (P_i(b) \longleftrightarrow P_i(c)) \wedge \neg(P_n(b) \longleftrightarrow P_n(c))$.

III. Otherwise.

As before, our choice is determined by which of the alternatives holds:

I. Fix such an i and $\underline{f}, \underline{g}$ and require that

$$S_j(b,c)^k \in D \text{ iff } S_j(b,a_i)^k \in \Delta_{\mathcal{B}}(S_j(c,a_i)^k \in \Delta_{\mathcal{C}}) \quad k = 0,1 \text{ and } j < \omega;$$

II. For that n fix $\underline{h} \in 2^n$ satisfying $P_i(b)^{\underline{h}(i)} \in \Delta_{\mathcal{B}}$ for $i < n$; obviously $\underline{h}^\frown\langle 0\rangle \in R_{s+1}$ for some $s < n$, so fix such an s and also $\underline{g} \in \mathrm{Ter}_2$, $\ell h(\underline{g}) = s$, then specify that

$$S_i(b,c)^{\underline{g}(i)}, \quad S_j(b,c) \in D \quad i < lh(\underline{g}) \leq j < \omega;$$

III. $S_i(b,c) \in D$, $i < \omega$.

The only check for consistency with T' that is not completely straightforward is with respect to defining condition I. We will check this and once again leave the remainder to the reader. The axioms in I-V present no difficulties. If an axiom in VI is violated then there would have to be an n and $\underline{h} \in F_2\text{-}\mathrm{Ter}_2$, $\ell h(\underline{h}) = H(n)$, $\underline{g}_1 \in F_1$, $\forall \underline{g}' \trianglelefteq \underline{g}_1 (\underline{g}' \notin R_{n+2})$ such that

$$D \vdash \mathrm{conj}(\underline{h}, b, c) \wedge \neg[\mathrm{conj}(\underline{g}_1, b) \longleftrightarrow \mathrm{conj}(\underline{g}_1, c)].$$

Let i, $\underline{f}$, and $\underline{g}$ be as in I. Now by the instructions, $\min_<(\underline{f}, \underline{g}) \triangleleft \underline{h}$ or $\underline{h} \triangleleft \min_<(\underline{f}, \underline{g})$. Thus by Lemma 1 (vi) and the axioms

$$\Delta_{\mathcal{B}} \vdash (\mathrm{conj}(\underline{g}_1, b) \longleftrightarrow \mathrm{conj}(\underline{g}_1, a_i)) \text{ and } \Delta_{\mathcal{C}} \vdash (\mathrm{conj}(\underline{g}_1, c) \longleftrightarrow \mathrm{conj}(\underline{g}_1, a_i)).$$

But this is a contradiction, since $\Delta_{\mathcal{B}} \cup \Delta_{\mathcal{C}} \vdash \neg(\mathrm{conj}(\underline{g}_1, b) \longleftrightarrow \mathrm{conj}(\underline{g}_1, c))$. Thus the axioms in VI are preserved.

A violation of an axiom in VII would imply that there was an n and an $\underline{h} \in \mathrm{Ter}_2$, $\ell h(h) = H(n)$, $\underline{g}_1 \in R_{n+1}$ such that

$$D \vdash \mathrm{conj}(\underline{h}, b, c) \wedge \mathrm{conj}(\underline{g}_1, b) \wedge \mathrm{conj}(\underline{g}_1, c).$$

But then for the i, $\underline{f}$, $\underline{g}$ as in I it must be that

$$\underline{h} < \max_<(\underline{f}, \underline{g}), \qquad \text{by the axioms in VIII}.$$

Therefore by Lemma 1 (vi) and the axioms,

$$\Delta_{\mathcal{B}} \cup \Delta_{\mathcal{C}} \vdash \mathrm{conj}(\underline{h}', a_i, b) \vee \mathrm{conj}(\underline{h}', a_i, c)$$

for some $\underline{h}' \in F\text{-}\mathrm{Ter}$, $\ell h(\underline{h}') = \ell h(\underline{h})$. Therefore by the axioms in VI

$$\Delta_{\mathcal{B}} \cup \Delta_{\mathcal{C}} \vdash (\mathrm{conj}(\underline{g}_1, a_i) \longleftrightarrow \mathrm{conj}(\underline{g}_1, b)) \vee (\mathrm{conj}(\underline{g}_1, a_i) \longleftrightarrow \mathrm{conj}(\underline{g}_1, c))$$

which is to say $\Delta_{\mathcal{B}} \cup \Delta_{\mathcal{C}} \vdash \mathrm{conj}(\underline{g}_1, a_i)$. But by the instructions for case II, it must be that

$$\Delta_{\mathcal{B}} \cup \Delta_{\mathcal{C}} \vdash \mathrm{conj}(\underline{h}, a_i, b) \vee \mathrm{conj}(\underline{h}, a_i, c),$$

and so an axiom from VII must fail in either $\mathcal{B}$ or $\mathcal{C}$, a contradiction. Thus the axioms in VII are preserved. The axioms in VIII are checked in a way similar to the verification in Lemma 3'.

References

1. Barwise, Jon, and J. Schlipf, "An Introduction to Recursively Saturated and Resplendent Models", Journal of Symbolic Logic, Vol. 41, No. 2, 1976, pp. 531-536.

2. Millar, T., "Some Results in Recursive Model Theory", Notices A. M. S., Vol. 24, 77T-E11.

3. Millar, T., "Vaught's Theorem Recursively Revisited", Journal of Symbolic Logic, to appear.

4. Sacks, G. E., Saturated Model Theory, W. H. Benjamin, Inc., 1972.

HIGH RECURSIVELY ENUMERABLE DEGREES AND THE ANTI-CUPPING PROPERTY[1]

by

David P. Miller

§1. Introduction

After seeing the Sacks Density Theorem [Sa2], Shoenfield conjectured [Sh2] that the recursively enumerable (r.e.) degrees **R** form a dense structure as an upper semi-lattice analogously as the rationals are a dense structure as a linearly ordered set, i.e., given a quantifier-free formula $\varphi(x_1, \ldots, x_n, y)$ in the language $L(\leq, \cup, \underset{\sim}{0}, \underset{\sim}{0}')$ and $\mathbf{a}_1, \ldots, \mathbf{a}_n \in \mathbf{R}$, there exists $\mathbf{b} \in \mathbf{R}$ such that $\varphi(\mathbf{a}_1, \ldots, \mathbf{a}_n, \mathbf{b})$ holds, unless the existence of **b** would lead to an "inconsistency." (The conjecture asserts that if $\vec{\underset{\sim}{a}} \in \underset{\sim}{R}$ satisfies the diagram $D(\vec{x})$ and $D_1(\vec{x}, y)$ is any consistent diagram in $L(\leq, \cup, \underset{\sim}{0}, \underset{\sim}{0}')$ extending $\underset{\sim}{D}$, then there exists $\underset{\sim}{b} \in \underset{\sim}{R}$ such that $D_1(\vec{\underset{\sim}{a}}, \underset{\sim}{b})$.)

Among the consequences of this conjecture Shoenfield listed:

(1.1) If $\mathbf{a}, \mathbf{b} \in \mathbf{R}$ are incomparable, then they have no greatest lower bound in **R**.

(1.2) Given r.e. degrees $\mathbf{0} < \mathbf{b} < \mathbf{a}$ there exists an r.e. degree $\mathbf{c} < \mathbf{a}$ such that $\mathbf{b} \cup \mathbf{c} = \mathbf{a}$.

(Consequence (1.2) was also conjectured by Sacks at the end of [Sa2].) Unfortunately, both consequences are false, but as Shoenfield anticipated [Sh2, p. 363], they led to the development of important new areas and techniques of proof.

We say that nonrecursive r.e. degrees **a** and **b** form a <u>minimal pair</u> if $\mathbf{a} \cap \mathbf{b} = \mathbf{0}$. Shoenfield's conjecture was first refuted by Lachlan [La1] and independently by Yates [Y], who disproved (1.1) by constructing a minimal pair. This led to the area of branching and nonbranching degrees, which has been summarized in [F] and [So2]. (An r.e. degree **a** is <u>branching</u> if there are r.e. degrees **b, c** both $> \mathbf{a}$ such that $\mathbf{a} = \mathbf{b} \cap \mathbf{c}$, and **a** is <u>nonbranching</u> otherwise. The minimal pair theorem of Lachlan and Yates asserts that $\mathbf{0}$ is branching.)

Consequence (1.2) led to results about cupping and anti-cupping. Given r.e. degrees $\mathbf{0} < \mathbf{b} < \mathbf{a}$, we say that **b** <u>cups to</u> **a** if there exists an r.e. degree $\mathbf{c} < \mathbf{a}$ such that $\mathbf{b} \cup \mathbf{c} = \mathbf{a}$; if no such degree **c** exists, we say that **b** is an <u>anti-cupping witness</u> for **a**. The r.e. degree **a** has the <u>anti-cupping (a.c.) property</u> if it has an anti-cupping witness. An r.e. degree **a** has the <u>strong a.c. property via witness</u> **b** if $\mathbf{0} < \mathbf{b} < \mathbf{a}$ and for no $\mathbf{c} \not\geq \mathbf{a}$ is $\mathbf{b} \cup \mathbf{c} \geq \mathbf{a}$. Consequence (1.2)

---

[1]This paper is derived from a talk based on some notes of Harrington given by the author in the logic seminar at the University of Connecticut in the fall of 1979, during which time the author was supported by the University of Connecticut. We would like to thank R. Soare for a number of helpful comments.

states that no nonrecursive $\mathbf{a} \in \mathbf{R}$ has the a.c. property. Lachlan [La2] constructed a counterexample to Consequence (1.2). Ladner and Sasso proved that r.e. degrees with a.c. property are abundant among the r.e. degrees $\mathbf{b}$ "close to" $\mathbf{0}$ (in the sense that $\mathbf{b}'' = 0''$). Their result [LdSs] states that every nonrecursive $\mathbf{a} \in \mathbf{R}$ has a nonrecursive r.e. predecessor $\mathbf{b}$ satisfying $\mathbf{b}'' = 0''$ with the a.c. property.

In this paper, we prove a result of Harrington (Theorem 3 below) which is analogous to, though stronger than, the Ladner-Sasso result, for the degrees close to $\mathbf{0}'$. An r.e. degree $\mathbf{a}$ is said to be <u>high</u> if $\mathbf{a}' = \mathbf{0}''$. The first in a series of theorems leading to Theorem 3 was announced by Yates:

<u>Theorem 1</u> (Cooper-Yates). $\mathbf{0}'$ has the a.c. property.

Cooper circulated a proof of Theorem 1, and Harrington followed with a proof based on Cooper's. A trivial modification of Harrington's proof makes the witness $\mathbf{b}$ constructed there satisfy $\mathbf{b}' = 0'$; i.e., the witness is "far" from $\mathbf{0}'$. Hence, the following theorem, also proved by Harrington, is a significant improvement.

<u>Theorem 2</u> (Harrington). $\mathbf{0}'$ has a high a.c. witness.

The proofs of Theorems 1 and 2 rely heavily on the fact that $0'$ contains a creative set. Following a suggestion of Soare, Harrington incorporated into his proof of Theorem 2 the method of permitting below a high r.e. degree introduced by Cooper [C], thereby removing this reliance and generalizing Theorem 2 from $\mathbf{0}'$ to all high r.e. degrees.

<u>Theorem 3</u> (Harrington). Every high r.e. degree $\mathbf{a}$ has the strong a.c. property via a high r.e. witness $\mathbf{b}$.

Our purpose is twofold. First we give a straightforward proof of Theorem 3 by modifying and rearranging some unpublished notes of Harrington on the proof of Theorem 3 (no proofs of Theorems 1, 2 and 3 have so far appeared in the literature). Second, we use the Cooper high-permitting method to prove a new result which will appear in [Mi], namely, for every high r.e. degree $\mathbf{a}$ there is a minimal pair of high r.e. degrees $\mathbf{b}_0, \mathbf{b}_1 < \mathbf{a}$. Moreover, it is clear that the proof of Theorem 3 may be incorporated into this proof so that the degrees $\mathbf{b}_0$ and $\mathbf{b}_1$ are both strong a.c. witnesses for $\mathbf{a}$.

A new method of proof is required for the Yates-Cooper-Harrington results. In the <u>priority method</u>, the desired r.e. set B is constructed by stages to meet a set of conditions. These conditions usually are split into <u>positive requirements</u> $\{P_e\}_{e\in N}$ and <u>negative requirements</u> $\{N_e\}_{e\in N}$, and given a priority ordering $N_0 < P_0 < N_1 < P_1 < \dots$; a requirement is said to have <u>higher priority</u> than the requirements appearing later in the ordering. The positive requirement P_e attempts to force certain numbers called followers into B. The negative requirement N_e attempts to prevent certain numbers from being enumerated into B. This is accomplished by associating with N_e the restraint function $r(e, s)$ and allowing a follower x of a positive requirement of lower priority than N_e to enter B at stage $s + 1$ only if $x > r(e, s)$. Of course, a higher priority positive requirement may cause a follower $x \leq r(e, s)$ to be enumerated in B at stage $s + 1$, thereby injuring N_e at stage $s + 1$. The priority method was introduced by Friedberg [Fr] and independently by Muchnik [Mu]. Their proofs have the property that the positive requirements are finitary, so that each negative requirement is injured at most finitely often, and that the restraint associated with each negative requirement is finitary (in fact, $\lim_s r(e, s) < \infty$ for all e), so that each positive requirement is satisfied.

Shoenfield [Sh1] and, independently, Sacks [Sa1], [Sa2], [Sa3] discovered a technique for handling a negative requirement which may be injured infinitely often. Sacks developed this technique into what he called the "infinite injury priority method." This method has been used to prove many important results on r.e. degrees, including the density theorem and the minimal pair construction. These constructions have the property that a single positive requirement P_e may contribute infinitely many elements to the set B, though in the simplest cases the set T_e of followers of P_e is recursive. The negative requirement N_e now can be injured infinitely often by the higher priority positive requirements, but the recursiveness of T_e enables the strategy for N_e to succeed. The main difficulty is that some P_i remains unsatisfied because the restraint function for N_e now may be unbounded in s (i.e., $\limsup_s r(e, s) = \infty$). This difficulty is surmounted by arranging that

$$\liminf_s R(e, s) < \infty , \tag{1.3}$$

where $R(e, s) = \max\{r(i, s) : i \leq e\}$, because then P_e has a "window" through the restraints at least infinitely often. More details and applications of the infinite injury method may be found in [So1].

One way of presenting priority method constructions is the pinball machine model, as introduced by Lerman in [Le]. In this model, the negative requirement N_e is associated with the gate G_e in the machine. Followers of P_e enter the machine by dropping from hole H_e to gate G_e. The follower x at gate G_e is allowed to pass to gate G_{e-1} at stage $s + 1$ just if $x > r(e, s)$; x is enumerated in B after it has passed all the gates. Thus, a follower of P_e is allowed to pass the higher priority negative requirements one at a time, instead of all at once as above. For P_e to be satisfied, one must arrange the construction so that each gate G_i, $i \leq e$, has only finitely many permanent residents, that is, followers which enter the gate and never leave. In most pinball machine constructions this follows from a property slightly easier to arrange than (1.3):

(1.4) $\liminf_s r(i, s) < \infty$, for $i \leq e$.

A pinball machine model is used to prove the Yates-Cooper-Harrington results. However, now it is possible that (1.4) will not be true--we may have $\liminf_s r(i, s) = \infty$ for some i. A new technique is required to guarantee that each gate has only finitely many permanent residents. The solution is to "spread out" the restraint associated with N_e over all the gates G_i, $i \geq e$. If we decide at stage $s + 1$ to increase the restraint for N_e $(r(e, s + 1) > r(e, s))$ the new restraint $r(e, s + 1)$ is applied at both G_e and at some gate $G_{e'}$, $e' > e$. Thus, no follower which might become a permanent resident of G_e due to this new restraint is allowed to pass $G_{e'}$. Therefore, any follower which does become a permanent resident of G_e due to the new restraint must have been in the "critical zone"--at a gate G_i or hole H_i, $e \leq i < e'$--at stage $s + 1$. New restraint is applied in such a way that if a follower from the critical zone reaches G_e, then only finitely much restraint will be needed to satisfy N_e: the construction has the property that either $\lim_s r(e, s) < \infty$ or gate G_e has <u>no</u> residents which are permanently restrained by N_e.

Any a.c. witness **b** for an r.e. degree **a** identifies infinitely many a.c. witnesses for **a**, namely, the "cone" $\{c: c \in \mathbf{R}$ and $c \leq \mathbf{b}\}$ of r.e. degrees below **b**. Harrington has constructed an r.e. degree **a** with exactly the opposite property of the degree **b** of Theorem 2.

<u>Theorem 4</u> (Plus-cupping Theorem - Harrington). There is an r.e. degree $\mathbf{a} > 0$ such that every nonrecursive r.e. $\mathbf{b} < \mathbf{a}$ can be cupped to every r.e. $\mathbf{d} \geq \mathbf{a}$.

Taking $\mathbf{d} = \mathbf{0}'$, the theorem yields an entire cone of r.e. degrees, $\{\mathbf{b}: \mathbf{b} \in \mathbf{R}$ and $\mathbf{b} \leq \mathbf{a}\}$, which are not a.c. witnesses for $\mathbf{0}'$. (The proof of the special case $\mathbf{d} = 0'$ is presented in a paper by Fejer and Soare in this volume [FeSo].)

Harrington and Shelah claim that the cupping/anti-cupping methods may be extended to show that any partially-ordered set with a $\mathbf{0}'$-recursive partial ordering is first-order definable from parameters in the language $L(\leq, \cup)$ for the r.e. degrees **R**, thus proving that the first-order theory of **R** is undecidable. Let $\Phi(\mathbf{a}, \mathbf{b}, \mathbf{m})$ be the formula "$\mathrm{b} \cup \mathbf{m} < \mathbf{a}$ and $(\forall c \in \mathbf{R})[\mathbf{a} \leq \mathrm{b} \cup \mathrm{c}$ or $\mathrm{c} \leq \mathrm{m}]$." Harrington and Shelah claim

<u>Theorem 5</u>. There exist $\mathbf{a}, \mathbf{b}, \mathbf{m} \in \mathbf{R}$ such that $\Phi(\mathbf{a}, \mathbf{b}, \mathbf{m})$.

The crucial point is that, for given parameters **a** and b, the r.e. degree **m** such that $\Phi(\mathbf{a}, \mathbf{b}, \mathbf{m})$ holds is uniquely determined; moreover, **m** is not obtainable from the parameters in a trivial algebraic way (i.e., from a combination of **a** and b using $\cap$ and $\cup$). They proceed to show:

Theorem 6. Fix a partial ordering $(P, \leq_p)$ where $\leq_p$ is $\mathbf{0}'$-presented. Then there are degrees $\mathbf{a}, \mathbf{b}, \mathbf{c} \in \mathbf{R}$ such that the set of degrees

$$\{\mathbf{d}: \mathbf{d} \in \mathbf{R} \text{ and } (\exists \mathbf{M})[\mathbf{m} \in \mathbf{R} \text{ and } \Phi(\mathbf{a}, \mathbf{b}, \mathbf{m}) \text{ and } \Phi(\mathbf{m}, \mathbf{c}, \mathbf{d})]\}$$

is isomorphic to $(P, \leq_p)$ under the ordering $\leq$.

Thus, the theory of partial orderings is interpretable in the theory of $(\mathbf{R}, \leq)$, so the latter is undecidable.

The remainder of this paper is devoted to Harrington's proof of Theorem 3. We present the requirements for the r.e. set B whose degree $\mathbf{b}$ is an a.c. witness for the fixed degree $\mathbf{a}$ and describe the pinball machine model in §2. The strategies for the different types of requirements are presented in §§3, 4, and 5. These sections also include proofs of the sufficiency of the strategies, using the crucial assumption that each gate in the machine has only finitely many permanent residents. In the final section, the new technique which guarantees that this hypothesis holds is presented and proved to work.

We follow the notation of [Rg], with a few minor changes and additions. We identify sets $A \subseteq N$ with their characteristic functions and let $A[n]$ denote the restriction of A to arguments $\leq n$. Let $A =^* B$ denote that $(A - B) \cup (B - A)$ is finite. Let $A^{(n)} = \{y: \langle n,y \rangle \in A\}$, where $\langle \ , \ \rangle: N \times N \to N$ is a fixed one-one, onto, recursive pairing function. Let $\{e\}_s(X; y)$ be the result, if any, after performing s steps in the eth Turing reduction with oracle X and input y. Let $\{e\} = \cup\{\{e\}_s : s \in N\}$. We write "$\{e\}_s(X; y)\downarrow$" if the computation converges, and "$\{e\}_s(X; y)\uparrow$" otherwise. Let $W_{e,s}$ (W_e) be the domain of $\{e\}_s$ ($\{e\}$). The <u>use function</u> u is defined by: $u(e, X, y, s)$ = the maximum element used in the computation $\{e\}_s(X; y)$ if $\{e\}_s(X; y)\downarrow$; otherwise, $u(e, X, y, s)$ is undefined. We adopt the convention that

$$(1.7) \qquad \{e\}_s(X; y)\downarrow \Rightarrow e, y, u(e, X, y, s) \leq s .$$

We say that the function f <u>dominates</u> the function g if $f(x) > g(x)$ for all but finitely many x.

§2. The Requirements and the Pinball Machine

Fix an r.e. degree $\mathbf{a}$ such that $\mathbf{a}' = \mathbf{0}''$. By a theorem of Robinson [Ro], we may choose an r.e. set A of degree $\mathbf{a}$ and an effective enumeration $\{A_s\}_{s \in N}$ of A so that the <u>computation function</u>

$$C_A(x) = (\mu s)[A_s[x] = A[x]]$$

dominates all recursive functions.

Define the increasing recursive sequence of finite sets $\{T_s\}_{s\in N}$ by

$$(2.1) \qquad \langle e, t\rangle \in T_s \Longleftrightarrow [t \leq s \text{ and } (\{e\}_t(A_t; e)\downarrow \Longrightarrow A_s[u] \neq A_t[u]),$$

$$\text{where } u = u(e, A_t, e, t)],$$

and let $T = \bigcup_s T_s$. Clearly, $T \leq_T A$ and

$$e \in A' \Longleftrightarrow T^{(e)} \text{ is finite} \Longleftrightarrow T^{(e)} \neq N^{(e)}.$$

Moreover, there is a recursive functional $\{\hat{e}\}$ satisfying

$$\{\hat{e}\}(A; \langle e, s\rangle)\downarrow \Longleftrightarrow T_s^{(e)} = T^{(e)} .$$

Let B be the r.e. set of degree **b** which has to be constructed. We make $\mathbf{b}' = 0''$ by satisfying the requirements

$$P_e \colon B^{(e)} =^* T^{(e)}, \quad e \in N .$$

Thus, by (2.1), $e \in A'$ if and only if $B^{(e)}$ is finite, so $\mathbf{b}' = \mathbf{a}' = 0''$. By satisfying the requirements

$$N_e \colon \{e_1\}(W_{e_0} \oplus B) = A \Longrightarrow A \leq_T W_{e_0}, \quad e = \langle e_0, e_1\rangle \in N ,$$

we guarantee that for no r.e. degree $\mathbf{c} < \mathbf{a}$ is $\mathbf{b} \cup \mathbf{c} \geq \mathbf{a}$. Additionally, we must make $B \leq_T A$; we refer to this as requirement R.

We use a pinball machine to describe the construction of B. The pinball machine M comprises <u>segments</u> G_e, $e \geq -1$, which form the <u>surface</u> of the machine, and <u>holes</u> H_e, $e \geq 0$, which we picture as leading down to the surface of the machine. The segment G_e is composed of <u>gates</u> G_e^0 and G_e^1. At each stage of the construction certain numbers, called <u>followers</u>, reside on the surface of M; a follower must be resident at some gate. A number $x = \langle e, y\rangle$ which is not on the surface of the machine is said to be <u>above hole</u> H_e.

The number x is said to be <u>above gate</u> G_e^0 if x is a resident of G_j^i, $j > e$ and $i = 0, 1$, or above hole H_j, $j > e$, and x is <u>above gate</u> G_e^1 if it is above G_e^0, a resident of G_e^0 or above H_e.

The number $x = \langle e, y\rangle$ first enters the machine by dropping from hole H_e to the gate G_e^1; when this happens we say x is <u>emitted from</u> H_e. After x has been emitted it moves down the machine--that is, from G_e^1 to G_{e-1}^0 to G_{e-1}^1, etc.-- until it reaches G_{-1}^1. A follower which reaches G_{-1}^1 stays there forever. We let B_s denote the followers which have reached G_{-1}^1 by the end of stage s of the construction, and we define $B = \bigcup_s B_s$.

The decision to emit a number is connected with the strategy for the positive requirements P_e. This strategy is discussed in §3. The movement of a follower down the machine is controlled by the strategy for the requirement R and the negative requirements N_e.

A follower which is e-restrained (respectively, e-frozen) at stage $s + 1$ may not pass gate G_e^1 (resp., G_e^0) during that stage. Additionally, a follower may not pass gate G_e^0 at stage $s + 1$ unless it is permitted by A to do so. Permitting is associated with the requirement R and will be described in §4. The requirement N_e imposes e-restraint using the restraint function $r(e, s)$; e-restraint corresponds to the conventional restriction of follower movement as described in §1. In contrast, e-freezing, imposed by all the requirements N_i, $i < e$, is the special restriction on follower movement which is the attempt to spread the restraint for N_i, $i < e$, to section G_e. Technical considerations force us to apply these restrictions separately at G_e; hence, G_e has been split into two gates. The definitions of e-restraint and e-freezing are given in §§5 and 6, respectively.

The following sequence of events takes place during stage $s + 1$ of the construction.

Step 1. A finite (possibly empty) set of holes will emit a finite number of followers at stage $s + 1$. For each e, place the followers emitted from H_e at G_e^1 one at a time in order of increasing size.

Step 2. For each e, place at gate G_e^1 all followers which are at gate G_e^0, are permitted by A at this stage, and are not e-frozen. As in Step 1, these followers should be placed at G_e^1 in order of increasing size.

Step 3. For each e, move to gate G_{e-1}^0 any follower which is currently resting at G_e^1 and which is not e-restrained at stage $s + 1$.

This completes the description of stage $s + 1$ of the construction.

§3. Emitting

The number x is emitted from hole H_e (becomes a follower of P_e) during step 1 of stage $s + 1$ if and only if $x = \langle e, y\rangle$ and $y \in T_{s+1}^{(e)} - T_s^{(e)}$. Thus, P_e will be satisfied if all but finitely many of the elements emitted from hole H_e are eventually enumerated in B. A follower x is said to be a permanent resident of gate G_e^j if x enters G_e^j at some stage and never leaves.

Lemma 1. Fix e. Suppose each gate G^0_i, $-1 \leq i < e$, and each gate G^1_i, $0 \leq i \leq e$, has only finitely many permanent residents. Then P_e is satisfied.

Proof. The gates mentioned in the hypothesis of the theorem are the only gates other than G^1_{-1} which followers of P_e enter. Hence, all but finitely many of the followers of P_e reach G^1_{-1} and are enumerated in B. □

§4. Permitting

We use the enumeration of A to control the movement of followers past the gates G^0_e, $e \geq -1$, so that we may A-recursively determine whether or not a follower which has been emitted ever enters B. The method employed here, introduced by Cooper in [Co], uses the highness of A to ensure that almost all followers which enter G^0_e are eventually permitted to leave (Lemma 6).

Let g_{e+1} enumerate by order of entry the followers which enter gate G^1_{e+1}. If x is at gate G^0_e or G^1_{e+1} by the end of Step 1 of Stage $s + 1$, the permitting number $p(x, s)$ associated with x at stage $s + 1$ is the unique m such that $g_{e+1}(m) = x$. The follower x is permitted by A at stage $s + 1$ if $A_s[p(x, s)] \neq A_{s+1}[p(x, s)]$.

Lemma 2. $B \leq_T A$.

Proof. Fix $x = \langle n, y\rangle$. We show how to A-recursively determine whether or not x enters B, that is, whether or not the follower x reaches gate G^1_{-1}. If $x \notin T$, then x never enters the machine and so cannot reach G^1_{-1}. Suppose $x \in T$. Then x reaches G^1_{-1} just if x enters each gate $G^1_{n-1}, G^1_{n-2}, \ldots, G^1_{-1}$. Suppose x enters G^1_{e+1} at stage s_0. The permitting number $p = p(x, s_0)$ remains associated with x until x enters G^1_e. Let s_1 be the least stage $s > s_0$ such that $A_s[p] = A[p]$. Then x enters G^1_e if and only if it enters G^1_e by the end of stage s_1. Since $T \leq_T A$ and s_1 may be found using an A-oracle, $B \leq_T A$. □

§5. Restraint

The strategy for N_e consists of attempting to protect certain computations of the form

$$\{e_1\}_s (W_{e_0,s} \oplus B_s;\ x)\downarrow \tag{5.1}$$

from being destroyed by followers entering B. We define e-restraint so as to prevent such injuries by followers which must pass gate G^1_e to enter B. Of course, followers which need not pass G_e^1 to enter B or followers which passed G_e^1 before restraint could be imposed on them may still injure computations which we have decided to protect, but these injuries to the strategy will be sufficiently well-behaved so that, under the assumption that $\{e_1\}(W_{e_0} \oplus B) = A$, we may W_{e_0}-recursively determine when a computation (5.1) is $W_{e_0} \oplus B$-correct, proving that $A \leq_T W_{e_0}$.

To describe which computations (5.1) are to be protected during stage $s + 1$, we define a sequence

$$\vec{b}_{e,s} = \langle b_{e,s}(e),\ b_{e,s}(e+1),\ \ldots,\ b_{e,s}(n)\rangle\ ,$$

where $n = n(e, s)$ can be determined effectively, and where

(5.2) $$-1 = b_{e,s}(e) < b_{e,s}(e+1) < \ldots < b_{e,s}(n), \quad \text{and}$$

(5.3) $$\{e_1\}_s(W_{e_0,s} \oplus B_s;\ x)\downarrow \quad \text{for all} \quad x \leq b_{e,s}(n).$$

Intuitively, we protect the computation (5.1) at stage $s + 1$ just if $x \leq b_{e,s}(n)$. The number $b_{e,s}(i)$, $e \leq i \leq n$, similarly describes which computations are being protected by i-freezing for N_e. The sequence $\vec{b}_{e,s}$ will be defined in §6. We conclude this section with a description of the properties of $\vec{b}_{e,s}$, and we prove that these properties suffice to guarantee that N_e is met.

We write "$b_{e,s}(i)\downarrow$" if $i \leq n(e, s)$, and "$b_{e,s}(i)\uparrow$" otherwise. Define the recursive functions q and r by

$$q(e, i, s) = \begin{cases} \max\{u(e, W_{e_0,s} \oplus B_s, x, s) : x \leq b_{e,s}(i)\} & \text{if } b_{e,s}(i)\downarrow\ , \\ -1 & \text{otherwise} \end{cases}$$

and

$$r(e, s) = q(e, n(e,s), s)\ .$$

Note that (5.3) guarantees that $q(e, i, s)$ is well-defined. The follower x is e-restrained at stage $s + 1$ if $x \leq r(e, s)$ and x is a resident of G_e^1 at the end of Step 2 at stage $s + 1$. It will follow immediately from the definition of $\vec{b}_{e,s}$ that

(5.4) $$\text{if } b_{e,s}(i)\downarrow, \text{ then} \quad (b_{e,s+1}(i)\uparrow \Longleftrightarrow (W_{e_0,s} \oplus B_s)[q(e,i,s)] \neq (W_{e_0,s+1} \oplus B_{s+1})[q(e,i,s)])\ .$$

We will prove later (Lemma 7) that

(5.5) $\{e_1\}(W_{e_0} \oplus B) = A \Rightarrow (\forall i)[\lim_s b_{e,s}(i) \text{ exists}]$.

Lemma 3. Fix e and suppose that the gates G_i^0, $-1 \leq i < e$, and G_i^1, $0 \leq i \leq e$, have only finitely many permanent residents. Then N_e is satisfied.

Proof. Let $C = \{i: i < e$ and $T^{(i)}$ is finite$\}$. Let s_0 be the least stage s such that $T_s^{(i)} = T^{(i)}$ for all $i \in C$. By the hypothesis of the lemma, we may assume we are supplied with a recursive oracle for F, the set of permanent residents of gates G_i^0, $-1 \leq i \leq e$, and G_i^1, $0 \leq i \leq e$.

Assume $\{e_1\}(W_{e_0} \oplus B) = A$ and fix x. We show how to W_{e_0}-recursively calculate $A(x)$. Let s_1 be the least stage $s > s_0$ such that

(5.6) $b_{e,s}(n(e, s)) > x$,

(5.7) $W_{e_0,s}[r(e, s)] = W_{e_0}[r(e, s)]$, and

(5.8) if $z = \langle i, y\rangle < r(e, s)$ and $z \notin B_s$, then either

(a) $z \in F$,

(b) $i \in C$ and $y \notin T_{s_0}^{(i)}$

(c) z is at a gate G_k^0 or G_k^1, or above hole H_k, for some $k > e$;

s_1 exists by (5.5) and the fact that $T^{(i)} = N^{(i)}$ for all $i < e$ which are not in C.

Claim: $\{e_1\}_{s_1}(W_{e_0,s_1} \oplus B_{s_1}; x) = \{e_1\}(W_{e_0} \oplus B; x) = A(x)$.

Proof of Claim: Suppose the claim is false. Then there is a stage $s > s_1$ such that $(W_{e_0,s} \oplus B_s)[r(e, s_1)] \neq (W_{e_0,s+1} \oplus B_{s+1})[r(e,s_1)]$; let s_2 be the least such stage and let $z = \langle i, y\rangle$ be the least number which enters $W_{e_0} \oplus B$ at stage $s_2 + 1$. Now $z < r(e, s_1)$, so (5.7) implies that z does not enter W_{e_0}; hence, $z \in B_{s_2+1} - B_{s_2}$. By (5.4) and the choice of s_2, we know that $r(e, t) \geq r(e, s_1)$ for all t, $s_1 \leq t \leq s_2$. Therefore, z was e-restrained and could not have passed G_e^1 during any stage t, $s_1 \leq t \leq s_2$. So z was either above a hole H_i, $0 \leq i < e$, or in some section G_i, $-1 \leq i < e$, at the end of stage s_1. But then (5.8) and the choice of $s_1 > s_0$ imply that either z is never emitted or z is the permanent resident of some gate, and so cannot enter B. This proves the claim.

Since s_1 can be computed recursively from x using a W_{e_0}-oracle, $A \leq_T W_{e_0}$, as required by N_e. □

§6. Freezing and the Finiteness Lemmas

The chief difficulty in the construction lies in the conflict between the hypothesis of Lemmas 1 and 3, namely, that each gate has only finitely many permanent residents, and property (5.5), which requires that $\liminf_s r(e, s) = \infty$ if $\{e_1\}(W_{e_0} \oplus B) = A$. The mechanism for the resolution of this conflict is freezing.

Suppose at the end of stage $s + 1$ we wish to protect additional computations for N_e, say the computations (5.1) for $x < b$. This increased protection will be indicated by adjoining b to $\vec{b}_{e,s}$, that is, we define $\vec{b}_{e,s+1} = \langle b_{e,s}(e),\ldots,b_{e,s}(n), b\rangle$, where $n = n(e, s)$. Then the restraint at G^1_e for the next stage is $r(e, s + 1)$. A follower x which could become a permanent resident of G^1_e due to this restraint falls into one of three categories according to its position at the end of stage $s + 1$:

(6.1) $\qquad x$ is above G^0_{n+1},

(6.2) $\qquad x$ is at a gate G^0_i or G^1_i, $e \leq i \leq n + 1$, or

(6.3) $\qquad x$ is above a hole H_i, $e \leq i \leq n + 1$.

Categories (6.2) and (6.3) correspond to the "critical zone" mentioned in §1.

A follower in the first category can be prevented from reaching G^1_e by $(n + 1)$-freezing it, thereby stopping it at G^0_{n+1} if it should reach that gate. A follower in the second (third) category will reach G^1_e only if it is A-permitted (emitted) at a later stage. We can use A_{s+1} and the permitter $p(x, s + 1)$ to predict if a follower in the second category will be permitted at a later stage. Similarly, A_{s+1} and the functional $\{\hat{e}\}$ can be used to predict if a hole in the critical zone is finite and, thus, whether or not a follower in the third category will be emitted. More precisely, we may be able to determine a number u such that a follower $x < r(e, s + 1)$ in the critical zone will move only if $A_{s+1}[u] \neq A[u]$. We will decide to protect the additional computations only if u can be found, $b > u$, and $\{e_1\}_{s+1}(W_{e_0,s+1} \oplus B_{s+1})[b] = A_{s+1}[b]$. Then either no followers become permanent residents of G^1_e due to the new restraint, or at some stage $t+1 > s+1$ we will have $A_{t+1}[b] \neq A_{s+1}[b]$, in which case the finite amount of restraint $r(e, s+1)$ will suffice to preserve a disagreement between $\{e_1\}_{s+1}(W_{e_0,s+1} \oplus B_{s+1})[b]$ and $A_{t+1}[b]$, thus satisfying N_e.

We must be careful to ensure that i-freezing does not combine with A-permitting at G^0_i in such a way that G^0_i ends up with infinitely many permanent residents. We say that N_e is <u>injured during stage</u> $s + 1$ if a number $x < r(e, s)$ enters $W_{e_0} \oplus B$ during stage $s + 1$. When N_e is injured we will drop the protection on certain computations. This will be indicated by defining $\vec{b}_{e,s+1} = \langle b_{e,s}(e), \ldots, b_{e,s}(i_0)\rangle$, where the computations (5.1) for $x < b_{e,s}(i_0)$ were not destroyed, and by i-thawing (that is, "un-i-freezing") all followers i-frozen by N_e, for $i > i_0$. Since N_e may be injured infinitely often, it is possible that a follower may be successively i-frozen and i-thawed by N_e infinitely many times. If we were to allow a follower which has reached G_i^0 to be i-frozen, that follower may be permanently held at G_i^0 even though it is not permanently i-frozen: the follower x may be i-frozen at precisely the stages when it is permitted, thus preventing its passage. Therefore, a condition on i-freezing necessary to prevent G_i^0 having infinitely many permanent residents is

(6.4) N_e is allowed to i-freeze the follower x at the end of stage $s + 1$ only if x is above G_i^0 at that time .

(It is this condition which prevents a simple implementation of i-freezing using a restraint function similar to that used to implement the conventional restraint at G_e^1.) The formal definitions of $\vec{b}_{e,s+1}$ and i-freezing follow.

Definition. Let $n = n(e, s)$. Let $G_{i,s+1}^j$ denote the set of followers which are residents of G_i^j at the end of stage $s + 1$. Note that the permitting number $p(x, s + 1)$ assigned to a follower x during stage $s + 2$ may be determined at this time for a follower x in $G_{i,s+1}^j$.

Case 1. N_e is injured during stage $s + 1$. Then $(W_{e_0,s} \oplus B_s)[r(e, s)] \neq (W_{e_0,s+1} \oplus B_{s+1})[r(e, s)]$. Let i_0 be the largest i such that $(W_{e_0,s+1} \oplus B_s)[q(e, i, s)] = (W_{e_0,s+1} \oplus B_{s+1})[q(e, i, s)]$. Define $\vec{b}_{e,s+1} = \vec{b}_{e,s}[i_0] = \langle b_{e,s}(e), \ldots, b_{e,s}(i_0)\rangle$, and i-thaw all followers which are i-frozen by N_e for all $i > i_0$.

Case 2. N_e is not injured during stage $s + 1$. Search for a number $b > b_{e,s}(n)$ such that

(6.5) $$(\forall x < b)[\{e_1\}_s(W_{e_0,s+1} \oplus B_{s+1}; x) = A_{s+1}(x)] ,$$

(6.6) $$(\forall x < r)(\forall i)[(x \in G_{i,s+1}^0 \cup G_{i,s+1}^1 \text{ and } e < i < n + 1) \Rightarrow p(x, s + 1) < b], \text{ and}$$

(6.7) for all i, $e < i < n + 1$, either

(a) $(\exists t < s)[\{\hat{e}\}_{s+1}(A_{s+1}; \langle i, t\rangle)\downarrow$ and $u(\hat{e}, A_{s+1}, \langle i, t\rangle, s + 1) < b]$, or

(b) $(\forall y)[\langle i, y\rangle < r \Rightarrow \langle i, y\rangle$ has been emitted$]$, where $r = \max\{u(e_1, W_{e_0,s+1} \oplus B_{s+1}, y, s+1): y \le b\}$. By the convention on use mentioned in §1, this search may be bounded above by $s + 1$. If no such b exists, define $\vec{b}_{e,s+1} = \vec{b}_{e,s}$. Otherwise, let b_0 be the least such b. Define $\vec{b}_{e,s+1} = \vec{b}_{e,s} * b = \langle b_{e,s}(e), \ldots, b_{e,s}(n), b\rangle$, and $(n + 1)$-freeze for N_e all followers $x \le r = r(e, s + 1)$ which are above G^0_{n+1}. □

Note that conditions (5.4) and (6.4) are automatically satisfied by the definition. It remains to verify the hypotheses of Lemmas 1 and 3 and to show that condition (5.5) is satisfied.

Lemma 4. For all $e \geq 0$, gate G^1_e has only finitely many permanent residents.

Proof. Fix e. Let $b_e(i) = \lim_s b_{e,s}(i)$ and $q(e, i) = \lim_s q(e, i, s)$, if the limits exist. First, suppose there is an i such that $b_e(i)$ does not exist. Let i_0 be the greatest i such that $b_e(i)$ does exist. Since $b_{e,s}(i)\uparrow$ implies $b_{e,s}(j)\uparrow$ for all $j \geq i$, there are infinitely many stages s such that $q(e, i, s) = -1$ for all $i > i_0$. Hence, $\liminf_s r(e, s) = \max\{q(e, i): i \le i_0\}$. The lemma follows immediately.

Now suppose that $b_e(i)$ exists for all i. Then $\{e_1\}(W_{e_0} \oplus B) = A$. We will show that G^1_e has no permanent residents. Assume x is a permanent resident of G^1_e. Let i_0 be the least i such that $x \le q(e, i_0)$, and let s_0 be the least s such that $q(e, i_0, t) = q(e, i_0)$ for all $t \geq s + 1$. (Intuitively x is permanently held at G^0_e by the permanent restraint associated with $b_e(i_0)$ established at stage $s_0 + 1$.) Note that $q(e, i_0, s_0) = -1$, $b_e(i_0) = b_{e,s_0+1}(i_0)$, and $A_{s_0+1}[b_e(i)] = \{e_1\}_{s_0+1}(W_{e_0,s_0+1} \oplus B_{s_0+1})[b_e(i)] = \{e_1\}(W_{e_0} \oplus B)[b_e(i)] = A[b_e(i)]$. Now x must have entered G^1_e after stage $s_0 + 1$, since $q(e, i_0, s_0) = -1$ and x would have moved to G^0_{e-1} during step 3 of stage $s_0 + 1$. Therefore, x must have been in one of the categories (6.1), (6.2) or (6.3) at the end of stage $s_0 + 1$. If x were in category (6.1), x would have been permanently i_0-frozen and so could not be a permanent resident of G^1_e. If x were in category (6.2), x must have been permitted after stage $s_0 + 1$. But then $A_{s_0+1}[p(x, s_0 + 1)] \neq A[p(x, s_0 + 1)]$ and $p(x, s_0 + 1) \le b_e(i)$, a contradiction. Therefore, x must have been in category (6.3). Now $x \le q(e, i_0)$ implies that (6.7b) does not apply. Therefore, there is a $t < s_0$ such that $\{\hat{e}\}_{s_0+1}(A_{s_0+1}; \langle i, t\rangle)\downarrow$ and $u = u(\hat{e}, A_{s_0+1}, \langle i, t\rangle, s_0 + 1) \le b_e(i_0)$. But x is emitted after stage $s_0 + 1$, so $T^{(i)}_{s_0} \neq T^{(i)}$. But then the computation $\{\hat{e}\}_{s_0+1}(A_{s_0+1}; \langle i, t\rangle)\downarrow$ must be incorrect, implying that $A_{s_0+1}[b] \neq A[b]$, a contradiction. Hence, no follower is a permanent resident of G^1_e. □

Lemma 5. For all $i \geq -1$, only finitely many permanent residents of G_i^0 are i-frozen for infinitely many stages.

Proof. Fix i and e, $0 \leq e \leq i$. By the definition of freezing either N_e permanently i-freezes only finitely many followers or it i-thaws the followers it has i-frozen infinitely often. If the former case holds, only finitely many permanent residents of G_i^0 are permanently i-frozen by N_e. If the latter case holds, no permanent resident x of G_i^0 is permanently i-frozen, since x must be i-thawed after it reaches G_i^0 and it cannot be i-frozen thereafter. Since only the requirements N_e, $0 \leq e \leq i$, may i-freeze followers, the lemma is true. □

Lemma 6. For all $i \geq -1$, gate G_i^0 has only finitely many permanent residents.

Proof. Fix i. We may assume that infinitely many followers enter G_i^0; otherwise, there is nothing to prove. Every follower that enters G_e^0 must first reside at G_{i+1}^1, so g_{i+1} is total. By Lemmas 4 and 5, there is an M such that for all $m \geq M$ the follower $g_{i+1}(m)$ eventually leaves G_{i+1}^1 and is eventually permanently i-thawed. Define the recursive function

$$f(m) = \begin{cases} 0 & \text{if } m < M, \\ (\mu s)[g_{i+1}(m) \in G_{i,s}^0 \text{ and is not i-frozen at stage } s] & \text{if } m \geq M. \end{cases}$$

Note that m is the permitting number associated with the follower $g_{i+1}(m)$ while it resides at G_e^0. Since C_A dominates f, there is an M_0 such that $C_A(m) > f(m)$ for all $m \geq M_0$. Then, for all $m \geq M_0$, $A_{f(m)+1}[m] \neq A[m]$ and follower $g_{i+1}(m)$ must be permitted after it has reached G_e^0. □

Lemma 7. For all $e \geq 0$, if $\{e_1\}(W_{e_0} \oplus B) = A$, then $\lim_s b_{e,s}(i)$ exists for all $i \geq e$.

Proof. Fix $e \geq 0$. The proof is by induction on i. The special case $i = e$ is trivial since $b_{e,s}(e) = -1$ for all s. Assume $\lim_s b_{e,s}(i)$ exists. Let F be the set of permanent residents of the gates G_j^k, for $e \leq j \leq i + 1$ and $k = 0, 1$. Let $C = \{j : T^{(j)}$ is finite and $e \leq j \leq i + 1\}$. For each $j \in C$ let t_j be such that $\{\hat{e}\}(A, \langle j, t_j \rangle)\downarrow$. Let s_0 be the least stage s such that

(6.8) $(\forall t \geq s)[b_{e,t}(i) = b_{e,t+1}(i)]$,

(6.9) $(\forall x \in F)$[x has arrived at its permanent residents by stage s], and

(6.10) $(\forall j \in C)[\{\hat{e}\}_s(A_s, \langle j, t_j\rangle)\downarrow$ and $A_s[u] = A[u]$,
where $u = u(\hat{e}, A_s, \langle j, t_j\rangle, s)]$.

Let $p = \max\{p(x, s_0): x \in F\}$, let $u_0 = \max\{u(\hat{e}, A_{s_0}, \langle i, t_i\rangle, s_0): i \in C\}$, and let $b = \max\{p, u_0, b_{e,s_0}(i) + 1\}$. Let s_1 be the least stage $s > s_0$ such that

(6.11) $$\{e_1\}_s(W_{e_0,s} \oplus B_s)[b] = A_s[b] = A[b] .$$

Let $u_1 = \max\{u(e_1, W_{e_0,s_1} \oplus B_{s_1}, y, s_1): y < b\}$ and let $r = \max\{b, u\}$. Let s_2 be the least stage $s > s_1$ such that

(6.12) $(\forall j)(\forall y)[(e < j < i$ and $j \in C$ and $\langle j, y\rangle < r) \Rightarrow$
$\langle j, y\rangle$ has been emitted by stage $s]$ and

(6.13) $$(W_{e_0,s} \oplus B_s)[r] = (W_{e_0} \oplus B)[r] .$$

The existence of s_0, s_1, and s_2 may be proved straightforwardly. Suppose there is a stage $s > s_2$ such that $b_{e,s}(i + 1)\uparrow$. Then b_0 satisfies (6.5), (6.6) and (6.7) at stage $s + 1$, so $b_{e,s+1}(i + 1)\downarrow < b_0$. But then $q(e, i + 1, s + 1) < r$ so, by (6.13), $b_{e,t+1}(i + 1) = b_{e,s+1}(i + 1)$ for all $t > s + 1$. □

<u>Lemma 8</u>. (i) For all $e > 0$, P_e is satisfied,
(ii) R is satisfied, and
(iii) For all $e > 0$, N_e is satisfied.

<u>Proof</u>. (i) follows from Lemmas 1, 4 and 6. (ii) is just a restatement of Lemma 2. (iii) follows from Lemmas 3, 4 and 7. □

This concludes the proof of Theorem 3. □

University of Chicago
Chicago, Illinois 60637

References

[C] S. B. Cooper, Minimal pairs and high recursively enumerable degrees, J. Symbolic Logic 39 (1974), 655-660.

[Fe] P. A. Fejer, The structure of definable subclasses of the recursively enumerable degrees, Ph.D. Dissertation, University of Chicago, 1980.

[FeSo] P. A. Fejer and R. I. Soare, The plus-cupping theorem for the recursively enumerable degrees, these Proceedings.

[Fr] R. M. Friedberg, Two recursively enumerable sets of incomparable degrees of unsolvability, Proc. Natl. Acad. Sciences, U.S.A. 43 (1957), 236-238.

[La1] A. H. Lachlan, Lower bounds for pairs of r.e. degrees, Proc. London Math. Soc. (3) 16 (1966), 537-569.

[La2] A. H. Lachlan, The impossibility of finding relative complements for recursively enumerable degrees, J. Symbolic Logic 31 (1966), 434-454.

[LdSs] R. E. Ladner and L. P. Sasso, The weak truth table degrees of recursively enumerable sets, Ann. Math. Logic 4 (1975), 429-448.

[Le] M. Lerman, Admissible ordinals and priority arguments, Proceedings of the Cambridge Summer School in Logic, 1971, Springer-Verlag Lecture Notes in Math., No. 337, 1973.

[Mi] D. P. Miller, Doctoral Dissertation, University of Chicago, 1981.

[Mu] A. A. Muchnik, On the unsolvability of the problem of reducibility in the theory of algorithms (Russ.), Doklady Academii Nauk SSSR, n.s., 108 (1956), 194-197.

[Ro] R. W. Robinson, A dichotomy of the recursively enumerable sets, Zeitschr. f. Math. Logik und Grundlagen d. Math. 14 (1968), 339-356.

[Rg] H. Rogers, Jr., Theory of recursive functions and effective computability, McGraw-Hill, N.Y., 1967.

[Sa1] G. E. Sacks, Recursive enumerability and the jump operator, Trans. Amer. Math. Soc. 108 (1963), 223-239.

[Sa2] G. E. Sacks, The recursively enumerable degrees are dense, Annals Math. (2) 80 (1964), 300-312.

[Sa3] G. E. Sacks, Degrees of Unsolvability, rev. ed., Annals of Math. Studies, No. 55, Princeton Univ. Press, Princeton, N.J., 1966.

[Sh1] J. R. Shoenfield, Undecidable and creative theories, Fundamenta Mathematicae 49 (1961), 171-179.

[Sh2] J. R. Shoenfield, Applications of model theory to degrees of unsolvability, 359-363, Symposium on the Theory of Models, North Holland, 1965.

[So1] R. I. Soare, The infinite injury priority method, J. Symbolic Logic 41 (1976), 513-530.

[So2] R. I. Soare, Fundamental methods for constructing recursively enumerable degrees, Recursion Theory, Its Generalizations and Applications, Logic Colloquium 79, Leeds, Cambridge University Press, to appear.

[Y] C. E. M. Yates, A minimal pair of r.e. degrees, J. Symbolic Logic 31 (1966), 159-168.

ON THE GRILLIOT-HARRINGTON-MacQUEEN THEOREM

Yiannis N. Moschovakis[(1)]

Department of Mathematics
University of California
Los Angeles, California 90024

One of the finest results in recursion in higher types is the Grilliot-Harrington-MacQueen Theorem, first discovered by Grilliot [1969] (who gave a wrong argument for it) and subsequently proved by Harrington and MacQueen [1976]. If, as usual

$$T_0 = \omega ,$$

$$T_{n+1} = {}^{(T_n)}\omega$$

and ${}^{j+2}E$ is the type-$(j + 2)$ object which embodies quantification over T_j, then the result says the following.

Theorem. If $k \geq j + 3$ and $R(x,\alpha^j)$ is semirecursive in ${}^kE$ with arguments of type $\leq j$, then the relation

$$P(x) \Leftrightarrow (\exists\alpha^j)R(x,\alpha^j)$$

is also semirecursive in ${}^kE$.

The bound $k \geq j + 3$ is best possible by Moschovakis [1967].

Our main purpose here is to give a proof of this result which appears to be new (at least in its details) and which is conceptually more direct than the Harrington-MacQueen argument.

The proof is best presented in an axiomatic setup and we will use the framework of functional induction developed in the first part of Kechris-Moschovakis [1977] which we will cite as KM; we will describe this briefly in §1 and we will assume no more knowledge of abstract recursion theory on the part of the reader.

After putting down the main argument in §2, we will refine it in several ways in §3 to derive some additional interesting selection theorems, including the somewhat surprising fact that on the structure $\langle V_{\omega+\omega}, \in\rangle$ (and others like it), recursion in E coincides with positive elementary induction.

---

(1) During the preparation of this paper the author was partially supported by NSF Grant #MCS 78-02989.

I want to thank Dag Normann with whom I discussed the contents of this paper during his visit to UCLA in March 1980. His stimulating comments made me realize that my old proof of Theorem A could be easily refined to yield the much more general Theorem 3 in section 3.

§1. Preliminaries. Fix an infinite set A such that $\omega \subseteq A$ and for each n, let $P^n(A)$ be the collection of all partial functions on A to A. A (partial, monotone) functional (on A with values in A) is any partial mapping

$$\Phi : A^n \times P^{k_1}(A) \times \cdots \times P^{k_m}(A) \to A$$

such that if $f_1 \subseteq g_1, \ldots, f_m \subseteq g_m$ and $\Phi(\bar{x}, f_1, \ldots, f_m) = w$, then $\Phi(\bar{x}, g_1, \ldots, g_m) = w$. The signature of a functional is the sequence of integers $(n, k_1, \ldots, k_m)$ which describes the kinds of arguments on which Φ acts. If the signature of Φ is of the form $(n, n, k_1, \ldots, k_m)$, we call Φ operative and we define its iterates by the recursion

$$\Phi^\xi(\bar{x}, \bar{g}) = \Phi(\bar{x}, \lambda\bar{x}'\, \Phi^{<\xi}(\bar{x}', \bar{g}), \bar{g}),$$

where

$$\Phi^{<\xi}(\bar{x}, \bar{g}) = w \Leftrightarrow (\exists \eta < \xi)\Phi^\eta(\bar{x}, \bar{g}) = w.$$

The union

$$\Phi^\infty = \bigcup_\xi \Phi^\xi$$

of all the iterates is the fixed point of Φ or the functional defined inductively by Φ.

If $\mathfrak{F}$ is a class of functionals on A, then a functional $\Psi(\bar{x}, \bar{g})$ is $\mathfrak{F}$-recursive if

$$\Psi(\bar{x}, \bar{g}) = \Phi^\infty(\bar{n}, \bar{x}, \bar{g})$$

for some operative Φ in $\mathfrak{F}$ and suitable integers $\bar{n} = n_1, \ldots, n_k$.

This is precisely the approach to abstract recursion theory developed in KM, except that there we dealt exclusively with partial functions and functionals on A to ω, while here it is convenient to allow values into A. The results in §1-§9 of KM go through word-for-word for these more general functionals with only one trivial change: in defining "$\Phi(\bar{x}, \bar{f})$ concentrates on $B \subseteq A$" in 4.1 of KM, we must add the condition

$$[\bar{x} \in B^n \;\&\; \Phi(\bar{x}, \bar{f}) = w] \Rightarrow w \in B.$$

From now on we assume the terminology and results of §1-§9 of KM with only this slight generalization of context.

A second-order structure (with a specified copy of ω) is a system

$$\mathfrak{A} = \langle A, \omega, R_1, \ldots, R_k, f_1, \ldots, f_\ell, c_1, \ldots, c_m, \Phi_1, \ldots, \Phi_n \rangle,$$

where $\omega \subseteq A$ and each R_i is a relation on A, each f_i is a function on A and each Φ_i is a functional on A. Given such a structure, put

$\mathrm{Exp}(\mathfrak{A})$ = the explicitly definable functionals on $\mathfrak{A}$
= the smallest suitable class of functionals on A which contains the characteristic functions $\chi_1,\ldots,\chi_k$ of $R_1,\ldots,R_k$, the functions $f_1,\ldots,f_\ell$, the constant functions $c_1,\ldots,c_m$ of any number of arguments and the functionals $\Phi_1,\ldots,\Phi_n$.

We call a functional Ψ recursive in $\mathfrak{A}$ if it is $\mathrm{Exp}(\mathfrak{A})$-recursive and we call a relation $P \subseteq A^n$ semirecursive in $\mathfrak{A}$ if there is a partial function $f : A^n \to A$, recursive in $\mathfrak{A}$, such that

$$P(\bar{x}) \Leftrightarrow f(\bar{x})\downarrow.$$

We collect these relations into the envelope of $\mathfrak{A}$,

$$\mathrm{Env}(\mathfrak{A}) = \{P \subseteq A^n : P \text{ is semirecursive in } \mathfrak{A}\}.$$

A quantifier-like, type-2 object on A is a functional $F(g)$ (with just one unary function argument, for simplicity), such that:

(i) $F(g)\downarrow \Leftrightarrow g$ is a total function on A.

(ii) If

$$g^*(t) = \begin{cases} 1 & \text{if } g(t) = 1 \\ 0 & \text{if } g(t)\downarrow \text{ and } g(t) \neq 1, \end{cases}$$

Then

$$F(g) = F(g^*).$$

Of course the type-2 object $E = E_A$ that represents quantification on A has these properties, where

$$E_A(g) = \begin{cases} 0 & \text{if } g \text{ is total } \& \ (\exists t)g(t) = 0, \\ 1 & \text{if } g \text{ is total } \& \ (\forall t)g(t) \neq 0. \end{cases}$$

To simplify matters, we will concentrate on good, type-2 structures $\mathfrak{A}$ which satisfy the following conditions.

(1) The equality relation = on A and the quantifier E_A are both recursive in $\mathfrak{A}$.

(2) All the functionals $\Phi_1,\ldots,\Phi_n$ in $\mathfrak{A}$ are quantifier-like, type-2 objects on A.

(3) There is a one-to-one function

$$\tau : A \times A \to A$$

(a pairing function) whose graph is $\mathfrak{A}$-recursive.

By 6.4 and 6.5 of KM, every good, type-2 structure $\mathfrak{A}$ is normal - i.e. all the functionals in $\mathrm{Exp}(\mathfrak{A})$ are normal and the Stage Comparison Theorem holds.

(This result essentially goes back to Moschovakis [1967].)

Kirousis [1978] proves a simple but very useful lemma which in particular implies the following: *if* $\mathfrak{A}$ *satisfies* (1) *and* (2) *above and if* $g : A^n \to A$ *has* $\mathfrak{A}$*-recursive graph, then*

$$\mathrm{Env}(\mathfrak{A}) = \mathrm{Env}(\mathfrak{A}, g).$$

Thus we can add to a good, type-2 structure the pairing function τ, its projection functions π and δ and the identity function id without enlarging the envelope; in the expanded structure $\mathfrak{A}^*$ then, *we have a full* $\mathfrak{A}^*$*-recursive coding scheme* in the sense of 8.2 of KM, since the relevant relations and functions $\mathrm{Seq}(x)$, $\ell h(x)$, $(x)_i$, etc. are easily defined by recursion from ω, the successor and predecessor functions, τ, π, δ and id.

From these remarks, it follows that in proving results about the envelope of a good, type-2 structure $\mathfrak{A}$, we may assume without loss of generality that we have an $\mathfrak{A}$-recursive coding scheme - we will do this without explicit mention.

There are two kinds of structures in which we are particularly interested here, although the selection theorems which we will prove have wider applicability.

Example 1. For each infinite ordinal λ, let V_λ be the set of sets of rank $< \lambda$ and put

$$\mathcal{V}_\lambda = \langle V_\lambda, \in, E\rangle.$$

When $\lambda = \omega + n$ $(n = 0,1,2,\ldots)$, this is essentially normal Kleene-recursion in higher types; i.e. *a relation* $P(x)$ *with arguments of type* $\leq n$ *is* $\mathcal{V}_\lambda$*-recursive exactly when it is Kleene-recursive in* ${}^{n+2}E$. This is tedious to verify, but it is easy and well-known.

It is also not hard to check that these structures $\mathcal{V}_\lambda$ are good, type-2 structures, as are their expansions by any relations, functions, constants and quantifier-like type-2 objects.

Example 2. For each infinite ordinal λ, let

$$\mathcal{J}_\lambda = \langle \lambda, \in, E\rangle.$$

These too are easily good, type-2 structures, as are their expansions by relations, functions, etc.

It is obvious that these examples are the extreme cases of structures of the form

$$\mathfrak{A} = \langle A, \in, E_A\rangle,$$

where A is a transitive set with a (recursive) pair on it.

We will end this preliminary section with a brief description of the *Kleene schemata* associated with recursion in a good, type-2 structure. We will need these for the proofs.

As usual, the idea is to define by recursion a relation

$$\{e\}(\bar{x}) = w,$$

where e varies over ω and $\bar{x} = x_1,\ldots,x_n$ and w vary over A, such that a partial function $f : A^n \to A$ is $\mathfrak{A}$-recursive exactly when for some $e \in \omega$,

$$f(\bar{x}) = \{e\}(\bar{x}).$$

Since the definition is by recursion, it also supplies us with an ordinal

$$|e,\bar{x}| = \text{the "length of the computation } \{e\}(\bar{x})\text{"}$$

whenever $\{e\}(\bar{x})\downarrow$.

If $\{e\}(\bar{x})$ is to be defined, then the argument e (the <u>code</u> or <u>index</u>) must satisfy one of a finite number of conditions (<u>schemata</u>); when e satisfies one of these conditions, we can extract from it recursively instructions for computing $\{e\}(\bar{x})$, either immediately or in terms of shorter computations.

This method of describing a given recursion theory by schemata is well-known and it is explained quite extensively in §14 and §15 of KM, so we will not elaborate on it further here. In fact, for once, we will not put down the schemata explicitly; it will suffice for our purposes to describe their nature and to group them according to their role.

<u>Group</u> A. <u>Explicitly given functions</u>. In these schemata we assign codes to several functions that we want to call recursive, directly.

For example, we may set

$$\{e\}(x_1,\ldots,x_n) = f_i(x_1,\ldots,x_n) \qquad e = \langle 2,n,i\rangle$$

if f_i is one of the functions in $\mathfrak{A}$. Here $e = \langle 2,n,i\rangle$ codes the information that we are in schema 2, that we are defining a function of n arguments and that f_i is the i'th function in $\mathfrak{A}$.

In this group we code the characteristic functions of the relations in $\mathfrak{A}$ and the functions and constants in $\mathfrak{A}$; we include the characteristic function of ω, the successor function and the characteristic function of $=$.

<u>Group</u> B. <u>Compositions with total functions</u>. There are three schemata in this group, <u>addition of variables</u>, <u>permutation of variables</u> and <u>definition by cases</u>, where (for example) the last of these is as follows, if this happens to be the b'th schema.

$$\{e\}(t,\bar{x}) = \begin{cases} \{m\}(\bar{x}) & \text{if } t = 0, \\ \{z\}(t,\bar{x}) & \text{if } t \neq 0 \end{cases} \qquad e = \langle b,n+1,m,z\rangle$$

It is well-known and easy to check that if we include this schema, we do not need to include definition by primitive recursion.

Group C. Functional substitution. These are schemata of the form

$$\{e\}(\bar{x}) = f(\lambda s\{m\}(\bar{x},s)) \qquad e = \langle c,n,i\rangle,$$

one for each quantifier-like functional in $\mathfrak{A}$. (The code $\langle c,n,i\rangle$ means that we are in schema c involving n variables and the i'th functional.) To compute $\{e\}(\bar{x})$ we must first compute all the values of

$$g_{\bar{x}}(s) = \{m\}(\bar{x},s),$$

and if this $g_{\bar{x}}$ is total, then set

$$\{e\}(\bar{x}) = f(g_{\bar{x}}).$$

Group D. Composition. This is the schema

$$\{e\}(\bar{x}) = \{m\}(\{z\}(\bar{x}),\bar{x}) \qquad e = \langle d,n,m,z\rangle,$$

assuming it is the d'th schema; to compute $\{e\}(\bar{x})$, we must first compute $\{z\}(\bar{x})$, and if this converges and has the value w, then we must compute $\{m\}(w,\bar{x})$.

Group E. Enumeration. This is the schema

$$\{e\}(m,\bar{x},\bar{y}) = \{m\}(\bar{x}) \qquad e = \langle f,n+k+1,n\rangle,$$

assuming it is the f'th schema and $\bar{x} = x_1,\ldots,x_n$, $\bar{y} = y_1,\ldots,y_k$ range over n-tuples and k-tuples respectively.

The fact that one can put down schemata for $\mathfrak{A}$-recursion that conform to these general conditions is not trivial, but can be established by a standard "index-transfer theorem," as in §15 of KM. We will omit putting down any details.

§2. The basic construction. We are now ready to state and prove the Grilliot-Harrington-MacQueen theorem. As usual, we let

$${}^{B}A = \text{all unary (total) functions on } B \text{ to } A$$

and if $X : A \times B \to C$ and $a \in A$, we define the fiber $X_a : B \to C$ by

$$X_a(t) = X(a,t).$$

Theorem A (Harrington-MacQueen [1976]). Suppose $\mathfrak{A}$ is a good, type-2 structure and B is an $\mathfrak{A}$-recursive subset of the domain A of $\mathfrak{A}$ such that the following two conditions hold:

(H1) There is a total function

$$\chi : A \times B \to A$$

with $\mathfrak{A}$-recursive graph, such that

$$^{B}A = \{\chi_a : a \in A\}.$$

(H2) There is an $\mathfrak{A}$-recursive prewellordering of A which is longer than every prewellordering of B.

Then $\mathrm{Env}(\mathfrak{A})$ is closed under existential quantification on B, $\exists^B$. ⊣

We will prove this (as in the Harrington-MacQueen argument) by showing first a selection lemma about computations, from which Theorem A follows very easily.

<u>Lemma</u> A.1. Assume the same hypotheses as in Theorem A; then there exists an $\mathfrak{A}$-recursive partial function $u(c,\bar{x},t)$, such that

$$\begin{aligned}(\exists t \in B)\{e\}(t,\bar{x})\downarrow \Rightarrow & (\forall t \in B)u(e,\bar{x},t)\downarrow \\ & \& \ (\exists t \in B)u(e,\bar{x},t) = 0 \\ & \& \ (\forall t \in B)[u(e,\bar{x},t) = 0 \Rightarrow \{e\}(t,\bar{x})\downarrow].\end{aligned}$$

<u>Proof of Theorem</u> A <u>from Lemma</u> A.1. Given an $\mathfrak{A}$-semirecursive relation

$$P(t,\bar{x}) \Leftrightarrow \{e\}(t,\bar{x})\downarrow,$$

verify that with the u of the lemma,

$$\begin{aligned}(\exists t \in B)P(t,\bar{x}) \Leftrightarrow & (\forall t \in B)u(e,\bar{x},t)\downarrow \\ & \& \ (\exists t \in B)u(e,\bar{x},t) = 0 \\ & \& \ (\forall t \in B)[u(e,\bar{x},t) = 0 \Rightarrow \{e\}(t,\bar{x})\downarrow];\end{aligned}$$

but this equivalence and the fact that E_A is $\mathfrak{A}$-recursive imply immediately that the relation

$$R(\bar{x}) \Leftrightarrow (\exists t \in B)P(t,\bar{x})$$

is $\mathfrak{A}$-semirecursive. ⊣

We would naturally like to define $u(e,\bar{x},t)$ by (effective) transfinite recursion on the ordinal $\mathrm{infimum}\{|e,t,x| : \{e\}(t,x)\downarrow \ \& \ t \in B\}$. To do this, however, we must reformulate the lemma so that we have a stronger and easier to use induction hypothesis. To put it briefly, Lemma A.1 deals with existential assertions of the form

$$(\exists t \in B)\{e\}(t,\bar{x})\downarrow;$$

we will have to deal with more complicated assertions of the form

$$(\exists t \in D \cap B)\{e\}(\varphi_1(t,\bar{y}),\ldots,\varphi_n(t,\bar{y}))\downarrow,$$

where D is a (suitably coded) subset of A, recursive in given parameters and $\varphi_1,\ldots,\varphi_n$ are given recursive functions.

To help simplify the notation, for each $m \in \omega$ and $\bar{z} = z_1,\ldots,z_\ell$, put

$$B(m,\bar{z}) = \{t \in B : \{m\}(t,\bar{z}) = 0\};$$

we will use these sets only in the case $\{m\}(t,z)\downarrow$ for each $t \in B$.

Given numbers $\hat{\varphi}_1,\ldots,\hat{\varphi}_n$, and $\bar{y} = y_1,\ldots,y_k \in A^k$, put

$$\varphi_i(t,\bar{y}) = \{\hat{\varphi}_i\}(t,\bar{y}) \qquad (i = 1,\ldots,n)$$

and collect these functions together into an n-tuple-valued function

$$\bar{\varphi}(t,\bar{y}) = (\varphi_1(t_1,\bar{y}),\ldots,\varphi_n(t,\bar{y}));$$

again, we will only be interested in the case each $\varphi_i(t,\bar{y})$ is defined for all $t \in B(m,\bar{z})$.

Now, we say that the tuple of parameters

$$m,\bar{z},\hat{\varphi} = \langle\hat{\varphi}_1,\ldots,\hat{\varphi}_n\rangle,\bar{y}$$

is <u>good</u> (for B) if for each $t \in B$, $\{m\}(t,\bar{z})\downarrow$ and if for each $t \in B(m,\bar{z})$, each $\varphi_i(t,\bar{y})$ is defined.

With these notation conventions we can now state the detailed generalization of Lemma A.1 which is easily amenable to proof. From the statement below we can get Lemma A.1 by choosing

$$B(m,\bar{z}) = B;\ \varphi_1(t,\bar{x}) = t;\ \varphi_2(t,\bar{x}) = x_1,\ldots,\varphi_{n+1}(t,\bar{x}) = x_n .$$

<u>Main Lemma</u>. As in Theorem A, assume that $\mathfrak{A}$ is a good, type-2 structure, B is a recursive subset of A and (H1), (H2) hold. Then, there is an $\mathfrak{A}$-recursive partial function

$$u(e,m,\hat{\varphi},z,y,t)$$

such that in the notation established above, if the tuple m, $\bar{z}$, $\hat{\varphi}$, $\bar{y}$ is good and

$$(\exists t \in B(m,\bar{z}))\{e\}(\bar{\varphi}(t,\bar{y}))\downarrow,$$

then

(i) $(\forall t \in B(m,z))u(e,m,\hat{\varphi},\langle\bar{z}\rangle,\langle\bar{y}\rangle,t)\downarrow$

and the set

$$C = C(e,m,\hat{\varphi},\bar{z},\bar{y}) = \{t \in B(m,\bar{z}) : u(e,m,\hat{\varphi},\langle\bar{z}\rangle,\langle\bar{y}\rangle,t) = 0\}$$

has the following additional properties:

(ii) $C \neq \emptyset$,

(iii) $t \in C \Rightarrow \{e\}(\overline{\varphi}(t,\overline{y}))\downarrow$.

<u>Proof</u>. By Kirousis' Lemma, we may assume that the function χ witnessing the truth of (H1) is $\mathfrak{A}$-recursive.

As usual, we will define u by the recursion theorem, i.e. we will define a recursive partial function

$$u(\hat{u},e,m,\hat{\varphi},z,y,t)$$

where $\hat{u}$ ranges over ω, we will choose some fixed $\hat{u}$ so that

$$u(\hat{u},e,m,\hat{\varphi},z,y,t) = \{\hat{u}\}(e,m,\hat{\varphi},z,y,t)$$

and we will then set

$$u(e,m,\hat{\varphi},z,y,t) = u(\hat{u},e,m,\hat{\varphi},z,y,t)$$

$$= \{\hat{u}\}(e,m,\hat{\varphi},z,y,t) \ .$$

After u is defined, the verification of (i)-(iii) will be by induction on

$$\text{infimum}\{|e,\varphi_1(t,\overline{y}),\ldots,\varphi_n(t,\overline{y})| : t \in B(m,\overline{z}) \ \& \ \{e\}(\overline{\varphi}(t,\overline{y}))\downarrow\}.$$

The definition of u is by cases on e, according to which recursive condition e satisfies as a code for a computation. It is important in this kind of definition to make sure that no induction hypotheses on the ordinals involved are essential in the definition of u before the recursion theorem is applied; nevertheless, we will follow well-established tradition and explain the proof <u>as if</u> we were defining $u(e,m,\hat{\varphi},z,y,t)$ by a direct, transfinite recursion, with an available induction hypothesis in each case. After this first informal definition of u in each case, we will make some remarks (when needed) to explain why no induction hypothesis was <u>actually used</u> in the definition.

In point of fact, we will avoid direct references to the partial function u in the informal explanations below in favor of the sets $C(e,m,\hat{\varphi},\overline{z},\overline{y})$ in the statement of the lemma which are what we really need. Of course u is nothing but the characteristic function of C on $B(m,\overline{z})$.

We now define C by cases on e, assuming that m, $\overline{z}$, $\hat{\varphi}$, $\overline{y}$ is a good sequence.

<u>Group</u> A. Now e defines a total function, so we can certainly take $C = B(m,\overline{z})$.

<u>Group</u> B. Here the definition of $\{e\}(\overline{\varphi}(t,\overline{y}))$ depends only on the convergence of certain fixed subcomputations and the definition of C is immediate. For example, if

$$\{e\}(\varphi_1(t,\bar{y}),\varphi_2(t,\bar{y}),\ldots,\varphi_n(t,\bar{y})) = \begin{cases} \{e_1\}(\varphi_2(t,\bar{y}),\ldots,\varphi_n(t,\bar{y})), & \text{if } \varphi_1(t,\bar{y}) = 0, \\ \{e_2\}(\varphi_1(t,\bar{y}),\ldots,\varphi_n(t,\bar{y})), & \text{otherwise,} \end{cases}$$

we put

$$C(e,m,\bar{z},\hat{\varphi},\bar{y}) = \{t \in B(m,\bar{z}) : [\varphi_1(t,\bar{y}) = 0 \;\&\; t \in C(e_1,m,\bar{z},\hat{\varphi}_2,\ldots,\hat{\varphi}_n,\bar{y})] \vee [\varphi_1(t,\bar{y}) \neq 0 \;\&\; t \in C(e_2,m,\bar{z},\hat{\varphi},\bar{y})]\}.$$

<u>Group</u> C, <u>functional</u> <u>substitution</u>. In this first interesting case, we have

$$\{e\}(\bar{\varphi}(t,\bar{y})) = F(\lambda s\{e_1\}(\bar{\varphi}(t,\bar{y}),s))$$

for some type-2 object F and some $e_1 \in \omega$. Computing from the hypothesis,

$$(*) \qquad (\exists t \in B(m,\bar{z}))[\{e\}(\bar{\varphi}(t,\bar{y}))\downarrow]$$
$$\Leftrightarrow (\exists t \in B(m,\bar{z}))(\forall s \in A)[\{e_1\}(\bar{\varphi}(t,\bar{y}),s)\downarrow]$$
$$\Leftrightarrow (\forall \pi : B(m,\bar{z}) \to A)(\exists t \in B(m,\bar{z}))[\{e_1\}(\bar{\varphi}(t,\bar{y}),\pi(t))\downarrow],$$

where we have used the dual of the axiom of choice. Using the hypothesis (H1), we can further simplify this to

$$\Leftrightarrow (\forall a \in A)(\exists t \in B(m,\bar{z}))[\{e_1\}(\bar{\varphi}(t,\bar{y}),\chi(a,t))\downarrow].$$

It is now clear from the general properties of the schemata that if one (and hence all) these conditions hold, then for each $a \in A$,

$$\text{infimum}\{|e_1,\bar{\varphi}(t,\bar{y}),\chi(a,t)| : t \in B(m,\bar{z}) \;\&\; \{e_1\}(\bar{\varphi}(t,\bar{y}),\chi(a,t))\downarrow\}$$
$$< \text{infimum}\{|e,\bar{\varphi}(t,\bar{y})| : t \in B(m,\bar{z}) \;\&\; \{e\}(\varphi(t,\bar{y}))\downarrow\},$$

so that if we have the induction hypothesis available, then for each $a \in A$ we have already defined a non-empty subset of $B(m,\bar{z})$

$$C(a) = C(e_1,m,\bar{z},\hat{\psi},\bar{y},a)$$

such that

$$t \in C(a) \Rightarrow \{e_1\}(\bar{\varphi}(t,\bar{y}),\chi(a,t))\downarrow;$$

here $\hat{\psi} = \langle\hat{\psi}_1,\ldots,\hat{\psi}_n,\hat{\psi}_{n+1}\rangle$, where

$$\psi_i(t,\bar{y},a) = \varphi_i(t,\bar{y}) \quad \text{for } i = 1,\ldots,n,$$
$$\psi_{n+1}(t,\bar{y},a) = \chi(a,t).$$

We put

$$C = \{t \in B(m,\bar{z}) : (\forall s \in A)(\exists a \in A)[s = \chi(a,t) \;\&\; t \in C(a)]\}.$$

To verify that C has the properties we want, assume first that $t \in C$ and s is any member of A; now choose $a \in A$ so that

$$s = X(a,t) \;\&\; t \in C(a),$$

so that by the property of $C(a)$,

$$\{e_1\}(\overline{\varphi}(t,\overline{y}), X(a,t)) = \{e_1\}(\overline{\varphi}(t,\overline{y}), s)\downarrow.$$

At the same time $C \neq \emptyset$, because if C were empty, then

$$\begin{aligned}(**)\quad & (\forall t \in B(m,\overline{z}))t \notin C \\ & \Rightarrow (\forall t \in B(m,\overline{z}))(\exists s \in A)(\forall a \in A)[s \neq X(a,t) \vee t \notin C(a)] \\ & \Rightarrow (\exists b \in A)(\forall t \in B(m,\overline{z}))(\forall a \in A)[X(b,t) \neq X(a,t) \vee t \notin C(a)] \\ & \quad\text{(by the axiom of choice and (H1))} \\ & \Rightarrow (\exists b \in A)(\forall t \in B(m,\overline{z}))[t \notin C(b)]\end{aligned}$$

which contradicts the assumed properties of each $C(b)$.

In this case it is very easy to verify that the formal definition of $\mathcal{U}(\hat{u},e,m,\hat{\varphi},z,y,t)$ which will yield this $C(e,m,\hat{\varphi},\overline{z},\overline{y})$ when the hypotheses hold is $\mathfrak{A}$-recursive. (The key fact of course is that E_A is $\mathfrak{A}$-recursive.)

Group D, composition. Now we have

$$\{e\}(\overline{\varphi}(t,\overline{y})) = \{e_1\}(\{e_2\}(\overline{\varphi}(t,\overline{y})), \overline{\varphi}(t,\overline{y})),$$

so that if

$$\text{infimum}\{|e,\overline{\varphi}(t,\overline{y})| : t \in B(m,\overline{z}) \;\&\; \{e\}(\overline{\varphi}(t,\overline{y}))\downarrow\} = \lambda,$$

then for some $t \in B(m,\overline{z})$ both $\{e_2\}(\overline{\varphi}(t,\overline{y})) = w$ and $\{e_2\}(w,\overline{\varphi}(t,\overline{y}))$ are defined with ordinal below λ and we may assume (at least heuristically) that we have defined C for these computations.

Let κ be the rank of the given $\mathfrak{A}$-recursive prewellordering of A which witnesses the truth of (H2). Keeping the parameters e, m, $\hat{\varphi}$, $\overline{z}$, $\overline{y}$ constant for a while (and suppressing them in the notation) we will define a sequence of sets

$$\{C_\xi : \xi < \kappa\}$$

such that for each ξ,

$$\emptyset \subsetneq C_\xi \subseteq B(m,\overline{z}), \qquad t \in C_\xi \Rightarrow \{e_2\}(\overline{\varphi}(t,\overline{y}))\downarrow.$$

The definition will assume the induction hypothesis and a few other things and will need to be made precise later on.

Step 0. Find C_0, using the induction hypothesis, so that

$$t \in C_0 \Rightarrow \{e_2\}(\overline{\varphi}(t,\overline{y}))\downarrow.$$

This is clearly possible, granting that

$$(\exists t \in B(m,\overline{z}))\{e\}(\overline{\varphi}(t,\overline{y}))\downarrow.$$

(This step can be omitted, and we will omit it in the precise definition below.)

Step $\xi \geq 0$. First set

$$D_\xi = \cup_{\eta<\xi} C_\eta$$

and then using the induction hypothesis and the stage comparison theorem do simultaneously two computations:

(a) Ask if $(\exists t \in D_\xi)\{e_1\}(\{e_2\}(\overline{\varphi}(t,\overline{y})),\overline{\varphi}(t,\overline{y}))\downarrow$.

(b) Ask if $(\exists t \in B(m,\overline{z}) - D_\xi)\{e_2\}(\overline{\varphi}(t,\overline{y}))\downarrow$.

Now by hypothesis, for some $t \in B(m,\overline{z})$ we have both

$$\{e_2\}(\overline{\varphi}(t,\overline{y}))\downarrow \ \& \ \{e_1\}(\{e_2\}(\overline{\varphi}(t,\overline{y})),\overline{\varphi}(t,\overline{y}))\downarrow;$$

if $t \in D_\xi$, then (a) holds while if $t \in B(m,\overline{z}) - D_\xi$, then (b) holds. In either case one of (a) or (b) holds (perhaps both), so one computation converges first.

If (a) terminates first, put $C_\xi = D_\xi$.

If (b) terminates first, use the induction hypothesis to find C_ξ such that

$$\emptyset \subsetneq C_\xi \subseteq B(m,\overline{z}) - D_\xi$$

and

$$t \in C_\xi \Rightarrow \{e_2\}(\overline{\varphi}(t,\overline{y}))\downarrow.$$

In a minute we will check that this is a legitimate definiton by recursion on $\xi < \kappa$. Granting that it is, we have the sequence

$$\{C_\xi : \xi < \kappa\}$$

and it cannot be the case that for each $\xi < \kappa$ case (b) of the definition applies, since then all the C_ξ would be pairwise disjoint and the relation on $B(m,\overline{z})$

$$x < y \Leftrightarrow (\exists\xi < \kappa)[x \in C_\xi \ \& \ (\forall\eta \leq \xi)y \notin C_\eta]$$

would be a prewellordering of this set of length κ, contradicting (H2).

Now set

$$C^* = \cup_{\xi<\kappa} C_\xi$$

and notice that by the construction, we know that

$$t \in C^* \Rightarrow \{e_2\}(\overline{\varphi}(t,\overline{y}))\downarrow,$$

and

$$(\exists t \in C^*)\{e_1\}(\{e_2\}(\overline{\varphi}(t,\overline{y})),\overline{\varphi}(t,\overline{y}))\downarrow.$$

Thus we can apply the induction hypothesis, replacing $B(m,\overline{z})$ by C^* and adding to the sequence of functions $\overline{\varphi}$, at the front, the function

$$\varphi_0(t,\overline{y}) = \{e_2\}(\overline{\varphi}(t,\overline{y})),$$

to find some set

$$C = C(e,m,\overline{z},\hat{\varphi},\overline{y})$$

with the required properties.

Since this is the most complicated case in the proof, we will take the space to outline the formal construction, as we need it before we apply the recursion theorem. (Those experienced in these arguments will surely want to skip to "Group E.")

Recall that what we are really doing, is defining an $\mathfrak{A}$-recursive partial function

$$u(\hat{u},e,m,\hat{\varphi},z,y,t)$$

by cases on e, and in this case e codes a composition, so we can get numbers e_1 and e_2 from it recursively.

We will first define an auxiliary $\mathfrak{A}$-recursive partial function

$$g(\xi,t) = g(\hat{u},e,m,y,z,\xi,t),$$

where ξ varies over A and denotes (heuristically) the ordinal which is its rank in the given prewellordering $<$ of A that witnesses (H2). When the hypotheses hold, then in the notation above (and identifying ξ with its rank), we will have

$$C_\xi = \{t \in B(m,\overline{z}) : g(\xi,t) = 0\}.$$

The partial function g will also be defined by the recursion theorem, i.e. we will put

$$\begin{aligned} g(\hat{u},e,m,y,z,\xi,t) &= \psi(\hat{g},\hat{u},e,m,y,z,\xi,t) \\ &= \{\hat{g}\}(\hat{u},e,m,y,z,\xi,t) \end{aligned}$$

for some $\mathfrak{A}$-recursive ψ and some $\hat{g} \in \omega$ which satisfies this equation. The partial function ψ is defined by a standard index-construction which we now outline; the main complication comes fromthe fact that we must switch parameters in applying the induction hypothesis.

In the formulas below, $<$ is the given $\mathfrak{A}$-recursive prewellordering of A

which witnesses (H2).

Step 1. Put

$$h_d(t,\hat{g},\hat{u},e,m,y,z,\xi) = \begin{cases} 0 & \text{if } \{m\}(t,\bar{z}) = 0 \\ & \quad \& \ (\forall\eta < \xi)\{\hat{g}\}(\hat{u},e,m,y,z,\eta,t)\downarrow \\ & \quad \& \ (\exists\eta < \xi)[\{\hat{g}\}(\hat{u},e,m,y,z,\eta,t) = 0], \\ 1 & \text{if } \{m\}(t,\bar{z}) = 0 \\ & \quad \& \ (\forall\eta < \xi)[\{\hat{g}\}(\hat{u},e,m,y,z,\eta,t)\downarrow \ \& \neq 0], \end{cases}$$

where we have omitted trivial conditions like

"z must code a tuple $\bar{z}$".

It is clear that if $\hat{h}_d$ is a code of h_d and the induction hypotheses hold, then we will have

$$B(\hat{h}_d,\hat{g},\hat{u},e,m,\bar{y},\bar{z},\xi) = D_\xi;$$

we will need to use the parameters $\hat{g}$, $\hat{u}$, e, m, $\bar{y}$, $\bar{z}$, ξ instead of just $\bar{z}$.

Step 2. Put

$$f_d(t,\hat{g},\hat{u},e,m,y,z,\xi) = \begin{cases} 0 & \text{if } \{m\}(t,\bar{z}) = 0 \ \& \ h_d(t,\hat{g},\hat{u},e,m,y,z,\xi) = 1, \\ 1 & \text{if } \{m\}(t,\bar{z}) = 0 \ \& \ h_d(t,\hat{g},\hat{u},e,m,y,z,\xi) = 0, \end{cases}$$

and let $\hat{f}_d$ be a code of f_d; now $\hat{f}_d$ codes $B(m,\bar{z}) - D_\xi$ in terms of the parameters.

Step 3. Let $\varphi_0(e_2,\hat{\varphi})$ be some fixed recursive function such that whenever $\hat{\varphi} = \langle\hat{\varphi}_1,\ldots,\hat{\varphi}_n\rangle$ is a sequence code, we have

$$\{\varphi_0(e_2,\hat{\varphi})\}(t,\bar{y}) = \{e_2\}(\varphi_1(t,\bar{y}),\ldots,\varphi_n(t,\bar{y})).$$

Now in the two cases (a) and (b) we want to use the induction hypothesis on the following two computations:

(a) $\{\hat{u}\}(e_1,\hat{h}_d,\langle\varphi_0(e_2,\hat{\varphi}),\hat{\varphi}_1,\ldots,\hat{\varphi}_n\rangle,\langle\hat{g},\hat{u},e,m,y,z,\xi\rangle,y,t)$

(b) $\{\hat{u}\}(e_2,\hat{f}_d,\hat{\varphi},\langle\hat{g},\hat{u},e,m,y,z,\xi\rangle,y,t)$

As in the proof of Theorem A from Lemma A.1, we want to execute these computations for all $t \in D_\xi$ or $B(m,\bar{z}) - D_\xi$ and ask if we get some correct values - and then we want to pick by stage-comparison one of these, where the verification that it works is shorter.

Step 4. Using fresh variables to avoid confusion, put

$$\omega(\hat{u}',e',m',\hat{\varphi}',z',y')\downarrow \Leftrightarrow [\forall t \in B(m,\bar{z}')][\{\hat{u}'\}(e',m',\hat{\varphi}',z',y',t)\downarrow]$$
$$\& \ [\exists t \in B(m,\bar{z}')][\{\hat{u}'\}(e',m',\hat{\varphi}',z',y',t) = 0]$$

and let $\hat{\omega}$ be a code of ω.

To determine whether we are in Case (a) or in Case (b), we must compare the ordinals

$$|\hat{\omega},\hat{u},e',m',\hat{\varphi}',z',y'|,$$

for the parameters indicated in (a) and (b) above. Thus the final step in defining the function ψ is as follows:

Step 5. Case (a). If

$$|\hat{\omega},\hat{u},e_1,\hat{h}_d,\langle\varphi_0(e_2,\hat{\varphi}),\hat{\varphi}_1,\ldots,\hat{\varphi}_n\rangle,\langle\hat{g},\hat{u},e,m,y,z,\xi\rangle,y|$$
$$\leq |\hat{\omega},\hat{u},e_2,\hat{f}_d,\hat{\varphi},\langle\hat{g},\hat{u},e,m,y,z,\xi\rangle,y|,$$

put

$$\psi(\hat{g},\hat{u},e,m,y,z,\xi,t) \simeq h_d(t,\hat{g},\hat{u},e,m,y,z,\xi).$$

Case (b). If the ordinals in Case (a) are ordered inversely, put

$$\psi(\hat{g},\hat{u},e,m,y,z,\xi,t) = \{\hat{u}\}(e_2,\hat{f}_d,\hat{\varphi},\langle\hat{g},\hat{u},e,m,y,z,\xi\rangle,y,t).$$

Of course we are comparing ordinals using the Stage Comparison Theorem and if one of the two computations in question converges, then we will get an answer.

Having defined ψ, we get a partial recursive function g with code $\hat{g}$ by the recursion theorem and then the remaining construction to define $u(\hat{u},e,m,\hat{\varphi},z,y,t)$ in this case is simple and we will omit it.

After $\hat{u}$ has been fixed by the recursion theorem again, we will be proving that it works by induction on the ordinal

$$\text{infimum}\{|e,\varphi_1(t,\bar{y}),\ldots,\varphi_n(t,\bar{y})| : t \in B(m,\bar{z}) \ \& \ \{e\}(\varphi_1(t,\bar{y}),\ldots,\varphi_n(t,\bar{y}))\downarrow\},$$

simultaneously for all values of the parameters. When e is a code for composition, in this case, we will need to check first that the auxiliary function

$$\{\hat{g}\}(\hat{u},e,m,y,z,\xi,t)$$

is defined for each ξ and each $t \in B(m,\bar{z})$ (by induction on ξ), so that each C_ξ is well-defined and the informal argument we gave in the beginning of this case works.

Group E, enumeration. In this case we have

$$\{e\}(\varphi_1(t,\bar{y}),\ldots,\varphi_n(t,\bar{y})) = \{\varphi_1(t,\bar{y})\}(\varphi_2(t,\bar{y}),\ldots,\varphi_k(t,\bar{y})),$$

for some $k \leq n$ recursively computable from e. (This function is only defined when $\varphi_1(t,\bar{y}) \in \omega$.) Computing as usual,

$$(\exists t \in B(m,\bar{z}))\{e\}(\varphi_1(t,\bar{y}),\ldots,\varphi_n(t,\bar{y}))\downarrow$$
$$\Rightarrow (\exists t \in B(m,\bar{z}))\{\varphi_1(t,\bar{y})\}(\varphi_2(t,\bar{y}),\ldots,\varphi_k(t,\bar{y}))\downarrow$$
$$\Rightarrow (\exists j)[j = \varphi_1(t,\bar{y}) \ \& \ (\exists t \in B(m,\bar{z}))\{j\}(\varphi_2(t,\bar{y}),\ldots,\varphi_k(t,\bar{y}))\downarrow].$$

Now for each $j \in \omega$, we can use the induction hypothesis and ask if

$$(*) \qquad [\exists t \in B(m,\bar{z}) \cap \{t : \varphi_1(t,\bar{y}) = j\}\{j\}(\varphi_2(t,\bar{y}),\ldots,\varphi_k(t,\bar{y}))\downarrow \ ;$$

we know that for at least one j this can be verified using computations shorter than

$$\text{infimum}\{|e,\bar{\varphi}(t,\bar{y})| : t \in B(m,\bar{z})\}.$$

Using Gandy-selection on ω, we can find one j for which $(*)$ holds recursively in the parameters and then plug it in $\{\hat{u}\}(j,\_\_)$ to define $\mathcal{U}(\hat{u},e,m,\hat{\varphi},z,y,t)$ in this case.

We will omit the formal details in this case which are somewhat simpler but similar to those of the composition case. (The complication comes from the fact that we must apply the induction hypothesis to a different $B(m,\bar{z})$ which is defined in terms of the longer list of parameters m, $\bar{z}$, $\hat{\varphi}_1$, $\bar{y}$, j, so that an index-construction is required.)

Treatment of this case completes the proof of the Main Lemma and hence the proof of Theorem A via Lemma A.1. ⊣

§3. Corollaries and refinements. Many of the selection theorems in which we are interested do not follow directly from Theorem A, but require going into - and sometimes refining - the proof of the Main Lemma.

First we look at the structures

$$\mathcal{V}_\lambda = \langle V_\lambda, \in, E\rangle$$

of Example 1.

Theorem 1. If λ is an infinite successor ordinal or an inaccessible cardinal, and if $\mathfrak{A}$ is any expansion of $\mathcal{V}_\lambda$ by relations, functions, constants and quantifier-like, type-2 objects, then $\mathrm{Env}(\mathfrak{A})$ is closed under restricted quantification.

Proof. The structure $\mathfrak{A} = \langle V_\lambda, \in, E, \_\_\rangle$ is a good, type-2 structure and so is the further expansion

$$\mathfrak{A}(B) = (\mathfrak{A}, \chi_B),$$

where we have added the characteristic function χ_B of any $B \in V_\lambda$. Moreover, B is a recursive subset of $\mathfrak{A}(B)$ and hypotheses (H1) and (H2) of Theorem A hold (easily), so that $\mathrm{Env}(\mathfrak{A}(B))$ is closed under $(\exists t \in B)$ - and it is also (trivially) closed under $(\forall t \in B)$.

The further assertion, that $\mathrm{Env}(\mathfrak{A})$ is closed under $(\exists t \in w)$ and $(\forall t \in w)$ follows from the observation that the proof of the main lemma is <u>uniform in</u> B - i.e. the choice function which we constructed in the proof depends recursively on the characteristic function of B and the obvious data that witness the truth of (H1) and (H2). We will omit the routine details. $\dashv$

This is the most elegant version of the Grilliot-Harrington-MacQueen Theorem and of course it includes the classical case when $\lambda = \omega + n$.

The next result implies that in many cases, $\mathrm{Env}(\mathcal{V}_\lambda)$ is closed under $\exists$, so that by 9A.3 of Moschovakis [1974] we have

$$\mathrm{Env}(\mathcal{V}_\lambda) = \mathrm{IND}(\mathcal{V}_\lambda)$$

= all relations on V_λ which are positive, elementary, absolutely inductive in $\langle V_\lambda, \in\rangle$.

<u>Theorem 2</u>. Let $\mathfrak{A} = \langle V_\lambda, \in, E, \_\rangle$ be an expansion of $\mathcal{V}_\lambda$ by relations, functions, constants and quantifier-like, type-2 objects, and suppose that λ is a limit ordinal and $\mu < \lambda$ is such that the constant function $t \mapsto \mu$ is $\mathfrak{A}$-recursive and there exists a function

$$\pi : \mu \to \lambda$$

which is cofinal in λ and has $\mathfrak{A}$-recursive graph. If $\mathrm{Env}(\mathfrak{A})$ is closed under $(\exists \xi < \mu)$, then $\mathrm{Env}(\mathfrak{A})$ is closed under $\exists$.

<u>Proof</u>. The function

$$\xi \mapsto V_\xi$$

clearly has $\mathfrak{A}$-recursive graph, so under the hypothesis, it is enough to prove that $\mathrm{Env}(\mathfrak{A})$ is closed under <u>restricted existential quantification</u>; because

$$\begin{aligned}(\exists x)P(y,x) &\Leftrightarrow (\exists \xi < \mu)(\exists x \in V_{\pi(\xi)})P(y,x)\\ &\Leftrightarrow (\exists \xi < \mu)(\forall z)[z = V_{\pi(\xi)} \Rightarrow (\exists x \in z)P(y,x)].\end{aligned}$$

We will show that the Main Lemma holds for this $\mathfrak{A}$, with any $B \in V_\lambda$ - and it will be obvious that the proof will be <u>uniform</u>, so that our argument will actually establish closure under $(\exists t \in B)$ with B a variable.

From the hypotheses of the Main Lemma, (H2) holds easily taking (for example) the prewellordering of all sets in V_λ which are prewellorderings. Of course (H1) need not hold, so we must modify the proof of the Main Lemma in the case of Group C, functional substitution - the only case where (H1) was used.

We may assume by Kirousis' Lemma that the functions π and $\xi \mapsto V_\xi$ are $\mathfrak{U}$-recursive, since adjoining them to $\mathfrak{U}$ does not alter the envelope. We will also use the following simple

Lemma. Suppose $P(\bar{x},\xi)$ is $\mathfrak{U}$-semirecursive; then there exists a partial function $\varphi(\bar{x})$ whose graph is $\mathfrak{U}$-semirecursive and such that

$$(\exists \xi < \mu)P(\bar{x},\xi) \Rightarrow \varphi(\bar{x})\downarrow \ \& \ \varphi(\bar{x}) < \mu$$
$$\& \ P(\bar{x},\varphi(\bar{x})).$$

Proof. For some $e \in \omega$, we have

$$P(\bar{x},\xi) \Leftrightarrow \{e\}(\bar{x},\xi)\downarrow.$$

Put

$$\varphi(\bar{x}) = \xi \Leftrightarrow \{e\}(\bar{x},\xi)\downarrow$$
$$\& \ (\forall \eta < \mu)[\,|e,\bar{x},\xi| \leq |e,\bar{x},\eta|\,]$$
$$\& \ (\forall \eta < \xi)[\,|e,\bar{x},\xi| < |e,\bar{x},\eta|\,],$$

where the ordinals are compared by the Stage Comparison Theorem.

New treatment of the case for Group C. We have

$$\{e\}(\bar{\varphi}(t,\bar{y})) = F(\lambda s\{e_1\}(\bar{\varphi}(t,\bar{y}),s))$$

and we compute as usual:

$$(\exists t \in B(m,\bar{z}))[\{e\}(\bar{\varphi}(t,\bar{y}))\downarrow]$$
$$\Leftrightarrow (\exists t \in B(m,\bar{z}))(\forall s \in V_\lambda)[\{e_1\}(\bar{\varphi}(t,\bar{y}),s)\downarrow]$$
$$\Leftrightarrow (\exists t \in B(m,\bar{z}))(\forall \xi < \mu)(\forall s \in V_{\pi(\xi)})[\{e_1\}(\bar{\varphi}(t,\bar{y}),s)\downarrow]$$
$$\Leftrightarrow (\forall f : B(m,\bar{z}) \to \mu)(\exists t \in B(m,\bar{z}))(\forall s \in V_{\pi(f(t))})[\{e_1\}(\bar{\varphi}(t,\bar{y}),s)\downarrow]$$
$$\Leftrightarrow (\forall f : B(m,\bar{z}) \to \mu)(\exists t \in B(m,\bar{z}))(\exists \xi < \mu)$$
$$[\xi = f(t) \ \& \ (\forall s \in V_{\pi(\xi)})[\{e_1\}(\bar{\varphi}(t,\bar{y}),s)\downarrow]]$$
$$\Leftrightarrow (\forall f : B(m,\bar{z}) \to \mu)(\exists \xi < \mu)(\exists t \in B(m,\bar{z}))(\forall s \in V_{\pi(\xi)})$$
$$[\xi = f(t) \ \& \ \{e_1\}(\bar{\varphi}(t,\bar{y}),s)\downarrow]$$

$$\Leftrightarrow (\forall f : B(m,\bar{z}) \to \mu)(\exists \xi < \mu)(\forall g : B(m,\bar{z}) \to V_{\pi(\xi)})$$
$$(\exists t \in B(m,\bar{z}))[\xi = f(t) \ \& \ \{e_1\}(\bar{\varphi}(t,\bar{y}),g(t))\downarrow].$$

Now fix any $f : B(m,\bar{z}) \to \mu$ and for every $\xi < \mu$, ask if

$$(\forall g : B(m,\bar{z}) \to V_{\pi(\xi)})(\exists t \in B(m,\bar{z}))[\xi = f(t) \ \& \ \{e_1\}(\bar{\varphi}(t,\bar{y}),g(t))\downarrow];$$

using the induction hypothesis, this can be translated into a question that is semirecursive in $\mathfrak{A}$, as in the treatment of questions (a) and (b) in the case of composition. Moreover, for at least one $\xi < \mu$ we will get a positive answer - and we can find one such ξ by the lemma, using a function, φ with $\mathfrak{A}$-semirecursive graph.

For each f then, find this ξ, then for each g find by the induction hypothesis some non-empty

$$C(f,g) \subseteq \{t : t \in B(m,\bar{z}) \ \& \ \xi = f(t)\}$$

such that

$$t \in C(f,g) \Rightarrow \{e_1\}(\bar{\varphi}(t,\bar{y}),g(t))\downarrow.$$

We can put these sets $C(f,g)$ together as in the treatment of this case in the main lemma to get the choice set C that we need. The whole computation is easily seen to be recursive, since the ordinal $\xi = \varphi(f,\_\_)$ is actually quantified out, i.e. symbolically

$$t \in C(f,g) \Leftrightarrow (\exists \xi < \mu)[\varphi(f,\_\_) = \xi \ \& \cdots \& \ \{\hat{u}\}(\ldots) = 0],$$
$$t \notin C(f,g) \Leftrightarrow (\exists \xi < \mu)[\varphi(f,\_\_) = \xi \ \& \cdots \& \ \{\hat{u}\}(\ldots) = 1].$$

We omit the details. ⊣

<u>Corollary</u> 2.1. If $\mathfrak{A} = \langle V_{\omega+\omega}, \in, E, \_\_\rangle$ is an expansion of $\mathcal{V}_{\omega+\omega}$ by relations, functions and constants, then

$$\mathrm{Env}(\mathfrak{A}) = \mathrm{IND}(\mathfrak{A}) = \Pi^1_1(\mathfrak{A}).$$

<u>Proof.</u> Take $\mu = \omega$ in the theorem for the first equality and apply Chang-Moschovakis [1970] for the second. ⊣

This result is somewhat surprising and it makes the universe of Zermelo $V_{\omega+\omega}$ seem that much more like ω. Notice that by Chang-Moschovakis [1970], for every ordinal λ of cofinality ω we have

$$\mathrm{IND}(\mathcal{V}_\lambda) = \Pi^1_1(\mathcal{V}_\lambda).$$

It is easy to see that this identification does not extend automatically to $\mathrm{Env}(\mathcal{V}_\lambda)$

as follows.

Choose first any κ of cofinality $> \omega$ and check easily that

$$\mathrm{Env}(\mathcal{U}_\kappa) \subsetneqq \mathrm{IND}(\mathcal{U}_\kappa).$$

This is well-known and follows easily from the abstract version of the representation theorem for recursion in type-2 objects, see Moschovakis [1967] or 18.2 of Kechris-Moschovakis [1977]. Recall now the infinitary language $\mathcal{L}_{\omega_1,G}$ introduced in 8D of Moschovakis [1974], where it is shown that it satisfies the Skolem-Löwenheim Theorem for countable sets of formulas; now check that $\mathrm{Env}(\mathcal{U}_\kappa) \subsetneqq \mathrm{IND}(\mathcal{U}_\kappa)$ is expressible in $\mathcal{L}_{\omega_1,G}$, and get a substructure of $\mathcal{U}_\kappa$ of the form $\mathcal{U}_\lambda$ with cofinality$(\lambda) = \omega$ where the same sentence is still true.

Theorem 2 also implies immediately that for ordinals like $\kappa = \aleph_{\aleph_1 \cdot 2}$, $\mathrm{Env}(\mathcal{U}_\kappa)$ is not closed under restricted quantification.

We now proceed to put down a strong refinement of Theorem A which follows easily from the proof we gave. To simplify the statement of this result, let us call any set-valued function with non-empty values

$$f : S \to \mathrm{Power}(T) - \{\emptyset\}$$

a <u>total, multiple-valued</u> function on S to T - we will obviously think of such f's as assigning possibly many values in T (but at least one) to each of their arguments in S. If $R \subseteq T$, then by definition,

$$R(f(t)) \Leftrightarrow (\forall s \in f(t))R(s).$$

<u>The representing relation</u> of a total multiple-valued function f is defined by

$$G_f(t,s) \Leftrightarrow s \in f(t).$$

Finally, if

$$X : A \times B \to A$$

is a total, multiple-valued function, then <u>the fiber</u> X_a $(a \in A)$ is defined as usual by

$$X_a(t) = X(a,t).$$

A relation $R(\bar{x})$ is $\mathfrak{A}$-recursive ($\mathfrak{A}$-recursive from parameters) if for some $\mathfrak{A}$-recursive partial function $g(a,\bar{x})$ and some fixed $a \in A$, $\lambda\bar{x}g(a,\bar{x})$ is total and

$$R(\bar{x}) \Leftrightarrow g(a,\bar{x}) = 0.$$

Theorem 3[(2)]. Suppose $\mathfrak{A}$ is a good, type-2 structure and B is an $\mathfrak{A}$-recursive subset of the domain A of $\mathfrak{A}$ such that the following two conditions hold:

$(H1)^*$ There is a total, multiple-valued function

$$X : A \times B \to A$$

with $\mathfrak{A}$-recursive representing relation such that for each $\mathfrak{A}$-recursive relation $R(t,s)$,

$$(\forall t \in B)(\exists s \in A)R(t,s) \Rightarrow (\exists a)(\forall s \in X(a,t))R(t,s).$$

$(H2)^*$ There is an $\mathfrak{A}$-recursive prewellordering of A which is longer than every $\mathfrak{A}$-recursive prewellordering of B.

Then $\mathrm{Env}(\mathfrak{A})$ is closed under existential quantification or B, $\exists^B$.

Proof. First we reformulate the Main Lemma to allow the functions $\varphi_1(t,\bar{y}),\ldots,\varphi_n(t,\bar{y})$ to be total, multiple-valued with $\mathfrak{A}$-recursive representing relations which have codes $\hat{\varphi}_1,\ldots,\hat{\varphi}_n$. The hypothesis of the lemma then is

$$(\exists t \in B(m,\bar{z}))(\forall s_1 \in \varphi_1(t,\bar{y})) \ldots (\forall s_n \in \varphi_n(t,\bar{y}))\{e\}(s_1,\ldots,s_n)\downarrow$$

and conclusion (iii) is interpreted similarly.

The definition of u is exactly as in the proof of the Main Lemma, except for the obvious (trivial) modifications which we must make to handle the many-valuedness. We only need make a few remarks on the proof in the cases of groups C and D where the hypotheses (H1) and (H2) were used.

For group D (composition), again, there is little to say: the only place where (H2) was used was in checking that for some ξ we must have case (a) apply, and this must be so under $(H2)^*$ also, else we would get a long prewellordering $\prec$ which is clearly $\mathfrak{A}$-recursive.

For group C (functional substitution), the definition of the choice set C is exactly as before, noting that in the crucial equivalences (*) we only need the direction $(\Rightarrow)$ which is true even without (H1). To verify that $C \neq \emptyset$, as in (**) in the proof of the Main Lemma, notice that C is surely $\mathfrak{A}$-recursive so that $(H2)^*$ suffices; here is how (**) looks with a multiple-valued X:

(2) During the final stage of the preparation of this manuscript, Peter Hinman announced a result in the Abstracts of the American Mathematical Society, volume 1, Number 6, October 1980, p. 549, #701E24 which is stated in the terminology of set recursion but is equivalent to a (slight) weakening of this theorem.

$(\forall t \in B(m,\bar{z}))t \notin C$

$$\Rightarrow (\forall t \in B(m,\bar{z}))(\exists s \in A)(\forall a \in A)[s \notin X(a,t) \vee t \notin C(a)]$$

$$\Rightarrow (\exists b \in A)(\forall t \in B(m,\bar{z}))(\forall s \in X(b,t))(\forall a \in A)[s \notin X(a,t) \vee t \notin C(a)]$$

$$\Rightarrow (\exists b \in A)(\forall t \in B(m,\bar{z}))(\forall s \in X(b,t))[s \notin X(b,t) \vee t \notin C(b)]$$

$$\Rightarrow (\exists b \in A)(\forall t \in B(m,\bar{z})[t \notin C(b)]. \qquad \dashv$$

In the case where we have an $\mathfrak{A}$-recursive wellordering of A, $(H2)^*$ may be easily reformulated to avoid reference to multiple-valued functions - in effect it asserts that the collection of all $\mathfrak{A}$-recursive functions on B to A can be parametrized on A by a single $\mathfrak{A}$-recursive function. In this form, Theorem 3 is relevant to recursion on ordinal structures of the form

$$\mathfrak{A} = \langle \kappa, \in, E \rangle$$

and in fact implies fairly easily the basic selection theorem of Kirousis [1978] about such structures. We will not pursue this here, since these examples are best discussed in the context of set recursion with which we have not concerned ourselves here.

References

C. C. Chang and Y. N. Moschovakis [1970], The Suslin-Kleene Theorem for V_κ with cofinality$(\kappa) = \omega$, Pacific J. Math. 35 (1970) 565-569.

T. J. Grilliot [1969], Selection functions for recursive functionals, Notre Dame J. of Formal Logic X (1969) 225-234.

L. Harrington and D. B. MacQueen [1976], Selection in abstract recursion theory, J. Symbolic Logic 41 (1976) 153-158.

A. S. Kechris and Y. N. Moschovakis [1977], Recursion in higher types, Handbook of Mathematical Logic (ed. J. Barwise), North Holland, 1977, 681-737.

L. Kirousis [1978], On abstract recursion theory and recursion in the universe of sets, doctoral dissertation, UCLA, 1978.

Y. N. Moschovakis [1967], Hyperanalytic predicates, Trans. Amer. Math. Soc. 129 (1967) 249-282.

Y. N. Moschovakis [1974], Elementary induction on abstract structures, North Holland, 1974.

RECURSIVELY SATURATED, RATHER CLASSLESS MODELS OF PEANO ARITHMETIC

James H. Schmerl*
Department of Mathematics
University of Connecticut
Storrs, Connecticut 06268

This paper is concerned with the construction of recursively saturated, rather classless models of Peano arithmetic. The term "rather classless", defined below, is taken from the title of Kaufmann's paper [5], where it was first shown that recursively saturated, rather classless models exist.

With regard to a model $\mathfrak{N}$ of PA we make the following basic definitions. A subset $X \subseteq N$ is <u>definable</u> if it is definable from parameters by a first-order formula. We let $\mathrm{Def}(\mathfrak{N})$ denote the set of definable subsets of N. If X is definable and bounded, then it is <u>$\mathfrak{N}$-finite</u>. For κ an infinite cardinal, $\mathfrak{N}$ is <u>κ-like</u> iff $|N| = \kappa$ yet for every $\mathfrak{N}$-finite X, $|X| < \kappa$. A subset $X \subseteq N$ is a <u>class</u> if $X \cap Y$ is $\mathfrak{N}$-finite for every $\mathfrak{N}$-finite Y. We say that $\mathfrak{N}$ is <u>rather classless</u> if each of its classes is definable.

The following theorem was proved by Kaufmann [5] under the assumption of the combinatorial principle $\diamondsuit$; subsequently, Shelah [11] eliminated this assumption by an absoluteness argument.

<u>Theorem 1.</u> (Kaufmann-Shelah). Every consistent extension of PA has an $\aleph_1$-like, recursively saturated, rather classless model.

This theorem will serve as the model for all the theorems proved in this paper. The main problem we investigate is that of determining which cardinals κ can replace $\aleph_1$ in Theorem 1. On the negative side, there is the obvious restriction that $\mathrm{cf}(\kappa) > \aleph_0$. In the case of singular κ, we have the following positive result.

<u>Theorem 2.</u> If $\kappa > \mathrm{cf}(\kappa) > \aleph_0$, then every consistent extension of PA has a κ-like, recursively saturated, rather classless model.

For regular κ there is the following negative result.

<u>Theorem 3.</u> If κ is regular and PA has a κ-like, recursively saturated, rather classless model, then there is an Aronszajn κ-tree.

---

*Research partially supported by NSF Grant MCS 79-05020.

In particular, it follows from the hypothesis that κ is not weakly compact, a result obtained independently by Kaufmann. It follows from a general result of Shelah (Theorem 10 of [11]) that, assuming V = L, if κ is the successor of a regular cardinal, then there are κ-like, recursively saturated, rather classless models of PA. This is improved in the next theorem.

Theorem 4. Assume V = L. If κ is regular, uncountable and not weakly compact, then every consistent extension of PA has a κ-like, recursively saturated, rather classless model.

Thus, in the constructible universe, there is a completely satisfactory solution to the problem we set out to solve.

Corollary 5. Assume V = L. Let κ be an infinite cardinal and let T be a consistent extension of PA. Then the follow are equivalent:

(1) there is a κ-like, recursively saturated, rather classless model of T;

(2) $cf(\kappa) > \aleph_0$ and κ is not weakly compact.

Kaufmann constructed the models of Theorem 1 in order to answer a question raised by Barwise. The significant features of the models were their recursive saturation and their rather classlessness; that the models were also $\aleph_1$-like was more of an accident of the construction. We now ask what the possibilities are if we drop requirements on the order type of the model but not on the cardinality. It follows from a general result of Shelah (Theorem 12 of [11]) that, if κ is the successor of a regular cardinal, then PA has a recursively saturated, rather classless model of cardinality κ. We improve that result with the last theorem of this introduction.

Theorem 6. If $\kappa > \aleph_0$, then every consistent extension of PA has a recursively saturated, rather classless model of cardinality κ.

This entire paper evolved out of an attempt to find a new proof of Kaufmann's Theorem (that is, Theorem 1 assuming $\diamondsuit$). Kaufmann's proof of his theorem uses admissible model theory, and requires familiarity with the deeper results of that theory. Our proof uses techniques (principally the MacDowell-Specker Theorem and satisfaction classes) which more readily lend themselves to generalization. It is in this manner that we were eventually led to the proofs of the above theorems.

The notion of recursive saturation has been of some recent interest. After several people used some very closely related ideas, Schlipf isolated the concept and characterized it in terms of the ordinal of the least admissible set above a structure

([9]). This characterization is the one used by Kaufmann [5]. The significance of recursive saturation of models of PA became clear when Barwise and Schlipf [1] showed that an arbitrary model of PA is recursively saturated just in case it is expandable to some rather weak fragments of second-order arithmetic.

The outline of this paper is as follows. In §1 we give some basic definitions and results, including the MacDowell-Specker Theorem and a generalization of Gaifman's Cofinality Theorem. Satisfaction classes are discussed in §2, and our alternate proof of Kaufmann's theorem is presented there. In §3 we prove Theorems 2 and 6. Theorem 3 is proved in §4, and finally in §5 we prove Theorem 4.

We would like to thank Matt Kaufmann and especially Steve Simpson for their continued interest in the results of this paper. We are grateful to Saharon Shelah who made a suggestion which eventually led us to the proofs of the current Corollaries 3.4 and 3.5.

§1. Basics.

Let PA be some usual first-order axiomatization of Peano arithmetic in the language L. It will at times be convenient to assume that PA is formalized so as to include Skolem terms. This is efficaciously done by introducing the μ-operator into the formalism. We shall assume that this has been done. Among the terms there is one which is the standard pairing function

$$\langle x,y\rangle = \frac{1}{2}[(x+y+1)(x+y)] + x.$$

We will let $\pi_1(z)$ and $\pi_2(z)$ be the first and second coordinates of z; that is $z = \langle \pi_1(z), \pi_2(z)\rangle$.

We will let $D(x,y)$ be the formula which canonically indexes finite sets. That is, $D(x,y)$ asserts: the y-th digit of the binary expansion of x is 1 . Thus, whenever $\mathfrak{N} \models PA$ and $a \in N$, then $\{b \in N : \mathfrak{N} \models D(a,b)\}$ is $\mathfrak{N}$-finite. Conversely, whenever X is $\mathfrak{N}$-finite, then there is a unique $a \in N$, called the canonical index of X, such that $X = \{b \in N : \mathfrak{N} \models D(a,b)\}$. We let D_x be the set whose canonical index is x.

We will often consider the language L^* which is L augmented by a unary predicate symbol. The theory PA^* is the extension of PA to the language L^* obtained by adjoining to PA all new instances of the induction scheme.

It will be useful to have available second-order variables which range over arbitrary subsets. These will be denoted by capital letters such as X. It will also be useful to have second-order terms. For example, we will make use of the terms

$$(X)_x = \{y : \langle x,y\rangle \in X\}$$

and

$$X|x = \{\langle y,z\rangle \in X : y < x\}.$$

There are several types of elementary extensions which will play key roles in this paper. Suppose $\mathfrak{M}$ and $\mathfrak{N}$ are models of PA (or PA^*) and $\mathfrak{M} \prec \mathfrak{N}$. Then $\mathfrak{N}$ is an end-extension of $\mathfrak{M}$ if whenever $a \in M$ and $b \in N-M$, then $\mathfrak{N} \models a<b$. The extension is conservative if whenever $X \in \mathrm{Def}(\mathfrak{N})$, then $X \cap M \in \mathrm{Def}(\mathfrak{M})$. The extension is finitely generated if there exists $b \in N$ such that $\mathfrak{N}$ has no proper elementary substructure containing $M \cup \{b\}$. The extension is cofinal if whenever $a \in N$ then there is $b \in M$ such that $\mathfrak{N} \models a<b$. It is easy to see that a conservative extension must be an end-extension.

We now state the very basic MacDowell-Specker Theorem [7].

Theorem 1.1 (MacDowell-Specker). Every model of PA^* has a proper, conservative elementary end-extension.

Of course this extension can be chosen to be finitely generated in which case it has the same cardinality as the original model. Or, by iterating the extensions sufficiently often, an extension can be constructed which has any desired cardinality greater than the cardinality of the original model.

Another basic theorem about extensions of models of PA is Gaifman's Cofinality Theorem [3], which asserts: If $\mathfrak{M}$ and $\mathfrak{N}$ are models of PA and $\mathfrak{M}$ is cofinal in $\mathfrak{N}$, then $\mathfrak{M} \prec \mathfrak{N}$. In our formalism, this theorem reduces to a triviality because we have included all Skolem functions. We need an extension of this theorem to models of PA^*. Suppose $(\mathfrak{M},X) \models PA^*$ and $(\mathfrak{M},X) \prec (\mathfrak{N},Y)$, this extension being cofinal. Then clearly

$$Y = \bigcup\{D_a^{\mathfrak{N}} : a \in M \text{ and } D_a^{\mathfrak{M}} \subseteq X\}.$$

We let $X^{\mathfrak{N}}$ be the set Y defined above.

Theorem 1.2.[1] Suppose that $(\mathfrak{M},X) \models PA^*$, $\mathfrak{N} \models PA$, and $\mathfrak{M}$ is a cofinal substructure of $\mathfrak{N}$. Then there is a unique $Y \subseteq N$ such that $(\mathfrak{M},X) \prec (\mathfrak{N},Y)$.

Proof. From the remark preceding the Theorem, the uniqueness is clear, and, in fact, if such a Y exists, then it must be that $Y = X^{\mathfrak{N}}$. An inspection of Gaifman's proof (of Theorem 3 in [3]) reveals that it suffices to prove: whenever $a \in M$ and $\phi(x,y)$ is a quantifier-free L^*-formula, then $(\mathfrak{M},X) \models \exists x \leq a\ \phi(x,a)$ implies $(\mathfrak{N},X^{\mathfrak{N}}) \models \exists x \leq a\ \phi(x,a)$. (In order to get this reduction, we relied heavily on the presence of Skolem functions.) Now let $a \in M$ and let $\phi(x,y)$ be such a formula. Then $\phi(x,y)$ is a Boolean combination of L-formulas and of formulas of the form $t(x,y) \in X$,

[1] This theorem also appears in a handwritten abstract by Henryk Kotlarski entitled "On Cofinal Extensions of Models of Arithmetic".

where t(x,y) is an L-term. Let $b \in M$ be such that whenever t(x,y) is a term occurring in $\phi(x,y)$, then $\mathfrak{M} \models \forall x \leq a\ (t(x,a) < b)$. There is $c \in M$ such that

$$D_c^{\mathfrak{M}} = \{x \in M : x \leq b\} \cap X.$$

Obtain an L-formula $\phi'(x,y,z)$ from $\phi(x,y)$ by replacing all occurrences of subformulas of the form $t(x,y) \in X$ by $D(z,t(x,y))$. Clearly $\mathfrak{M} \models \forall x \leq a\ \phi'(x,a,c)$, so that also $\mathfrak{N} \models \forall x \leq a\ \phi'(x,a,c)$. From the definition of $X^{\mathfrak{N}}$ it is clear that $(\mathfrak{N}, X^{\mathfrak{N}}) \models \forall x \leq a\ \phi'(x,a,c) \longleftrightarrow \forall x \leq a\ \phi(x,a)$. Thus, we get that $(\mathfrak{N}, X^{\mathfrak{N}}) \models \forall x \leq a\ \phi(x,a)$. □

§2. Satisfaction Classes.

One of the standard effective Gödel numberings of L^*-formulas will be used; if ϕ is an L^*-formula, then let $\#(\phi)$ be its Gödel number. There is a Σ_1 formula $W(X,x,y,z)$ with the following property: whenever $\phi(X,x,y)$ is a Σ_1 formula, $\mathfrak{N} \models PA$ and $X \subseteq N$, then

$$(\mathfrak{N},X) \models \forall xy(\phi(X,x,y) \longleftrightarrow W(X,x,y,\#(\phi))).$$

Let $j(X)$ be the second-order term denoting $\{x : W(X,x,\pi_1(x),\pi_2(x))\}$. For $n < \omega$ define $j^n(X)$ by $j^0(X) = X$ and $j^{n+1}(X) = j(j^n(X))$. Thus, for $1 \leq n < \omega$, the term $j^n(\emptyset)$ formally denotes the complete Σ_n set. We will occasionally refer to $\{x : W(j^n(\emptyset),x,0,b)\}$ as the b-th Σ_n set.

For a model $\mathfrak{N} \models PA$ we will call a set $S \subseteq N$ a <u>satisfaction class</u> if each of the following holds:

(1) $(\mathfrak{N},S) \models (S)_n = j^n(\emptyset)$, for $n < \omega$;

(2) $(\mathfrak{N},S) \models PA^*$.

In case $\mathfrak{N}$ is nonstandard and $S \subseteq N$ satisfies (2), then (1) is equivalent to

(3) $(\mathfrak{N},S) \models \forall x < b\ ((S)_{x+1} = j((S)_x))$, for some nonstandard $b \in N$.

It should be observed that our definition of satisfaction class is stronger than other definitions which do not require the full strength of condition (2).

In Lemmas 2.1 - 2.4 we state the basic properties of satisfaction classes needed for the proof of Kaufmann's lemma (Lemma 2.5 below). The first two lemmas indicate the relationship that exists between recursive saturation and the existence of satisfaction classes. Indeed, a necessary and sufficient condition for a countable model of PA to be recursively saturated is that it be nonstandard and have a satisfaction class.

<u>Lemma 2.1.</u> If $\mathfrak{N}$ is recursively saturated and countable, then $\mathfrak{N}$ has a satisfaction class.

Lemma 2.2. If $\mathfrak{N}$ is nonstandard and has a satisfaction class, then $\mathfrak{N}$ is recursively saturated.

The proofs of Lemmas 2.1 and 2.2 are essentially given in Kotlarski [6]. The proof of Lemma 2.1 uses the fact that countable recursively saturated models are resplendent, and the proof of Lemma 2.2 uses that there is a nonstandard $b \in M$ such that for any L-formula $\phi(x)$,

$$(\mathfrak{N},X) \models \forall x(\phi(x) \longleftrightarrow W((X)_b,x,0,\#(\phi)),$$

together with the representability of recursive sets.

The next lemma is obvious.

Lemma 2.3. If S is a satisfaction class for $\mathfrak{N}$ and $b \in N$ is nonstandard, then $S|b$ is also a satisfaction class.

Finally we get to Lemma 2.4 which is the weapon that will be used to kill undefinable classes.

Lemma 2.4. If S is a satisfaction class for nonstandard $\mathfrak{N}$, and if $X \in \mathrm{Def}((\mathfrak{N},S|a))$ for each nonstandard $a \in N$, then $X \in \mathrm{Def}(\mathfrak{N})$.

Proof. Let $a \in N$ be nonstandard such that $(\mathfrak{N},S) \models \forall x \leq a\ (j((S)_x) = (S)_{x+1})$. Notice that if $n < \omega$, $b < a$, and X is definable in $(\mathfrak{N},S|b)$ by a Σ_{n+1} formula, then X is definable in $(\mathfrak{N},S|(b+1))$ by a Σ_n formula. Assume that $X \in \mathrm{Def}((\mathfrak{N},S|b))$ for each nonstandard $b < a$, and let $b' < b$ be such that both b' and $b-b'$ are nonstandard. Then, since X is definable in $(\mathfrak{N},S|b')$, it follows that X is definable in $(\mathfrak{N},S|b)$ by a Σ_1 formula. So for each nonstandard $b \in N$, X is definable in $(\mathfrak{N},S|b)$ by a Σ_1 formula. Hence, by underspill, there is some standard $b \in N$ such that X is definable in $(\mathfrak{N},S|b)$. But $S|b \in \mathrm{Def}(\mathfrak{N})$, so that $X \in \mathrm{Def}(\mathfrak{N})$. □

The proof of Theorem 1 is divided into three parts. The first part consists of Lemma 2.5 below, which was proved by Kaufmann [5] using admissible model theory. We shall show how it follows easily from Lemmas 2.1 - 2.4 and the MacDowell-Specker Theorem. The second part of the proof of Theorem 1, which is also in [5], consists of a standard implementation of the combinatorial principle $\diamondsuit$ to build the desired model. The third part, Shelah's contribution, is to show that Theorem 1 is absolute; this is a consequence of the absoluteness theorem of [11]. Shelah's proof uses a modification of the Baumgartner-Malitz-Reinhardt [2] forcing construction which transforms an Aronszajn tree into a special one.

Lemma 2.5 (Kaufmann). Let $\mathfrak{M}$ be a countable, recursively saturated model of PA, and let $X \subseteq M$ be an undefinable class. Then $\mathfrak{M}$ has a countable, recursively saturated end-extension $\mathfrak{N}$ such that if $Y \in \mathrm{Def}(\mathfrak{N})$ is a class, then $Y \cap M \neq X$.

Proof. Let S be a satisfaction class for $\mathfrak{M}$, the existence of which is guaranteed by Lemma 2.1. By Lemmas 2.3 and 2.4 we can choose S so that $X \notin \mathrm{Def}((\mathfrak{M},S))$. Use Theorem 1.1 to obtain a proper, conservative elementary end-extension $(\mathfrak{N},S')$ which, moreover, is countable. By Lemma 2.2, $\mathfrak{N}$ is recursively saturated, so that clearly $\mathfrak{N}$ is as desired. □

In §5 we will use a generalization of Lemma 2.1 from the countable case to the countable cofinality case.

Lemma 2.6. If $\mathfrak{N}$ is recursively saturated and $\mathrm{cf}(\mathfrak{N}) = \aleph_0$, then $\mathfrak{N}$ has a satisfaction class.

Proof. Let $\mathfrak{M} \prec \mathfrak{N}$ be a countable, cofinal, recursively saturated elementary substructure. By Lemma 2.1 there is a satisfaction class for $\mathfrak{M}$. Theorem 1.2 asserts the existence of $Y \subseteq N$ such that $(\mathfrak{M},S) \prec (\mathfrak{N},Y)$. Clearly Y is a satisfaction class for $\mathfrak{N}$.

Lemma 2.6 implies a corresponding strengthening of Lemma 2.5. Note also that the same method can be used to prove the theorem of Smorynski and Stavi [12] that a cofinal extension of a recursively saturated model of PA is itself recursively saturated.

The next lemma is another extension of Lemma 2.1. It is the key model-theoretic fact which will be used in proving Theorem 4 in §5.

Lemma 2.7. Suppose $\mathfrak{M}_0$, $\mathfrak{M}_1$, $\mathfrak{N}$ are recursively saturated such that $\mathrm{cf}(\mathfrak{N}) = \aleph_0$ and $\mathfrak{M}_1, \mathfrak{N}$ are proper elementary end-extensions of $\mathfrak{M}_0$, $\mathfrak{M}_1$ respectively. Furthermore, suppose that S_0 and S_1 are satisfaction classes for $\mathfrak{M}_0$ and $\mathfrak{M}_1$ such that $(\mathfrak{M}_0,S_0) \prec (\mathfrak{M}_1, S_1)$. Then there is a satisfaction class S for $\mathfrak{N}$ such that $(\mathfrak{M}_0,S_0) \prec (\mathfrak{N},S)$.

Proof. We can assume that $\mathrm{cf}(\mathfrak{M}_1) = \aleph_0$. (If that is not the case then replace $(\mathfrak{M}_1,S_1)$ by an appropriate elementary substructure of itself.) Let $(\mathfrak{M}_1',S_1') \prec (\mathfrak{M}_1,S_1)$ be a countable cofinal substructure. Let $a \in M_1' - M_0$.

We will obtain a cofinal, elementary embedding $f: \mathfrak{M}_1' \longrightarrow \mathfrak{N}$, which is the identity on $\{x \in M_1' : x \leq a\}$, in the following manner. Let $\{a_n : n < \omega\} = M_1'$, and let $\langle b_n : n < \omega\rangle$ be an increasing sequence cofinal in N. Define f inductively. Suppose that $f(a_o),\ldots, f(a_{n-1})$ have already been defined, and then let $f(a_n) = b$ be such that:

(1) $(\mathcal{N}, f(a_o),\ldots, f(a_{n-1}), b, x)_{x\leq a} \equiv (\mathcal{N}, a_o,\ldots, a_n, x)_{x\leq a}$;

(2) if possible, $b > b_n$.

There always exists b satisfying (1) by the recursive saturation of $\mathcal{N}$. Thus, f is elementary. Also, there are arbitrarily large n for which b exists with $b > b_n$, since, by the recursive saturation of $\mathcal{M}_1'$, there are arbitrarily large n for which a_n is not in the substructure generated by $\{x \in M_1' : x < a_i \text{ and } i < n\}$. Thus f is cofinal.

Now let $\mathcal{M}_1''$ be the substructure of $\mathcal{N}$ generated by $M_1' \cup \{x \in N : x \leq a\}$. $\mathcal{M}_1''$ is a cofinal extension of $\mathcal{M}_1'$, so by Theorem 1.2 there is $S_1'' \subseteq M_1''$ such that $(\mathcal{M}_1', S_1') \prec (\mathcal{M}_1'', S_1'')$, so that S_1'' is a satisfaction class for $\mathcal{M}_1''$. It is clear that $S_1'' = M_1'' \cap S_1$. Thus from Theorem 1.2 it easily follows that $(\mathcal{M}_1'', S_1'') \prec (\mathcal{M}_1, S_1)$. Then, since $(\mathcal{M}_o, S_o) \prec (\mathcal{M}_1, S_1)$, we get that $(\mathcal{M}_o, S_o) \prec (\mathcal{M}_1'', S_1'')$.

There is a unique elementary embedding, call it $g: \mathcal{M}_1'' \longrightarrow \mathcal{N}$, which extends f and which is the identity on $\{x \in M_1'' : x \leq a\}$. Let $(\mathcal{N}', S')$ be the image of $(\mathcal{M}_1'', S_1'')$ under this embedding. Since g is the identity on M_o, it follows that $(\mathcal{M}_o, S_o) \prec (\mathcal{N}', S')$. But using Theorem 1.2 again, there is $S \subseteq N$ such that $(\mathcal{N}', S') \prec (\mathcal{N}, S)$, so that $(\mathcal{M}_o, S_o) \prec (\mathcal{N}, S)$, as required. □

§3. The Proofs of Theorem 2 and 6.

Before giving the proofs of these theorems, we extend an observation we had previously made in [10]. There we observed that by iterating conservative extensions through an uncountable regular cardinal κ so as to obtain a κ-like model, then this model is rather classless. We extend this observation here so as to obtain κ-like, rather classless models where κ need only have uncountable cofinality. A similar argument will occur in the proofs of Theorems 2 and 6.

Lemma 3.1. Suppose that $\mathrm{cf}(\alpha) > \aleph_o$ for limit ordinal α, and that $\langle \mathcal{N}_\nu : \nu \leq \alpha \rangle$ is a continuous chain of conservative extensions. Then $\mathcal{N}_\alpha$ is rather classless.

Proof. Suppose X is a class of $\mathcal{N}_\alpha$. Our object is to prove that $X \in \mathrm{Def}(\mathcal{N}_\alpha)$. Each $X \cap N_\nu$ is a class of $\mathcal{N}_\nu$ for $\nu < \alpha$. In fact, since $\mathcal{N}_{\nu+1}$ is a conservative extension of $\mathcal{N}_\nu$, we get that $X \cap N_\nu \in \mathrm{Def}(\mathcal{N}_\nu)$. Let $f : \alpha \longrightarrow \alpha$ be such that

$$f(\nu) = \mu\beta[X \cap N_\nu \text{ is definable in } \mathcal{N}_\nu \text{ from a parameter in } N_\beta].$$

Then for limit ordinal $\nu < \alpha$, $f(\nu) < \nu$, so that by Fodor's Theorem there is some $\beta < \alpha$ such that $f^{-1}(\beta)$ is unbounded. In addition, since $\mathrm{cf}(\alpha) > \aleph_0$, there is some $n < \omega$ such that for some unbounded $I \subseteq f^{-1}(\beta)$, $X \cap N_\nu$ is definable by a Σ_n formula using only parameters from N_α, for each $\nu \in I$. Let

$$a_\nu = \mu a[X \cap N_\nu \text{ is the } a\text{-th } \Sigma_n \text{ subset of } \mathcal{N}_\nu].$$

We now claim that for some $a \in N_\beta$, $a_\nu = a$ for each $\nu \in I$. Clearly, this will finish the proof for it will show that X is the a-th Σ_n subset of $\mathcal{N}_\alpha$.

To prove the claim, suppose $\nu < \mu$ are both in I. Then $X \cap N_\mu$ is the a_μ-th Σ_n subset of $\mathcal{N}_\mu$. Since $\mathcal{N}_\nu \prec \mathcal{N}_\mu$ and $a_\mu \in N_\nu$, then $X \cap N_\nu$ is the a_μ-th Σ_n subset of $\mathcal{N}_\nu$. Then, in $\mathcal{N}_\nu$, the a_ν-th and a_μ-th Σ_n subsets are the same, so this is also true in $\mathcal{N}_\mu$. Hence, $X \cap N_\mu$ is the a_ν-th Σ_n subset of $\mathcal{N}_\mu$. Thus $a_\nu = \min(a_\nu, a_\mu) = a_\mu$, proving the claim. □

The following two corollaries are immediate consequences of Lemma 3.1 using, of course, Theorem 1.1.

Corollary 3.2. If $\mathrm{cf}(\kappa) > \aleph_0$, then every model of PA of cardinality $< \kappa$ has a κ-like, rather classless elementary end-extension.

Proof. Let $\mathcal{M} \models$ PA such that $|M| < \kappa$. Let $\langle \mathcal{N}_\nu : \nu \leq \kappa \rangle$ be a continuous chain of finitely generated, conservative extensions, where $\mathcal{N}_0 = \mathcal{M}$. Then $\mathcal{N}_\kappa$ is as required. □

Corollary 3.3. If $\kappa > \aleph_0$, then evèry model of PA of cardinality $\leq \kappa$ has a rather classless elementary end-extension of cardinality κ.

Proof. Proceed as in the proof of Corollary 3.2 with a chain indexed by $\kappa \cdot \omega_1$. □

We now turn to the proofs of Theorems 2 and 6. We will first prove Theorem 2, and then just indicate the necessary modifications to make to get a proof of Theorem 6.

Proof of Theorem 2. Let $cf(\kappa) = \lambda > \aleph_0$, and let $\langle \kappa_\alpha : \alpha < \lambda \rangle$ be a continuous, increasing sequence of cardinals converging to κ such that $\kappa_0 = \lambda$. Let T be a consistent extension of PA. We easily obtain $(\mathcal{N}_0, S_0)$, where $\mathcal{N}_0 \models T$ is such that $|N_0| = \kappa_0$, S_0 is a satisfaction class for $\mathcal{N}_0$, and there is a decreasing sequence $\langle b_\alpha : \alpha < \lambda \rangle$ of elements of N_0 with no nonstandard lower bound. Let $\mathcal{N}_0^* = (\mathcal{N}_0, S_0)$. We use Theorem 1.1 to build a chain $\langle \mathcal{N}_\nu^* : \nu < \kappa \rangle$, where $\mathcal{N}_\nu^* = (\mathcal{N}_\nu, S_\nu)$, as follows. Having $\mathcal{N}_\nu^* = (\mathcal{N}_\nu, S_\nu)$, let $\mathcal{N}_{\nu+1}^* = (\mathcal{N}_{\nu+1}, S_{\nu+1})$ be a finitely generated, conservative extension of $\mathcal{N}_\nu^*$. If ν is a limit ordinal and $\kappa_\alpha \leq \nu < \kappa_{\alpha+1}$, then let $\mathcal{N}_\nu^* = \bigcup_{\beta<\nu} (\mathcal{N}_\beta, S_\beta | b_\alpha)$. We claim that $\mathcal{N} = \bigcup_{\nu<\kappa} \mathcal{N}_\nu$ is as required.

Clearly, $\mathcal{N} \models T$, is κ-like, and also is recursively saturated since each of the $\mathcal{N}_\nu$ is by Lemmas 2.2 and 2.3. It remains to verify that $\mathcal{N}$ is rather classless, so let $X \subseteq N$ be a class, and suppose $X \in \mathrm{Def}(\mathcal{N})$. As in the proof of Lemma 3.1, there is some $\nu < \kappa$ such that $X \cap N_\nu \notin \mathrm{Def}(\mathcal{N}_\nu)$. By Lemma 2.4, there is $\alpha < \lambda$ such that $\kappa_\alpha > \nu$ and $X \cap N_\nu \notin \mathrm{Def}((\mathcal{N}_\nu, S_\nu | b_\alpha))$. But $(\mathcal{N}_\nu, S_\nu | b_\alpha) \prec \mathcal{N}_{\kappa_\alpha}^*$, so that $X \cap N_{\kappa_\alpha} \notin \mathrm{Def}(\mathcal{N}_{\kappa_\alpha}^*)$, and therefore $X \cap N_{\kappa_\alpha+1}$ is not a class of $\mathcal{N}_{\kappa_\alpha+1}$. This contradicts X being a class of $\mathcal{N}$, and completes the proof. □

Proof of Theorem 6. Start off with a model $\mathcal{N}_0^* = (\mathcal{N}_0, S_0)$ as in the proof of Theorem 2, only with $\lambda = \aleph_1$. The model $\mathcal{N}_0$ has a decreasing sequence $\langle b_\alpha : \alpha < \omega_1 \rangle$ with no nonstandard lower bound. Let κ_α be the ordinal $\kappa \cdot \alpha$. Now form the sequence $\langle \mathcal{N}_\nu^* : \nu < \kappa \cdot \omega_1 \rangle$ as in the proof of Theorem 2. The model $\mathcal{N} = \bigcup \{\mathcal{N}_\nu : \nu < \kappa \cdot \omega_1\}$ will have the desired properties. □

If the model $\mathcal{N}_0$ which occurs in the proofs of Theorem 2 and 6 is $\aleph_0$-saturated, then so will be the resulting model. Such an $\mathcal{N}_0$ can always be found provided that $|N_0| \geq 2^{\aleph_0}$. Thus, we get the following corollaries.

Corollary 3.4. If $\kappa > cf(\kappa) > \aleph_0$ and $\kappa > 2^{\aleph_0}$, then every consistent extension of PA has a κ-like, $\aleph_0$-saturated, rather classless model.

Corollary 3.5. If $\kappa \geq 2^{\aleph_0}$, then every consistent extension of PA has an $\aleph_0$-saturated, rather classless model of cardinality κ.

§4. The Proof of Theorem 3.

There are many properties of cardinals which are known to be equivalent to weak compactness. Let us say that a regular cardinal κ has the tree property if there do not exist any Aronszajn κ-trees. Then, assuming GCH, a cardinal is weakly compact if it is regular and has the tree property. (Some assumption such as GCH is needed as has been shown by Mitchell and Silver [8]). To prove Theorem 3, we will show that for regular κ, a κ-like recursively saturated, rather classless model of PA generates in a natural way an Aronszajn κ-tree.

Since nonstandard models of PA are not well-ordered, it will be convenient, but not essential, to weaken the notion of κ-tree to allow non-well-founded trees whose elements have ranks occurring in some κ-like linearly ordered set.

Proof of Theorem 3. Let κ be a regular cardinal, and let $\mathfrak{N}$ be a κ-like, recursively saturated, rather classless model of PA. Let $c \in N$ be some nonstandard element, and then define A to be the set of those elements $a \in N$ such that:

(1) if $X = D_a$, then $(X)_n$ is an initial segment of $j^n(\emptyset)$ for each $n < \omega$;

(2) $(X)_x = \emptyset$ whenever $x \geq c$.

Define $a \prec b$ for elements of A iff for some $d \in N$,

$$(D_a)_x = (D_b)_x \cap \{y \in N : y \leq d\},$$

for each $x < c$. Clearly $(A, \prec)$ is a tree. There is a natural way to assign ranks; we describe it in the following anthropomorphic way: the rank of a is just what $\mathfrak{N}$ thinks the cardinality of the set $\{x \in N : x \prec a\}$ is. To see that there are elements of arbitrary rank, just note that recursive saturation of $\mathfrak{N}$ implies that for any $d \in N$ there is some $a \in N$ such that for all $n < \omega$, $(D_a)_n = j^n(\emptyset) \cap \{x \in N : x \leq d\}$. Therefore, it is clear that $(A, \prec)$ is a κ-tree.

It remains to show that $(A, \prec)$ is an Aronszajn tree. So suppose to the contrary that it is not and that B is a branch through the tree. Let $X = \bigcup\{D_a : a \in B\}$. Clearly, X is a class, and $(X)_n = j^n(\emptyset)$ for each $n < \omega$; thus X is not definable, contradicting the rather classlessness of $\mathfrak{N}$. This completes the proof of Theorem 3. □

What we actually showed in the preceding proof is that if $\mathfrak{N}$ is κ-like and recursively saturated, where κ is a regular cardinal which has the tree property, then there is a class X such that $(X)_n = j^n(\emptyset)$ for each $n < \omega$. Of course, X need not be a satisfaction class. However, we can show that $\mathfrak{N}$ does have a satisfaction class. This is a consequence of the following lemma.

Lemma 4.1. Suppose that $\mathcal{N}$ is κ-like for some regular, uncountable κ. If $X \subseteq N$ is a class such that $(X)_n = j^n(\emptyset)$ for each $n < \omega$, then there is $b \in N$ such that $X|b$ is a satisfaction class.

Proof. First we will show that there is $b_o \in N$ such that

$$(\mathcal{N},X) \models \forall x{<}b_o(j((X)_x) = (X)_{x+1}).$$

For each $a \in N$ there is $c \in N$ such that

$$(\mathcal{N},X) \models \forall x{<}a(x\varepsilon(X)_{n+1} \longleftrightarrow x\varepsilon j(\{y\varepsilon(X)_n : y \leq c\})),$$

for each $n \in \omega$. By overspill, there is some nonstandard $i \in N$ such that

$$(*) \qquad (\mathcal{N},X) \models \forall z{<}i\,\forall x{<}a(x\varepsilon(X)_{z+1} \longleftrightarrow x\varepsilon j(\{y\varepsilon(X)_z : y \leq c\})).$$

Thus, for each $a \in N$, if we let I_a be the set of all $i \in N$ such that (*) holds for some $c \in N$, then I_a is an initial segment of N which contains some nonstandard element. Clearly, if $a < a'$, then $I_a \supseteq I_{a'}$. Since $\mathcal{N}$ is κ-like, there is a nonstandard $b_o \in \bigcap_{a \in N} I_a$. This b_o is as required.

In a very similar manner we can show that there is a nonstandard $b_1 \in N$ such that $X|x \in \mathrm{Def}((\mathcal{N},(X)_x))$ for each $x < b_1$.

Now let $b < b_1$ be nonstandard such that $b_1 - b$ is also nonstandard. We claim that $X|b$ is a satisfaction class. If it is not, then there is $Y \in \mathrm{Def}\ ((\mathcal{N},X|b))$ which is bounded but not $\mathcal{N}$-finite. By the properties of b, Y is definable in $(\mathcal{N},(X)_{b_o})$ by a Σ_1 formula. But again using that $\mathcal{N}$ is κ-like for regular κ, it follows that $Y \in \mathrm{Def}(\mathcal{N})$, and this is a contradiction. ◻

Corollary 4.2. Suppose that $\mathcal{N}$ is κ-like and recursively saturated for some regular cardinal κ which has the tree property. Then $\mathcal{N}$ has a satisfaction class.

§5. The Proof of Theorem 4.

Jensen [4] proved that V = L settles Souslin's hypothesis at all regular cardinals. Specifically, he showed that, assuming V = L, whenever κ is an uncountable regular cardinal, then the existence of a Souslin κ-tree is equivalent to κ not being weakly compact. Our proof of Theorem 4 is similar in outline to his proof.

We will assume V = L throughout this section, and we will assume that κ is some fixed uncountable regular cardinal.

Jensen showed (Theorem 6.1 of [4]) that there is a stationary $E \subseteq \kappa$ and a sequence $\langle C'_\alpha : \alpha < \kappa \text{ a limit ordinal}\rangle$ such that each C'_α is a closed, unbounded subset of

α, and such that whenever $\beta \in C'_\alpha$ is a limit point of C'_α, then $\beta \notin E$ and $C'_\beta = \beta \cap C'_\alpha$. Let $\langle C_\alpha : \alpha < \kappa \rangle$ be the sequence such that if α is a limit ordinal, then C_α is the set of limit points of C'_α, and $C_\alpha = \emptyset$ otherwise.

The following properties are all obvious for each $\alpha < \kappa$:

(i) C_α is a closed subset of α;

(ii) if $cf(\alpha) > \aleph_0$, then C_α is an unbounded subset of α;

(iii) $C_\alpha \cap E = \emptyset$;

(iv) if $\beta \in C_\alpha$, then β is a limit ordinal and $\beta \cap C_\alpha = C_\beta$.

We will also need the combinatorial principle $\Diamond_\kappa(E)$: there is a sequence $\langle D_\alpha : \alpha < \kappa \rangle$ such that whenever $X \subseteq \kappa$ then $\{\alpha \in E : X \cap \alpha = D_\alpha\}$ is a stationary subset of κ.

Let $(\mathcal{N}_0, S_0)$ be such that $\mathcal{N}_0$ is a model of the given extension of PA, S_0 a satisfaction class, and, without loss of generality, let N_0 be some ordinal $< \kappa$. We will construct an elementary end-extension of $\mathcal{N}_0$ with the required properties. To do so we will construct a sequence $\langle (\mathcal{N}_\alpha, S_\alpha) : \alpha < \kappa \rangle$ which has the following properties:

(1) $\langle \mathcal{N}_\alpha : \alpha < \kappa \rangle$ is a continuous chain of proper elementary end-extension;

(2) each S_α is a satisfaction class for $\mathcal{N}_\alpha$;

(3) N_α is some ordinal $< \kappa$;

(4) if $\alpha \in E$, $D_\alpha \subseteq N_\alpha$, and $D_\alpha \notin \mathrm{Def}(\mathcal{N}_\alpha)$ then $D_\alpha \neq Y \cap N_\alpha$ for any $Y \in \mathrm{Def}(\mathcal{N}_{\alpha+1})$.

Having constructed such a sequence, we then let $\mathcal{N} = \bigcup\{\mathcal{N}_\alpha : \alpha < \kappa\}$. Properties (1) and (3) imply that $\mathcal{N}$ is a κ-like elementary end-extension of $\mathcal{N}_0$; property (2) and Lemma 2.2 imply that $\mathcal{N}$ is recursively saturated. To see that $\mathcal{N}$ is rather classless, let $X \subseteq N$ be a class. Since $\mathcal{N}$ is κ-like and κ is regular and uncountable, there is $\alpha \in E$ such that $N_\alpha = \alpha$, $(\mathcal{N}_\alpha, X \cap N_\alpha) \prec (\mathcal{N}, X)$ and $X \cap \alpha = D_\alpha$. Then, from (4), if $X \cap N_\alpha \notin \mathrm{Def}(\mathcal{N}_\alpha)$, then $X \cap N_\alpha \neq Y \cap N_\alpha$ for any $Y \in \mathrm{Def}(\mathcal{N}_{\alpha+1})$, so that X cannot be a class. Thus, $X \cap N_\alpha \in \mathrm{Def}(\mathcal{N}_\alpha)$, so that also $X \in \mathrm{Def}(\mathcal{N})$.

In order to construct the sequence $\langle (\mathcal{N}_\alpha, S_\alpha) : \alpha < \kappa \rangle$ we will require of it two more properties:

(5) if $\alpha \notin E$, then $(\mathcal{N}_\alpha, S_\alpha) \prec (\mathcal{N}_{\alpha+1}, S_{\alpha+1})$;

(6) if $\beta \in C_\alpha$, then $(\mathfrak{N}_\beta, S_\beta) \prec (\mathfrak{N}_\alpha, S_\alpha)$.

There is no problem in constructing this sequence by induction on the ordinals $\alpha < \kappa$ if α is not a limit ordinal or if C_α is unbounded in α: just use Theorem 1.1 and when necessary also Lemmas 2.3 and 2.4. Now let's consider the problematic case: α is a limit ordinal and C_α is not unbounded in α. Condition (1) forces $\mathfrak{N}_\alpha = \bigcup\{\mathfrak{N}_\beta : \beta < \alpha\}$. To get S_α, let γ be the maximum ordinal in C_α; γ exists since C_α is closed and bounded. Notice that $\mathfrak{N}_\alpha$ is recursively saturated and that $cf(\mathfrak{N}_\alpha) = \aleph_0$ since $cf(\alpha) = \aleph_0$ by condition (ii) above. From (5), $(\mathfrak{N}_\gamma, S_\gamma) \prec (\mathfrak{N}_{\gamma+1}, S_{\gamma+1})$, so all the hypotheses of Lemma 2.7 are met. Consequently, there is a satisfaction class S_α for $\mathfrak{N}_\alpha$ such that $(\mathfrak{N}_\gamma, S_\gamma) \prec (\mathfrak{N}_\alpha, S_\alpha)$.

It is an easy verification that $\langle (\mathfrak{N}_\alpha, S_\alpha) : \alpha < \kappa \rangle$ satisfies properties (1) - (6). This completes the proof of Theorem 4.

The proof that we have just presented actually yields additional information. The model $\mathfrak{N}$ that was constructed is an elementary end-extension of $\mathfrak{N}_0$. Since both models are recursively saturated, they are $\aleph_0$-homogeneous and realize the same types, so that $\mathfrak{N} \equiv_{\infty,\omega} \mathfrak{N}_0$. If we start with any recursively saturated $\mathfrak{M}$, then there is a recursively saturated $\mathfrak{N}_0 \equiv_{\infty,\omega} \mathfrak{M}$ such that $|N_0| \leq \aleph_1$ and $cf(\mathfrak{N}_0) = \aleph_0$. Following Lemma 2.6, this model has a satisfaction class S_0. Thus we get the following corollary to the proof.

Corollary 5.1. Assume V = L. Suppose $\kappa \geq \aleph_2$ is a regular cardinal which is not weakly compact, and $\mathfrak{M} \models PA$ is recursively saturated. Then there is a κ-like, recursively saturated, rather classless $\mathfrak{N} \equiv_{\infty,\omega} \mathfrak{M}$.

By starting out with an $\aleph_0$-saturated model, and combining the previous corollary with Corollary 3.4, we get the following.

Corollary 5.2. Assume V = L. If $\kappa \geq \aleph_2$, $cf(\kappa) > \aleph_0$, and κ is not weakly compact, then every consistent extension of PA has a κ-like, $\aleph_0$-saturated, rather classless model.

References

[1] K. J. Barwise and J. Schlipf, On recursively saturated models of arithmetic, Model Theory and Algebra, Lecture Notes in Mathematics 498, Springer-Verlag, Berlin, 1976, pp. 42-55.

[2] J. E. Baumgartner, J. I. Malitz, and W. Reinhardt, Embedding trees in the rationals, Proc. Nat. Acad. Sci. USA. 67 (1970), 1748-1753.

[3] H. Gaifman, A note on models and submodels of arithmetic, Conference in Mathematical Logic-London '70, Lecture Notes in Mathematics 255, Springer-Verlag, Heidelberg, 1972, pp. 128-144.

[4] R. B. Jensen, The fine structure of the constructible hierarchy, Annals Math. Logic 4 (1972), 229-308.

[5] M. Kaufmann, A rather classless model, Proc. A.M.S. 62 (1977), 330-333.

[6] H. Kotlarski, On elementary cuts in models of arithmetic, II, (preprint).

[7] R. MacDowell and E. Specker, Modelle der Arithmetik, Infinitistic Methods, Proceedings of the Symposium on the Foundations of Mathematics (Warsaw, 1959), Pergammon, 1961, pp. 257-263.

[8] W. Mitchell, Aronszajn trees and the independence of the transfer property, Annals Math. Logic 5 (1972), 21-46.

[9] J. S. Schlipf, A guide to the identification of admissible sets above structures, Annals Math. Logic 12 (1977), 151-192.

[10] J. H. Schmerl, Peano models with many generic classes, Pacific J. Math. 46 (1973), 523-536.

[11] S. Shelah, Models with second order properties II. Trees with no undefined branches, Annals Math. Logic 14 (1978), 73-87.

[12] C. Smorynski and J. Stavi, Cofinal extension preserves recursive saturation, (preprint).

THE DEGREES OF UNSOLVABILITY: GLOBAL RESULTS

Richard A. Shore
Cornell University
Ithaca, NY 14853/USA

0. INTRODUCTION

Our goal in this paper is, with the benefit of hindsight, to give a coherent straight line development of our work on the global structure of $\mathcal{D}$, the degrees of unsolvability ordered by Turing reducibility.

The results that we will present fall into three or four basic categories: the characterization of the first order theories of $\mathcal{D}$ and of various substructures of $\mathcal{D}$, the definability of certain degrees and relations on degrees in these structures, restrictions on automorphisms of and isomorphisms between them and in particular non-homogeneity results. Problems of these types were first raised in the context of degrees of unsolvability by Hartley Rogers Jr. in 1965 (see Rogers [1967a]). They have since then been repeated with many variations by various other workers such as Sacks [1966], Yates [1970] and Simpson [1977].

For some time it seemed that no progress was being made toward the solutions of these problems for $\mathcal{D}$. On the other hand there were some successes for the structure $\mathcal{D}'$, the degrees with ordering and the jump operator. Feiner [1970] disproved Rogers' strong homogeneity conjecture by showing that $\mathcal{D}' \not\cong \mathcal{D}'(\geq \underset{\sim}{0}^{(6)})$. ($\mathcal{D}(\geq \underset{\sim}{a})$ means the structure of degrees greater than or equal to $\underset{\sim}{a}$ under $\leq_T$. $\mathcal{D}'(\geq \underset{\sim}{a})$, $\mathcal{D}(\leq \underset{\sim}{a})$ and $\mathcal{D}[\underset{\sim}{a},\underset{\sim}{b}]$ are all defined analogously.) Yates [1972] gave another proof of this result which was improved upon by Jockusch and Solovay [1977]. They used it to show that every automorphism of $\mathcal{D}'$ is the identity on $\mathcal{D}'(\geq \underset{\sim}{0}^{4})$. Richter [1979] improves this result by one more jump to show that the base of the fixed cone can be taken to be $\underset{\sim}{0}^{(3)}$ in place of $\underset{\sim}{0}^{(4)}$.

As for definability questions, one can view a number of early theorems of hierarchy theory as giving results of this kind. This view becomes explicit in Jockusch and Simpson [1976] where some very nice results on definability by natural methods in $\mathcal{D}'$ are given. A

Preliminary versions of this paper were given in a series of seminars at both the University of Conn. and UCLA. We would like to thank the participants in those seminars and the institutions for their support, help and hospitality. The preparation of this paper was partially supported by NSF Grant MCS 80-03016.

much more general approach to these questions is presented in Simpson [1977]. In this paper Simpson first shows that the first order theory of $\mathcal{D}'$ is recursively isomorphic to that of true second order arithmetic. Then using the methods developed for this result he shows, for example, that all relations on degrees $\geq \underset{\sim}{0}^{(\omega)}$ definable in second order arithmetic are in fact definable in $\mathcal{D}'$. He also makes the first successful attack on any of these problems for $\mathcal{D}$ by showing that its theory is also recursively isomorphic to true second order arithmetic. In this paper we will present our approach to this last result and apply it to get results in all the other areas both for $\mathcal{D}$ and many of its substructures.

As far as substructures are concerned we will deal mainly with jump ideals, i.e., subsets closed downward and under jump and join, and cones, i.e., ones the form $\mathcal{D}(\geq \underset{\sim}{a})$. Thus we will only mention in passing results on segments such as $\mathcal{D}(\leq \underset{\sim}{0}')$ and $\mathcal{D}[\underset{\sim}{0}', \underset{\sim}{0}'']$ which can be found in Epstein [1979] and Shore [1981a] or ones on the r.e. degrees as in Lerman, Shore and Soare [1981] and Shore [1980]. The theorems we will prove are drawn from the following papers: Nerode and Shore [1979] and [1980], Shore [1979] and [1980], and Harrington and Shore [1981]. As a short preview we list some typical results:

Theorem 2.7. If $\mathbf{C} \subseteq \mathcal{D}$ is a jump ideal, then the first order theory of $\mathbf{C}$ is recursively isomorphic to that of the two sorted structure of true arithmetic with quantification over sets whose degrees are in $\mathbf{C}$.

Theorem 3.5. Every degree above those of all the hyperarithmetic sets which is definable in second order arithmetic is definable in $\mathcal{D}$.

Theorem 4.1. If $\mathcal{D}(\geq \underset{\sim}{a}) \cong \mathcal{D}(\geq \underset{\sim}{b})$ then $\underset{\sim}{a}$ and $\underset{\sim}{b}$ are contained in the same hyperdegree.

Theorem 4.5. Every automorphism of $\mathcal{D}$ is the identity on a cone with a hyperarithmetic base. Thus every degree above those of all the hyperarithmetic sets is fixed by every automorphism of $\mathcal{D}$.

Modulo citing a few basic facts from the literature we hope to give a self-contained presentation of the proofs of these and other related theorems. Our plan is to explain the coding apparatus in section one and the results on theories in section two. The theorems on definability will be proved in section three while those on automorphisms and homogeneity problems will appear in section four. Note, however, that the main results of this last section are essentially independent of the rest of the paper and except for the results on elementary equivalences it can be read on its own.

1. THE CODING APPARATUS

The starting point of our coding procedure is the observation that any countable sequence P_i of m_i-ary relations can be coded by a single graph, i.e., a symmetric irreflexive binary (s.i.b.) relation. In the setting of arithmetic this idea first appears in Church and Quine [1952]. Its general formulation for first order logic is given in Rabin and Scott [n.d.] and Lavrov [1963] (or see Rabin [1965] or Ershov et. al. [1965]). A fairly detailed version of this coding covering the second order case as well is also given in Nerode and Shore [1979, §1] where it is used to show that the theories of many reducibility orderings including $\mathfrak{D}$ are recursively isomorphic to that of true second order arithmetic.

The point of these codings in the first order case is that given any countable first order language L there is an effective transformation $\varphi \to \varphi^F$ taking formulas of L to ones in the language of a single s.i.b. relation such that φ is valid iff φ^F is. This is accomplished by actually giving an effective transformation $\mathfrak{A} \to \mathfrak{A}^{\mathcal{F}}$ from structures for L to graphs such that $\mathfrak{A} \models \varphi$ iff $\mathfrak{A}^{\mathcal{F}} \models \varphi^F$. We will never have to consider any of the particularities of the various possible coding schemes but details can be found in any of the above references. (The best pictures are unfortunately in Rabin and Scott [n.d.] but most are reproduced in Nerode and Shore [1979, §1].)

Our next step is to code graphs as ideals in a distributive lattice L with 0 and 1. The atoms of our lattice will be the domain of the s.i.b. relation to be coded. As an s.i.b. relation is simply a set of unordered pairs we first want a way of coding unordered pairs of distinct atoms. We specify such a coding procedure by requiring our lattice to satisfy the following condition:

(1.1) For every pair of distinct atoms $\{a_1, a_2\}$ there is a unique element which is strictly above $a_1 \vee a_2$ but not above any elements other than 0, a_1, a_2 and $a_1 \vee a_2$. (We call this element the code for $\{a_1, a_2\}$ and designate it by $C(a_1, a_2)$.)

Now any s.i.b. relation R on A_L, the atoms of L, determines an ideal I_R = the ideal generated by $\{C(a_1, a_2) \mid R(a_1, a_2)\}$. Conversely any ideal J determines an s.i.b. relation S_J on A_L by $S_J(a_1, a_2) \leftrightarrow C(a_1, a_2) \in J$. To see that we can really code every s.i.b. relation on A_L this way one must show that $S_{(I_R)} = R$ for every such relation R. The point to be verified is that no code $C(a_1, a_2)$ with (a_1, a_2) not in R is in the ideal generated by the codes of pairs that are in

R. In other words $C(a_1, a_2)$ is not below any finite join of other codes. In our original presentation we went to some trouble to guarantee this form of independence in a first order way. J. Schmerl pointed out to us, however, that their independence follows automatically from the fact that they are join irreducibles in a distributive lattice (see Balbes and Dwinger [1974, III.2]). Thus quantification over ideals of L can code quantification over relations on A_L. With a bit of care to fix a particular domain this shows that the theory of distributive lattice with quantification over ideals is equivalent to full second order logic. The details of the transformation are in Nerode and Shore [1979] along with some other similar examples.

For our purposes here we only need a single relation on A_L to code the relations of $\leq$, + and x of arithmetic. In terms of structures, we can effectively go from a graph $\mathfrak{L} = \langle B, R \rangle$ to a lattice L with an ideal I such that $\langle A_L, S_I \rangle \cong \langle B, R \rangle$. Indeed for a single relation we can just as well make the ideal principal. That is, we can construct L with an element r such that $\langle A_L, S_r \rangle \cong \langle B, R \rangle$ where $S_r = S_{I(r)}$ and $I_r = \{x | x \leq r\}$ is the principal ideal generated by r.

As the set of atoms of a lattice and the property of being a code for a pair of atoms are clearly first order definable in the language of lattices we can by the usual procedure of relativization effectively say in this language that the structure $\langle A_L, S_r \rangle$ satisfies any given sentence of the language of graphs. The only point to make is that $S_r(a_1, a_2)$ is translated by $C(a_1, a_2) \leq r$. We can now combine this translation with our effective transformation F of the language of arithmetic to that of graphs to say that a lattice codes a model of arithmetic.

<u>Definition 1.2. A structure L with distinguished element r codes a model of arithmetic</u> if

1) L is a distributive lattice with 0 and 1 satisfying (1.1) and

2) The structure $\langle A_L, S_r \rangle$ satisfies the translation of the following axioms for arithmetic:

(i) $<$ is an ω-like ordering (discrete with first but no last element) and (ii) + and x satisfy the usual inductive definitions over this ordering.

If the ω-like ordering of $\langle A_L, S_r \rangle$ is in fact an ω-ordering we say that <u>L codes a standard model of arithmetic</u>.

Now a model for second order arithmetic is a two sorted model $\langle B, \mathcal{P}, \leq, +, x; \varepsilon \rangle$ with elements in B and sets in $\mathcal{P}$. For our lattices this corresponds to adding on a class of ideals $\mathcal{I}$. The interpretation of course is that an atom a of L is in the set coded by an ideal $I \in \mathcal{I}$ iff $a \in I$. As the atoms are also join irreducible, distinct sets of atoms generate distinct ideals and so quantification over all ideals of L corresponds to quantification over all subsets of the model.

2. THEORIES OF DEGREE STRUCTURES

We will now consider our coding apparatus in the context of the degrees of unsolvability. Our lattice will be given by initial segments $\mathcal{D}(\leq \underset{\sim}{d})$ of $\mathcal{D}$. This suffices by Lachlan [1968]: Every countable distributive lattice with 0 and 1 is isomorphic to an initial segment of $\mathcal{D}$. Quantification over ideals of $\mathcal{D}(\leq \underset{\sim}{d})$ will be given by first order quantification over pairs of degrees. That this entails no loss is assured by Spector [1956]: Every countable ideal I of $\mathcal{D}$ has an exact pair, i.e., a pair $\underset{\sim}{x}$, $\underset{\sim}{y}$ such that I $= \{\underset{\sim}{d} \mid \underset{\sim}{d} \leq \underset{\sim}{x} \;\&\; \underset{\sim}{d} \leq \underset{\sim}{y}\}$.

Definition 2.1. The pair of degrees $\langle \underset{\sim}{d}, \underset{\sim}{r} \rangle$ codes a (standard) model of arithmetic if the structure $\mathcal{D}(\leq \underset{\sim}{d})$ with $\underset{\sim}{r} < \underset{\sim}{d}$ as a distinguished element codes a (standard) model of arithmetic.

We can now, of course, say in the first order language of $\mathcal{D}$ that $\langle \underset{\sim}{d}, \underset{\sim}{r} \rangle$ codes a model of arithmetic. The crucial point however is that we can also say it is a standard model. Spector's theorem says that quantification over pairs of degrees gives quantification over all countable ideals and so over all subsets of the model. With quantification over all subsets it is of course trivial to guarantee that the ω-like ordering is well founded. Section one now supplies us with a translation of arithmetic into the language of $\mathcal{D}$. As we can pick out the standard model in $\mathcal{D}$ and then quantify over all subsets in a first order way, we have reduced the theory of true second order arithmetic, $\mathrm{Th}^2(N) = \mathrm{Th}(\langle N, 2^N, \leq, +, \times; \epsilon \rangle)$, to the first order theory of $\mathcal{D}$. The reduction in the other direction is trivial and so $\mathrm{Th}(\mathcal{D})$ is recursively isomorphic to $\mathrm{Th}^2(N)$. As the analogs for the theorems of Lachlan and Spector hold for most reducibilities one also sees that the first order theories of 1-1, m-1, tt, wtt, and arithmetic degrees are all recursively isomorphic to $\mathrm{Th}^2(N)$. The sources for the initial segments results are Lachlan [1969] and [1970], Nerode and Shore [1979] and Harding [1974]. The versions of Spector's theorem needed are presented in Nerode and Shore [1979, §3].

Now on its face the above analysis required quantification over all subsets of a model of arithmetic (and so over all ideals) to guarantee that it was a standard model. In fact we only really need one subset (or ideal) - the one containing (generated by) the standard integers. To be precise if $\mathcal{L} = \langle B, \mathcal{P}, \leq +, \times; \epsilon\rangle$ is a model of second order arithmetic and the set of standard integers is an element of $\mathcal{P}$ then we can guarantee that $\mathcal{L}$ is a standard model by saying that

(2.2) Every proper initial segment of $\leq$ has a maximum element.

For our translation of arithmetic into the language of $\mathcal{D}$ we thus need only exact pairs for the ideals generated by the degrees representing standard integers in the models of arithmetic coded by any $\langle \underset{\sim}{d}, \underset{\sim}{r}\rangle$.

Definition 2.3. If $\langle \underset{\sim}{d}, \underset{\sim}{r}\rangle$ codes a model of arithmetic we let $\underset{\sim}{d}_n$, for $n \in N$, be the degree representing the integer n in this model. If $D \in \underset{\sim}{d}$ we let f_D be a function such that $\{f_D(n)\}^D \in \underset{\sim}{d}_n$.

Lemma 2.4. f_D is arithmetic in D, i.e., $f_D \leq_T D^{(n)}$ for some n. (We write this as $f_D \leq_a D$.)

Proof: First note that the ordering of Turing reducibility on representatives $\{i\}^D$ of $\mathcal{D}(\leq \underset{\sim}{d})$ is arithmetic in D. (In fact, recursive in $D^{(3)}$ by a quantifier counting argument.) Next consider the formula $\varphi(x,y)$ of arithmetic which asserts that x is the immediate successor of y. The translation of φ into the language of $\mathcal{D}(\leq \underset{\sim}{d})$ has some fixed number of quantifiers and so to see if it is satisfied by some pair is recursive in the corresponding number of jumps of the ordering relation. As this is itself arithmetic in D the relation $R(i,j)$ which says that $\deg(\{i\}^D)$ is the immediate successor of $\deg(\{j\}^D)$ is arithmetic in D. Thus beginning with an index i_0 with $\{i_0\}^D \in \underset{\sim}{d}_0$. We can recursively in a fixed number of jumps successively calculate i_n such that $\{i_n\}^D \in \underset{\sim}{d}_n$.

The point of this result is that once we have D and f_D Spector [1956] says that the exact pair for the ideal generated by $\{\deg\{f_D(n)\}^D | n \in N\}$ is recursive in just one more jump and so also arithmetic in D. Thus if $\mathbf{C} \subseteq \mathcal{D}$ is a jump ideal (and so closed under "arithmetic in") and some $\langle \underset{\sim}{d}, \underset{\sim}{r}\rangle$ in $\mathbf{C}$ codes a model of arithmetic, quantification over pairs in $\mathbf{C}$ suffices to guarantee that $\langle \underset{\sim}{d}, \underset{\sim}{r}\rangle$ codes a standard model of arithmetic. Now Lachlan [1968] shows that all recursively presented distributive lattices can be embedded as initial segments below $\underset{\sim}{0}^{(2)}$. As there is clearly one such which codes a standard model of arithmetic, every jump ideal $\mathbf{C}$ contains a $\langle \underset{\sim}{d}, \underset{\sim}{r}\rangle$ coding such a model. Since we can by the above computation always pick out the standard models in a jump ideal $\mathbf{C}$, we can, via our effective translation of arithmetic, reduce Th(N), the first order theory of true arithmetic, to Th($\mathbf{C}$) for any jump ideal $\mathbf{C}$. The question then is to see if there is a second order standard model of arithmetic whose theory is actually recursively isomorphic to Th($\mathbf{C}$) the way the full model was seen to be to that of $\mathfrak{D}$. To answer this question we must see what sets are coded in models $\langle \underset{\sim}{d}, \underset{\sim}{r} \rangle$ in $\mathbf{C}$ by ideals determined by exact pairs which are also in $\mathbf{C}$.

Definition 2.5. Suppose $\langle \underset{\sim}{d}, \underset{\sim}{r} \rangle$ codes a standard model of arithmetic. We say that the pair $\langle \underset{\sim}{x}, \underset{\sim}{y} \rangle$ codes the set $W \subseteq N$ for $\underset{\sim}{d}$ if $\forall n\ (n \in W \iff \underset{\sim}{d}_n \leq \underset{\sim}{x}\ \&\ \underset{\sim}{d}_n \leq \underset{\sim}{y})$.

Lemma 2.6. Let $\langle \underset{\sim}{d}, \underset{\sim}{r} \rangle$ code a standard model of arithmetic.

a) If $\langle \underset{\sim}{x}, \underset{\sim}{y} \rangle$ codes W for $\underset{\sim}{d}$ then W is arithmetic in $\underset{\sim}{x} \vee \underset{\sim}{y} \vee \underset{\sim}{d}$.

b) There is an $\underset{\sim}{e}$ arithmetic in $\underset{\sim}{d}$ such that for any set $W \in \underset{\sim}{w}$ there is a pair $\langle \underset{\sim}{x}, \underset{\sim}{y} \rangle$ recursive in $\underset{\sim}{w} \vee \underset{\sim}{e}$ coding W for $\underset{\sim}{d}$. Indeed the proof will show that there is an n independent of $\underset{\sim}{d}$ such that in part (a), $W \leq (\underset{\sim}{x} \vee \underset{\sim}{y} \vee \underset{\sim}{d})^{(n)}$ and in part (b) $\underset{\sim}{e} \leq \underset{\sim}{d}^{(n)}$.

Proof: Let $D \in \underset{\sim}{d}$, $X \in \underset{\sim}{x}$, $Y \in \underset{\sim}{y}$.

a) $n \in W \iff \{f_D(n)\}^D \leq_T X, Y$. As $f_D \leq_a D$ by Lemma 2.4 this relation is clearly arithmetic in $D \vee X \vee Y$.

b) We need an exact pair for the ideal generated by $\{\{f_D(n)\}^D | n \in W\}$. Let $\underset{\sim}{e} = \deg(f_D \vee D)'$. It is arithmetic in $\underset{\sim}{d}$ by Lemma 2.4. The sequence $S_i = \bigvee\{(\{f_D(n)\}^D)' | n \leq i\ \&\ n \in W\}$ is then uniformly recursive in $\underset{\sim}{w} \vee \underset{\sim}{e}$ and it is all that Spector [1956] needs to build the required pair.

With this lemma we can now characterize the theories of jump ideals in $\mathfrak{D}$.

Theorem 2.7. If $\mathbf{C} \subseteq \mathfrak{D}$ is a jump ideal then Th($\mathbf{C}$) is recursively isomorphic to $\mathrm{Th}(N, \mathbf{C}^*) = \mathrm{Th}(\langle N, \mathbf{C}^*, <, +, ;\, \in \rangle)$ where $\mathbf{C}^*$ is the class of all subsets of N whose degrees are in $\mathbf{C}$.

Proof: That $\mathrm{Th}(\mathbf{C}) \leq_{1-1} \mathrm{Th}(N, \mathbf{C}^*)$ is obvious. Consider then any sentence φ of second order arithmetic. The transformation of section one give us an equivalent formula φ^G in the language of lattices with a distinguished element and quantification over ideals. We translate the language of lattices into that of partial orderings by defining $\vee$ and $\wedge$ in the usual way. If $\langle \underset{\sim}{d}, \underset{\sim}{r} \rangle \in \mathbf{C}$ codes a standard model of arithmetic we say it satisfies the formula if $\mathfrak{D}(\leq \underset{\sim}{d})$ with r as distinguished element does. To finish the conversion into a formula in the language of partial orderings we now replace the quantification over ideals I of $\mathcal{D}(\leq \underset{\sim}{d})$ by quantification over pairs $\underset{\sim}{x}, \underset{\sim}{y}$ and the associated formula $\underset{\sim}{a} \in I$ by $\underset{\sim}{a} \leq \underset{\sim}{x}$ & $a \leq y$. This gives us a new formula $\varphi^{H(\underset{\sim}{d},\underset{\sim}{r})}$. (It depends on $\underset{\sim}{d}$ and $\underset{\sim}{r}$ because they are used as parameter to say that $\mathcal{D}(\leq \underset{\sim}{d})$ with $\underset{\sim}{r}$ satisfies φ^G.) We now claim that for any formula φ of second order arithmetic $(N, \mathbf{C}^*) \models \varphi$ iff $\mathbf{C} \models \exists \underset{\sim}{d}, \underset{\sim}{r}(\langle \underset{\sim}{d}, \underset{\sim}{r}\rangle$ codes a standard model of arithmetic & $\varphi^{H(\underset{\sim}{d},\underset{\sim}{r})})$. (We write this as $\mathbf{C} \models \varphi^H$.)

Suppose $(N, \mathbf{C}^*) \models \varphi$. We surely have a pair $\langle \underset{\sim}{d}, \underset{\sim}{r}\rangle$ in $\mathbf{C}$ that codes a standard model of arithmetic and so one that satisfies the appropriate requirements in $\mathbf{C}$. Now φ^G translates faithfully from the language of arithmetic to that of lattices. Thus the only worry is to determine the range of the second order quantifiers over this model when they are given by exact pairs in $\mathbf{C}$. As $\mathbf{C}$ is closed under "arithmetic in", Lemma 2.6 says that this range is precisely $\mathbf{C}^*$. Thus $\mathbf{C} \models \varphi^{H(\underset{\sim}{d},\underset{\sim}{r})}$.

For the converse let $\underset{\sim}{d}, \underset{\sim}{r}$ be some appropriate witnesses. By the remarks to Lemma 2.4, $\langle \underset{\sim}{d}, \underset{\sim}{r}\rangle$ does in fact code a standard model of arithmetic. Again the question is what is the class of the sets coded for $\underset{\sim}{d}$ by pairs in $\mathbf{C}$. Once again Lemma 2.6 says that it is precisely $\mathbf{C}^*$. Thus $\mathbf{C} \models \varphi^H$ implies that $(N, \mathbf{C}^*) \models \varphi$.

This theorem characterizes the theories of most of the familiar substructures of $\mathcal{D}$. By relativizing the initial segment results and Spector's theorem it also covers the corresponding substructures of $\mathcal{D}(\geq \underset{\sim}{a})$ as well. (Note that $\mathbf{C}^*$ must here be closed under $\leq_T$.) The major structures left unanalyzed are $\mathcal{D}(\leq \underset{\sim}{0}^{\omega})$ and RED, the r.e. degrees, $\mathcal{D}(\leq 0^{\omega})$ was first proved undecidable in Epstein [1979] using methods like those of Simpson [1977] but below $\underset{\sim}{0}'$. Lerman [1978] gives an initial segment result also giving its undecidability. Th($\mathcal{D}(\leq \underset{\sim}{0}^{\omega})$) was then shown to be recursively isomorphic to true first order arithmetic by a procedure similar to, but technically more difficult than, the one above in Shore [1981a] by using Lerman [1978] and new strengthened versions of Spector's theorem. The theory of RED remains fairly obscure although Lerman, Shore and Soare [1981] shows that it is not $\aleph_0$-categorical.

An important point to notice about Theorem 2.7 is that it gives a uniformly effective translation $\varphi \mapsto \varphi^H$ such that $\langle N, \mathbf{C}^*\rangle \models \varphi$ iff $\mathbf{C} \models \varphi^H$. Thus we can prove that two jump ideals $\mathbf{C}_1$, and $\mathbf{C}_2$ are not elementary equivalent simply by showing that $\langle N, \mathbf{C}_1^*\rangle \not\equiv \langle N, \mathbf{C}_2^*\rangle$.

Corollary 2.8. Let $\mathcal{D}_n$ be the degrees of the Δ^1_n sets for $n \geq 0$.
1) $\mathcal{D} \not\equiv \mathcal{D}_n$ for any $n \in N$.

2) $\mathcal{D}_0 \not\equiv \mathcal{D}_1$

3) If $V = L$ or AD holds and $n < m$ then $\mathcal{D}_n \not\equiv \mathcal{D}_m$.

Proof: The only sets explicitly definable in $(N, \mathcal{D}_n^*)$ are obviously recursive in $E_n^{(\omega)}$ where E_n is the complete Σ_n^1 set.

1) In $(N, \mathcal{D}^*) = (N, 2^N)$ one can of course explicitly define every E_{n+1} by some $\varphi_{n+1}(x)$. Thus it and $(N, \mathcal{D}_n^*)$ must differ on some instance of this formula $\varphi_{n+1}(\underline{k})$. As each integer is definable, the structures cannot be elementary equivalent.

2) In $(N, \mathcal{D}_1^*)$ one can define E_1 as Π_1 over Δ_1^1 is equivalent to Σ_1^1 by the Spector-Gandy theorem (see Rogers [1967], §16.7). Now argue as in (1).

3) If $V = L$ then Δ_m^1 is a basis for Σ_m^1 by Addison [1959] and so E_m is explicitly definable in $(N, \mathcal{D}_m^*)$. If AD holds then Δ_{2k}^1 is a basis for Σ_{2k}^1 by Moschovakis [1971]. Thus the only new worry is when $n = 2k$ and $m = 2k+1$. Here E_{2k} is explicitly definable in both structures but only in $(N, \mathcal{D}_{2k+1}^*)$ does the sentence which asserts that there is a set satisfying this definition hold.

Note: A very different proof of (1) for $n = 0$ appears in Jockusch [1973]. We give one last application:

Corollary 2.9. Let $\mathcal{D}_n(A)$ be the degrees of the set $\Delta_n^1(A)$ for $n = 0,1$.

$$\mathcal{D}_n \equiv \mathcal{D}_n(A) \text{ iff } A \in \Delta_n^1 .$$

Proof: The if direction is trivial so suppose $A \notin \Delta_n^1$. Then $(N, \mathcal{D}_n(A)^*)$ satisfies the sentence saying there is a set which is not Δ_n^1 but $(N, \mathcal{D}_n^*)$ does not. (We can say that S is not Δ_0^1 in such a two sorted model as follows:

$$\forall n \; \exists X(X^{[0]} = \phi \;\&\; \forall i \leq n(X^{[i]'} = X^{[i+1]}) \;\&\; S \not\leq_T X^{[n]}) .$$

We can say that S is not Δ_1^1 by saying that it is not recursive in any set which is implicitly Π_2^0 definable.) As a final remark we note that all of the results of this section hold for 1-1, m-1, tt, and wtt degrees as well as by essentially the same proofs.

3. DEFINABILITY

In this section we will consider questions of definability in $\mathcal{D}$ and its substructures. Major results on definability in $\mathcal{D}'$ appear in Jockusch and Simpson [1976] and Simpson [1977]. In Nerode and Shore [1979, §2] we showed how to replace the jump operator by either a parameter such as $\mathbf{0}^{\omega\omega}$ or a predicate for any jump ideal and still get the major definability results of Simpson [1977]. The missing ingredient is then supplied by Harrington and Shore [1981] where a particular jump ideal is proved to be definable in $\mathcal{D}$. Thus we get that almost all possible definability results actually are true. (Almost all means roughly on a cone and possible refers to the obvious restriction that anything definable in $\mathcal{D}$ is definable in second order arithmetic.) We first restate the essential lemma which is proved in Harrington and Shore [1981].

<u>Lemma 3.1</u>. If $\mathbf{x} \notin \mathfrak{H}$ (the degrees of hyperarithmetic sets) and $\mathbf{x}, \mathcal{O} \leq \mathbf{t}$, then there is a degree $\mathbf{s} \not\geq \mathbf{x}$ such that $\mathbf{t}$ is a minimal cover of $\mathbf{s}$, i.e., $\mathbf{t} > \mathbf{s}$ and there is no degree strictly between $\mathbf{t}$ and $\mathbf{s}$. ($\mathcal{O}$ is the degree of the complete Π^1_1 set.)

Now we producc our definable jump ideal.

<u>Lemma 3.2</u>. $\mathbf{C} = \{\mathbf{x} \mid \exists \mathbf{y}[\mathbf{x} \leq \mathbf{y} \;\&\; \forall \mathbf{z}(\mathbf{z} \vee \mathbf{y}$ is not a minimal cover of $\mathbf{z})]\}$ is a jump ideal with $\mathfrak{A} \subseteq \mathbf{C} \subseteq \mathfrak{H}$. ($\mathfrak{A}$ is the class of degrees of arithmetic sets.)

<u>Proof</u>: $\mathbf{C}$ is obviously closed downward. Suppose $\mathbf{x}_1, \mathbf{x}_2 \in \mathbf{C}$. Let $\mathbf{y}_1$ and $\mathbf{y}_2$ be the appropriate witnesses and consider any degree $\mathbf{z}$. As neither $\mathbf{z} \vee \mathbf{y}_1$ nor $\mathbf{z} \vee \mathbf{y}_2$ is a minimal cover of $\mathbf{z}$ neither is $\mathbf{z} \vee \mathbf{y}_1 \vee \mathbf{y}_2$. Thus $\mathbf{y}_1 \vee \mathbf{y}_2$, witnesses that $\mathbf{x}_1 \vee \mathbf{x}_2 \in \mathbf{C}$. Next consider $\mathbf{z} \vee \mathbf{y}_1$. It is r.e. in $\mathbf{z} \vee \mathbf{y}_1 \geq \mathbf{z}$ and so by the relativized version of the non minimality of r.e. degrees (Friedberg [1957] or see Sacks [1966]) it could not be a minimal cover of $\mathbf{z}$ unless it equaled $\mathbf{z} \vee \mathbf{y}_1$. As $\mathbf{z} \vee \mathbf{y}_1$ is not a minimal cover of $\mathbf{z}$ by assumption $\mathbf{y}_1'$ witnesses that $\mathbf{x}_1' \in \mathbf{C}$.

Thus $\mathbf{C}$ is a jump ideal and so contains $\mathfrak{A}$. The point of Lemma 3.1 is that $\mathbf{C} \subseteq \mathfrak{H}$ as the requirement there that $\mathbf{s} \not\geq \mathbf{x}$ implies that $\mathbf{t} = \mathbf{s} \vee \mathbf{x}$ is the witness that $\mathbf{x} \notin \mathbf{C}$ if $\mathbf{x} \notin \mathfrak{H}$.

We note that the proof Lemma 3.1 is based on that of Harrington and Kechris [1975] of which it is a strengthening. The proof of Lemma 3.2 follows that of Jockusch and Soare [1970]. These earlier results were the ones actually used in Shore [1979] and [1981]. The versions given here lead to sharper and simpler results as well as the new results of this section on definability. As a sample of what can now be defined rather simply we prove the following:

<u>Theorem 3.3</u>. $\mathfrak{H}$ is definable in $\mathcal{D}$.

Proof: Let $\varphi(S)$ be the formula of second order arithmetic which says that S is not hyperarithmetic. We claim that $\underset{\sim}{z} \notin \mathbf{H}$ iff

$$\mathcal{D} \models (\exists \underset{\sim}{d}, \underset{\sim}{r}, \underset{\sim}{e} \in \mathbf{C})(\exists \underset{\sim}{x}, \underset{\sim}{y} \leq \underset{\sim}{z} \vee \underset{\sim}{e})[\langle \underset{\sim}{d}, \underset{\sim}{r} \rangle \text{ codes a standard}$$

$$\text{model of arithmetic } \& \ \varphi^{H(\underset{\sim}{d},\underset{\sim}{r})}(\langle \underset{\sim}{x}, \underset{\sim}{y} \rangle)].$$

(We use $\varphi^{H(\underset{\sim}{d},\underset{\sim}{r})}(\langle \underset{\sim}{x}, \underset{\sim}{y} \rangle)$ to mean that we substitute the set coded for $\underset{\sim}{d}$ by the pair $\langle \underset{\sim}{x}, \underset{\sim}{y} \rangle$ for the free set variable in φ in the translation given by $H(\underset{\sim}{d}, \underset{\sim}{r})$.)

If $\underset{\sim}{z} \notin \mathbf{H}$ we choose any $\underset{\sim}{d}, \underset{\sim}{r} \in \mathbf{C}$ coding a standard model. We then let $\underset{\sim}{e}$ be given by Lemma 2.6(b) and choose $\underset{\sim}{x}, \underset{\sim}{y}$ to code a set Z of degree $\underset{\sim}{z}$. As $(N, \mathcal{D}^*) \models \varphi(Z)$, $\mathcal{D} \models \varphi^{H(\underset{\sim}{d},\underset{\sim}{r})}(\langle \underset{\sim}{x}, \underset{\sim}{y} \rangle)$ as required. On the other hand if $\underset{\sim}{d}, \underset{\sim}{r}, \underset{\sim}{e}, \underset{\sim}{x}$ and $\underset{\sim}{y}$ are witnesses as required and $\mathcal{D} \models \varphi^{H(\underset{\sim}{d},\underset{\sim}{r})}(\langle \underset{\sim}{x}, \underset{\sim}{y} \rangle)$ then $(N, \mathcal{D}^*) \models \varphi(W)$ where W is the set coded by $\langle \underset{\sim}{x}, \underset{\sim}{y} \rangle$ for $\underset{\sim}{d}$. Thus W is not hyperarithmetic. By Lemma 2.6(a) however W is arithmetic in $\underset{\sim}{d} \vee \underset{\sim}{z}$. As $\underset{\sim}{d} \in \mathbf{C} \subseteq \mathbf{H}$ we conclude that $\underset{\sim}{z} \notin \mathbf{H}$.

Essentially this proof shows that any relation on degrees closed under both arithmetic in and joining with an arbitrary degree in $\mathbf{C}$ is definable in $\mathcal{D}$ iff it is definable in second order arithmetic. Other examples include all the other $\mathcal{D}_n$'s for $n > 1$ and the class of degrees of constructible sets. To get finer relations and even particular degrees to be definable we need one more fact and a bit more work.

Lemma 3.4. (Selman [1972]): For each $n \geq 1$ and any degree $\underset{\sim}{c} \geq \underset{\sim}{d}^{(n)}$ there are $\underset{\sim}{a}, \underset{\sim}{b} \geq \underset{\sim}{d}$ such that $\underset{\sim}{a} \vee \underset{\sim}{b} = \underset{\sim}{c} = \underset{\sim}{a}^{(n)} = \underset{\sim}{b}^{(n)}$. (For another proof see Jockusch [1974].)

Theorem 3.5. If $\underset{\sim}{s}$ is above every degree in $\mathbf{C}$ (e.g. above every one in $\mathbf{H}$) then $\underset{\sim}{s}$ is definable in $\mathcal{D}$ iff it is definable in second order arithmetic.

Proof: Let $\varphi(T)$ be the formula of second order arithmetic which says that $\deg(T^{(n)}) \leq \underset{\sim}{s}$ where n is the integer given in Lemma 2.6. (We are assuming of course that $\underset{\sim}{s}$ is definable in $(N, 2^N)$.) We claim that $\underset{\sim}{s}$ is the unique solution of $\psi(\underset{\sim}{w}) \equiv \forall \underset{\sim}{d}, \underset{\sim}{r}, \underset{\sim}{a} \in \mathbf{C}\ [\underset{\sim}{a} \geq \underset{\sim}{d}$ and $\langle \underset{\sim}{d}, \underset{\sim}{r} \rangle$ codes a standard model of arithmetic $\rightarrow \exists \underset{\sim}{b} \in \mathbf{C}(\underset{\sim}{b} \geq \underset{\sim}{c}\ \&\ \underset{\sim}{w} = \mathrm{LUB}\{\underset{\sim}{z} \geq \underset{\sim}{b} \mid (\forall \underset{\sim}{x}, \underset{\sim}{y} < \underset{\sim}{z})\, \varphi^{H(\underset{\sim}{d},\underset{\sim}{r})}(\langle \underset{\sim}{x}, \underset{\sim}{y} \rangle)\})]$.

Suppose $\mathcal{D} \models \psi(\underset{\sim}{w})$. Let $\underset{\sim}{d}, \underset{\sim}{r} \in \mathbf{C}$ code a standard model and let $\underset{\sim}{a}$ be the degree $\underset{\sim}{e}$ given by Lemma 2.6(b). Let $\underset{\sim}{b}$ be a witness as required in ψ. Consider now any $\mathbf{z} \geq \mathbf{b}$. If $\mathbf{z}^{(2n)} \leq \mathbf{s}$ and $\mathbf{x}, \mathbf{y} \leq \mathbf{z}$ then the set T coded for $\mathbf{d}$ by $\langle \mathbf{x}, \mathbf{y} \rangle$ is recursive in $\mathbf{z}^{(n)}$ by Lemma 2.6(a) and so its nth jump is recursive in $\mathbf{s}$. Thus $\varphi(T)$ holds and so therefore does $\varphi^{H(\mathbf{d},\mathbf{r})}(\langle \mathbf{x}, \mathbf{y} \rangle)$. Thus any such $\mathbf{z}$ is recursive in $\mathbf{w}$. As $\mathbf{b} \in \mathbf{C}$, $\mathbf{b}^{(2n)} \in \mathbf{C}$ and so $\mathbf{b}^{(n)} \leq \mathbf{s}$. Thus by Lemma 3.4 there are $\mathbf{z}_1, \mathbf{z}_2 \geq \mathbf{b}$ with $\mathbf{z}_1 \vee \mathbf{z}_2 = \mathbf{s} = \mathbf{z}_1^{(2n)} = \mathbf{z}_2^{(2n)}$. As $\mathbf{z}_1, \mathbf{z}_2 \leq \mathbf{w}$ by this argument $\mathbf{s} \leq \mathbf{w}$. On the other hand, if $\mathbf{z} \geq \mathbf{b}$ but $\mathbf{z} \nleq \mathbf{s}$ then by Lemma 2.6(a) there are $\mathbf{x}, \mathbf{y} \leq \mathbf{z}$ coding a set $Z \in \mathbf{z}$ for $\mathbf{d}$. As $\mathbf{z} \nleq \mathbf{s}$, $\deg(Z^{(n)}) \nleq \mathbf{s}$ and so $\varphi^{H(\mathbf{d},\mathbf{r})}(\langle \mathbf{x}, \mathbf{y} \rangle)$ fails. Thus $\mathbf{s}$ is an upper bound for $\{\mathbf{z} \geq \mathbf{b} \mid (\forall \mathbf{x}, \mathbf{y} < \mathbf{z})\, \varphi^{H(\mathbf{d},\mathbf{r})}(\langle \mathbf{x}, \mathbf{y} \rangle)\}$ and so $\mathbf{w} \leq \mathbf{s}$ and indeed $\mathbf{w} = \mathbf{s}$ as required. Finally we must show that $\mathcal{D} \models \psi(\mathbf{s})$. Let $\mathbf{d}$ $\mathbf{r}$, $\mathbf{a} \in \mathbf{C}$ be as assumed in ψ. Choose $\mathbf{b} \in \mathbf{C}$ to be $\mathbf{a} \vee \mathbf{e}$ where $\mathbf{e}$ is given by Lemma 2.6(b). The above argument (once all of these degrees are fixed) now shows that $\mathbf{s} = \mathrm{LUB}\{\mathbf{z} \geq \mathbf{b} \mid (\forall \mathbf{x}, \mathbf{y} < \mathbf{y})\, \varphi^{H(\mathbf{d},\mathbf{r})}(\langle \mathbf{x}, \mathbf{y} \rangle)\}$ as required.

We can now elaborate on this proof a bit to cover relations and at the same time relax the assumption that the degrees be above all those of $\mathbf{C}$.

<u>Theorem 3.6</u>. If $R(\mathbf{s}_1, \mathbf{s}_n)$ is a relation on degrees such that $R(\mathbf{s}_1, \ldots, \mathbf{s}_n)$ and $\mathbf{c}_1, \ldots, \mathbf{c}_n \in \mathbf{C}$ implies that $R(\mathbf{s}_1 \vee \mathbf{c}_n, \ldots, \mathbf{s}_n \vee \mathbf{c}_n)$ then R is definable in $\mathcal{D}$ iff it is definable in second order arithmetic. (In particular this holds if R is a relation only on degrees above all those in $\mathbf{C}$.)

<u>Proof</u>: Let $\varphi(S_1, \ldots, S_n)$ be the formula of second order arithmetic which says that $R(\deg(S_1), \ldots, \deg(S_n))$ holds. Let $\theta(A, B)$ say that $A^n \leq_T B$. We claim that $R(\mathbf{s}_1, \ldots, \mathbf{s}_n) \Longleftrightarrow \forall \mathbf{d}, \mathbf{r}, \mathbf{a} \in \mathbf{C}(\mathbf{a} \geq \mathbf{d}$ & $\langle \mathbf{d}, \mathbf{r} \rangle$ codes a standard model of arithmetic $\rightarrow \exists \mathbf{b}, \mathbf{x}_1, \mathbf{y}_1, \ldots, \mathbf{x}_n, \mathbf{y}_n$ $[\mathop{\&}_{i \leq n} (\mathbf{s}_i = \mathrm{LUB}\{\mathbf{z} \geq \mathbf{b} \mid (\forall \mathbf{x}, \mathbf{y} \leq \mathbf{z})\, \theta^{H(\mathbf{d},\mathbf{r})}(\langle \mathbf{x}, \mathbf{y} \rangle, \langle \mathbf{x}_i, \mathbf{y}_i \rangle)\})$ & $\varphi^{H(\mathbf{d},\mathbf{r})}(\langle \mathbf{x}_1, \mathbf{y}_1 \rangle, \ldots, \langle \mathbf{x}_n, \mathbf{y}_n \rangle)]$.

Suppose the indicated formula holds for $\mathbf{s}_1, \ldots, \mathbf{s}_n$. Choose any $\mathbf{d}$, $\mathbf{r} \in \mathbf{C}$ coding a standard model of arithmetic. Let $\mathbf{a}$ be the degree $\mathbf{e}$ given by Lemma 2.6(b) and $\mathbf{b}, \mathbf{x}_1, \mathbf{y}_1, \ldots, \mathbf{x}_n, \mathbf{y}_n$ the assumed witnesses. The proof of the last theorem shows that if S_i is the set coded by $\langle \mathbf{x}_i, \mathbf{y}_i \rangle$ for $\mathbf{d}$ then $S_i \in \mathbf{s}_i$. As $\varphi^{H(\mathbf{d},\mathbf{r})}(\langle \mathbf{x}_1, \mathbf{y}_1 \rangle, \ldots, \langle \mathbf{x}_n, \mathbf{y}_n \rangle)$ holds in $\mathcal{D}$, $\varphi(S_1, \ldots, S_n)$ is true and so $R(\mathbf{s}_1, \ldots, \mathbf{s}_n)$ holds as required.

Conversely if $R(\mathbf{s}_1, \ldots, \mathbf{s}_n)$ holds and $\langle \mathbf{d}, \mathbf{r} \rangle$ codes a standard model with $\mathbf{d} \leq \mathbf{a} \in \mathbf{C}$, we let $\mathbf{b} = \mathbf{a} \vee \mathbf{e}$ where $\mathbf{e}$ is again given by

Lemma 2.6(b). We can now by Lemma 2.6(a) choose $\underset{\sim}{x}_1, \underset{\sim}{y}_1, \ldots, \underset{\sim}{x}_n, \underset{\sim}{y}_n$ so that $\langle \underset{\sim}{x}_i, \underset{\sim}{y}_i \rangle$ codes a set $S_i \in \underset{\sim}{s}_i$ for $\underset{\sim}{d}$. Again the argument for Theorem 3.5 shows that $\underset{\sim}{s}_i = \mathrm{LUB}\{\underset{\sim}{z} \geq \underset{\sim}{b} \mid (\forall \underset{\sim}{x}, \underset{\sim}{y} \leq \underset{\sim}{z})\, \theta^{H(\underset{\sim}{d}, \underset{\sim}{r})}(\langle \underset{\sim}{x}, \underset{\sim}{y} \rangle, \langle \underset{\sim}{x}_i, \underset{\sim}{y}_i \rangle)\}$. As $\varphi(S_1, \ldots, S_n)$ is true we have then that $\varphi^{H(\underset{\sim}{d}, \underset{\sim}{r})}(\langle \underset{\sim}{x}_1, \underset{\sim}{y}_1, \rangle, \ldots, \langle \underset{\sim}{x}_n, \underset{\sim}{y}_n \rangle)$ holds and the formula on the right hand side is verified.

We now just list some examples of such definable relations.

Corollary 3.7. The following relations on $\mathcal{D}$ are definable in $\mathcal{D}$.

1) $\underset{\sim}{a}$ is Δ^1_n in $\underset{\sim}{b}$ for $n \geq 1$.

2) $\underset{\sim}{a} \equiv_T \mathcal{O}^{\underset{\sim}{b}}$ i.e., $\underset{\sim}{a}$ is the Turing degree of the hyperjump of $\underset{\sim}{b}$.

3) $\underset{\sim}{a}$ is constructible from $\underset{\sim}{b}$.

4) $\underset{\sim}{a} \equiv_T \underset{\sim}{b}^{\#}$.

(Of course we may definably set $\underset{\sim}{b} = \underset{\sim}{0}$ to get the definability of the corresponding classes and degrees.)

It is possible to reach inside $\mathcal{C}$ or more widely to its side if one has some appropriate starting point. Thus for example Nerode and Shore [1979, Theorem 2.8], the model for Theorem 3.5, shows that any degree $\geq \underset{\sim}{0}^{(7)}$ definable in $(N, 2^N)$ is definable in $\mathcal{D}$ from a parameter for $\underset{\sim}{0}^{(2)}$. Relations on such degrees are similarly definable by Shore [1981] Theorem 3.4 whose proof is the model for Theorem 3.6. Further technical improvements are also possible as described therein. The natural question that remains open is if the jump operator is actually definable in $\mathcal{D}$.

4. AUTOMORPHISMS AND HOMOGENEITY

One of the most pervasive phenomena in recursion theory is relativization. Given a proof of some structural fact about $\mathcal{D}$ one can almost invariably relativize it, i.e., switch to functions recursive in $\underset{\sim}{a}$ and degrees above $\underset{\sim}{a}$, to get the same fact about $\mathcal{D}(\geq \underset{\sim}{a})$ for any $\underset{\sim}{a}$.

This process led Rogers [1967, p.261] to the homogeneity problems: Is $\mathcal{D}(\geq \underset{\sim}{a}) \cong \mathcal{D}$ for every $\underset{\sim}{a}$ or even is $\mathcal{D}'(\geq \underset{\sim}{a}) \cong \mathcal{D}'$ for every $\underset{\sim}{a}$. It is however by applications of relativization itself that one proves these conjectures false. Thus for example the relativization of Lachlan [1968] says that every $\underset{\sim}{a}$-presentable countable distributive lattice is isomorphic to an initial segment of $\mathcal{D}[\underset{\sim}{a}, \underset{\sim}{a}'']$. Now it is easy to construct an $\underset{\sim}{a}$-presentable distributive lattice $L_{\underset{\sim}{a}}$ with 0 and 1 such that $\underset{\sim}{a}$ is recursive in the jump of any presentation of

$L_{\mathbf{a}}$ as a p.o. (As Yates [1972] suggests we can use a "chain of lines and diamonds" i.e., an increasing string of either successive elements or copies of $2^{\{0,1\}}$ with a copy of $2^{\{0,1\}}$ in the nth position iff $n \in A \in \mathbf{a}$. A largest element is then put at the top of the chain.) On the other hand, $\mathcal{D}[\mathbf{a}, \mathbf{a}'']$ and so every one of its initial segments is $\mathbf{a}^{(5)}$-presentable since Turing reducibility on indices $\{i\}^X$ is recursive in $X^{(3)}$. Thus if $\mathcal{D}[\mathbf{a}, \mathbf{a}''] \cong \mathcal{D}[\mathbf{b}, \mathbf{b}'']$, $L_{\mathbf{a}}$ is an initial segment of $\mathcal{D}[\mathbf{b}, \mathbf{b}'']$ and so $\mathbf{a} \leq \mathbf{b}^{(6)}$. So in particular if $\mathcal{D}'(\geq \mathbf{a}) = \mathcal{D}'(\leq \mathbf{b})$ then $\mathbf{a} \leq \mathbf{b}^{(6)}$. These ideas and results on $\mathcal{D}'$ come from Feiner [1970] and Yates [1972]. The sharpest known version is that if $\mathcal{D}'(\geq \mathbf{a}) \cong \mathcal{D}'(\geq \mathbf{b})$ then $\mathbf{a}^{(3)} \equiv \mathbf{b}^{(3)}$ (see Richter [1979] and Nerode and Shore [1979], Theorem 5.3). The failure of the homogeneity conjecture for $\mathcal{D}$ was first established in Shore [1979] by somewhat more indirect means. We are now, however, in a position to give just as simple a proof for $\mathcal{D}$ as for $\mathcal{D}'$.

<u>Theorem 4.1</u>. If $\mathcal{D}(\geq \mathbf{a}) \cong \mathcal{D}(\geq \mathbf{b})$ then $\mathbf{a} \equiv_h \mathbf{b}$ i.e., $\mathbf{a}$ and $\mathbf{b}$ are contained in the same hyperarithmetic degree.

<u>Proof</u>: Let $\mathbf{C}^{\mathbf{a}}$ be the substructure of $\mathcal{D}(\geq \mathbf{a})$ determined by interpreting the definition of $\mathbf{C}$ in $\mathcal{D}(\geq \mathbf{a})$. The relativizations of Lemmas 3.1 and 3.2 show that $\mathbf{C}^{\mathbf{a}}$ is a jump ideal in $\mathcal{D}(\geq \mathbf{a})$ with $\mathbf{A}^{\mathbf{a}} \subseteq \mathbf{C}^{\mathbf{a}} \subseteq \mathbf{H}^{\mathbf{a}}$. (Here $\mathbf{A}^{\mathbf{a}}$ and $\mathbf{H}^{\mathbf{a}}$ are the degrees in $\mathcal{D}(\geq \mathbf{a})$ which contain sets arithmetic and hyperarithmetic in $\mathbf{a}$ respectively.) As $\mathbf{A}^{\mathbf{a}} \subseteq \mathbf{C}^{\mathbf{a}}$, $L_{\mathbf{a}}$ is isomorphic to some initial segment of $\mathbf{C}^{\mathbf{a}}$. As $\mathbf{C}^{\mathbf{a}}$ and $\mathbf{C}^{\mathbf{b}}$ are defined by the same formulas in the isomorphic structures $\mathcal{D}(\geq \mathbf{a})$ and $\mathcal{D}(\geq \mathbf{b})$, $L_{\mathbf{a}}$ is isomorphic to some initial segment of $\mathbf{C}^{\mathbf{b}}$. Suppose the top element of this segment is $\mathbf{x} \in \mathbf{C}^{\mathbf{b}}$. Now $\mathcal{D}[\mathbf{b}, \mathbf{x}]$ is $\mathbf{x}^{(3)}$ presentable and so by the choice of $L_{\mathbf{a}}$, $\mathbf{a} \leq \mathbf{x}^{(4)}$. As $\mathbf{x} \in \mathbf{C}^{\mathbf{b}}$ $\mathbf{x}$ is hyperarithmetic in $\mathbf{b}$. Thus $\mathbf{a} \leq_h \mathbf{b}$. By symmetry $\mathbf{a} \equiv_h \mathbf{b}$ as required.

The natural weakening of the homogeneity conjectures suggested in Yates [1970] is the possibility that the structures $\mathcal{D}(\geq \mathbf{a})$ and $\mathcal{D}(\geq \mathbf{b})$ (or even $\mathcal{D}'(\geq \mathbf{a})$ and $\mathcal{D}'(\geq \mathbf{b})$) might be elementarily equivalent. One must tread carefully here for if Projective Determinary holds then indeed there is an $\mathbf{a}$ such that $\forall \mathbf{b} \geq \mathbf{a}$ $(\mathcal{D}'(\geq \mathbf{a}) \equiv \mathcal{D}'(\geq \mathbf{b}))$. This degree $\mathbf{a}$ however is not readily definable. Indeed Simpson [1977] refuted the conjecture for $\mathcal{D}'$ by showing that there is an $\mathbf{a}$ with $\mathcal{D}'(\geq \mathbf{a}) \not\equiv \mathcal{D}'$. The sharpest result here is in Shore [1981]: $\mathcal{D}'(\geq \mathbf{a}) \equiv \mathcal{D}' \Rightarrow \mathbf{a}^{(3)} = \mathbf{0}^{(3)}$. In the same paper we also gave a fairly complicated argument that the elementary equivalence version also fails for

$\mathcal{D}$: If $\underset{\sim}{a} \geq \mathbf{0}$ then $\mathcal{D}(\geq \underset{\sim}{a}) \not\equiv \mathcal{D}$. Once again Lemmas 3.1 and 3.2 give much simpler proofs.

Theorem 4.2. If $\underset{\sim}{a}$ is definable in second order arithmetic and $\mathcal{D}(\geq \underset{\sim}{a}) \equiv \mathcal{D}(\geq \underset{\sim}{b})$ then $\underset{\sim}{a} \equiv_h \underset{\sim}{b}$. Thus in particular if $\mathcal{D}(\geq \underset{\sim}{a}) \equiv \mathcal{D}$ then $\underset{\sim}{a}$ is hyperarithmetic.

Proof: Let $\psi(S) \iff S \in \underset{\sim}{a}$ and $\theta(S) \iff S \leq_h \underset{\sim}{a}$. We, of course, begin with $\mathbf{C}^{\underset{\sim}{a}}$ and $\mathbf{C}^{\underset{\sim}{b}}$ defined by the same formulas in the appropriate structures. We write $\underset{\sim}{x} \in \mathbf{C}$ for this formula. Now our initial segment facts and results on ideals all relativize and so we can correctly say in $\mathcal{D}(\geq \underset{\sim}{a})$ that some $\langle \underset{\sim}{d}, \underset{\sim}{r} \rangle$ codes a standard model of arithmetic and our translations $\varphi^{H(\underset{\sim}{d}, \underset{\sim}{r})}$ remain faithful. (Even though we are in $\mathcal{D}(\geq \underset{\sim}{a})$, quantifunction over all ideals is still quantification over all sets.) Thus $\mathcal{D}(\geq \underset{\sim}{a}) \models (\exists \underset{\sim}{d}, \underset{\sim}{r}, \underset{\sim}{x}, \underset{\sim}{y} \in \mathbf{C})(\langle \underset{\sim}{d}, \underset{\sim}{r} \rangle$ codes a standard model of arithmetic & $\psi^{H(\underset{\sim}{d}, \underset{\sim}{r})}(\langle \underset{\sim}{x}, \underset{\sim}{y} \rangle))$. By elementary equivalence $\mathcal{D}(\geq \underset{\sim}{b})$ also satisfies this sentence. Thus there is a pair $\langle \underset{\sim}{x}, \underset{\sim}{y} \rangle$ in $\mathbf{C}^{\underset{\sim}{b}}$ coding a set S of degree $\underset{\sim}{a}$ in a model $\underset{\sim}{d} \in \mathbf{C}^{\underset{\sim}{b}}$. By Lemmas 2.6(a) and 3.2 (relativized) $\underset{\sim}{a}$ is hyperarithmetic in $\underset{\sim}{b}$.

On the other hand we have that $\mathcal{D}(\geq \underset{\sim}{a}) \models (\forall \underset{\sim}{d}, \underset{\sim}{r}, \underset{\sim}{x}, \underset{\sim}{y} \in \mathbf{C})(\langle \underset{\sim}{d}, \underset{\sim}{r} \rangle$ codes a standard model of arithmetic $\rightarrow \theta^{H(\underset{\sim}{d}, \underset{\sim}{r})}(\langle \underset{\sim}{x}, \underset{\sim}{y} \rangle))$. Now in $\mathcal{D}(\geq \underset{\sim}{b})$ there is a $\langle \underset{\sim}{d}, \underset{\sim}{r} \rangle \in \mathbf{C}$ coding a standard model and $\underset{\sim}{x}, \underset{\sim}{y} \in \mathbf{C}^{\underset{\sim}{b}}$ coding a set $B \in \underset{\sim}{b}$ for $\underset{\sim}{d}$. Again by elementary equivalence and the faithfulness of the translation we have that $\theta(B)$ holds i.e., $\underset{\sim}{b} \leq_h \underset{\sim}{a}$.

Although we cannot in general remove the definability assumption on $\underset{\sim}{a}$ in this result (without, of course, contradicting PD) we can do so under suitable set theoretic hypotheses.

Theorem 4.3. If there is a well ordering of 2^N definable in second order arithmetic (e.g. if $V = L$) and $\mathcal{D}(\geq \underset{\sim}{a}) \equiv \mathcal{D}(\geq \underset{\sim}{b})$ then $\underset{\sim}{a} \equiv_h \underset{\sim}{b}$.

Proof: By the faithfulness of our usual translations and the elementary equivalence assumption the least pair of sets forming an exact pair for all the degrees hyperarithmetic in any set coded in $\mathbf{C}^{\underset{\sim}{a}}$ is the same as the one for $\mathbf{C}^{\underset{\sim}{b}}$. Thus the degrees hyperarithmetic in $\underset{\sim}{a}$ are the same as those in $\underset{\sim}{b}$.

Note that if we consider such questions for substructures of $\mathcal{D}$ we get similar results. Indeed for jump ideals the proof of Corollary 2.9 already gives results such as $\mathfrak{a}^{\underset{\sim}{a}} \equiv \mathfrak{a}$ or $\mathfrak{H}^{\underset{\sim}{a}} \equiv \mathfrak{H}$ implies that $\underset{\sim}{a}$ is arithmetic or hyperarithmetic respectively. Moving inside jump ideals Shore [1981a] shows that if $\mathcal{D}[\underset{\sim}{a}, \underset{\sim}{a}^{\omega}] \equiv \mathcal{D}[\underset{\sim}{0}, \underset{\sim}{0}^{\omega}]$ or $\mathcal{D}(\leq \underset{\sim}{a}') \equiv \mathcal{D}(\leq \underset{\sim}{0}^{\omega})$

then $\underset{\sim}{a}^{(4)} = \underset{\sim}{0}^{(4)}$. In the same vein if the degrees r.e. in and above $\underset{\sim}{a}$ are isomorphic to the r.e. degrees then $\underset{\sim}{a}$ is arithmetic by Shore [1980]. This isomorphism result and the isomorphism version of the other elementary equivalence results mentioned all generalize to pairs $\underset{\sim}{a}$, $\underset{\sim}{b}$, but of course the elementary ones cannot so relativize by arithmetic or analytic determinacy. The isomorphism versions of these results for $\mathcal{D}$ as well as its substructures can also be derived without any appeal to initial segment results. If one uses the finitely generated coding schemes of Shore [1980] then Kleene-Post type extension methods are all that one needs.

Now in the setting of $\mathcal{D}'$ the methods of Feiner [1970] and Yates [1972] used to disprove the homogeneity conjecture were turned by Jockusch and Solovay [1977] into a proof that all automorphisms of $\mathcal{D}'$ are the identity on the cone $\mathcal{D}'(\geq \underset{\sim}{0}^{(4)})$. (This was improved to $\mathcal{D}'(\geq \underset{\sim}{0}^{3})$ by Richter [1979].) For $\mathcal{D}$ however Nerode and Shore [1980, Theorem 4.1] first proved that every automorphism of $\mathcal{D}$ is the identity in some cone and then this was used in Shore [1979] to refute the homogeneity conjecture. We can now bring the analogy with $\mathcal{D}'$ back into line by reversing the order (of quantifiers) again.

Theorem 4.4. If $\varphi\colon \mathcal{D}(\geq \underset{\sim}{a}) \to \mathcal{D}(\geq \underset{\sim}{b})$ is an isomorphism then φ is the identity on a cone with base the join of one element in $\mathcal{C}^{\underset{\sim}{a}}$ and one in $\mathcal{C}^{\underset{\sim}{b}}$.

Proof: If $\varphi(\underset{\sim}{x}) \geq \underset{\sim}{b}^{(2)}$ i.e., $x \geq \varphi^{-1}(b^{2}) = \underset{\sim}{c}$ then $L_{\varphi(\underset{\sim}{x})}$ is isomorphic to a segment of $\mathcal{D}[\underset{\sim}{b}, \varphi(\underset{\sim}{x})]$ and so by the isomorphism to one of $\mathcal{D}[\underset{\sim}{a}, \underset{\sim}{x}]$. As $\mathcal{D}[\underset{\sim}{a}, \underset{\sim}{x}]$ is $\underset{\sim}{x}^{(3)}$ presentable $\varphi(x) \leq x^{(4)}$. Now by the completeness theorem of Friedberg [1957a] if $x \geq \underset{\sim}{c}^{(4)}$ there is a $\underset{\sim}{y} \geq \underset{\sim}{c}$ such that $\underset{\sim}{x} = \underset{\sim}{y}^{(4)} = \underset{\sim}{y} \vee \underset{\sim}{c}^{(4)}$ and so $\varphi(\underset{\sim}{x}) = \varphi(\underset{\sim}{y}) \vee \varphi(\underset{\sim}{c}^{(4)})$. Thus if $\underset{\sim}{x} \geq \varphi(\underset{\sim}{c}^{(4)})$ as well as $\underset{\sim}{c}^{(4)}$ then $\varphi(\underset{\sim}{x}) \leq \underset{\sim}{x}$. Applying the same argument to φ^{-1} in place of φ shows that if $x \geq \underset{\sim}{d}^{(4)} \vee \varphi^{-1}(d^{(4)})$ then $\varphi^{-1}(\underset{\sim}{x}) \leq \underset{\sim}{x}$ where $\underset{\sim}{d} = \varphi(\underset{\sim}{a}^{(2)})$. Combining these results shows that $\varphi(\underset{\sim}{x}) = \underset{\sim}{x}$ if $\underset{\sim}{x} \geq \underset{\sim}{c}^{(4)} \vee \varphi(\underset{\sim}{c}^{(4)}) \vee \underset{\sim}{d}^{(4)} \vee \varphi^{-1}(\underset{\sim}{d}^{(4)}) = \underset{\sim}{e}$. Now $\underset{\sim}{a}^{(2)}$ and $\underset{\sim}{b}^{(2)}$ are in $\mathcal{C}^{\underset{\sim}{a}}$ and $\mathcal{C}^{\underset{\sim}{b}}$ respectively and so $\underset{\sim}{d} = \varphi(\underset{\sim}{a}^{(2)})$ and $\underset{\sim}{c} = \varphi^{-1}(\underset{\sim}{b}^{(2)})$ are in $\mathcal{C}^{\underset{\sim}{b}}$ and $\mathcal{C}^{\underset{\sim}{a}}$ respectively. As these are jump ideals $\underset{\sim}{d}^{(4)} \in \mathcal{C}^{\underset{\sim}{b}}$ and $\underset{\sim}{c}^{(4)} \in \mathcal{C}^{\underset{\sim}{a}}$ and so $\varphi^{-1}(\underset{\sim}{d}^{(4)}) \in \mathcal{C}^{\underset{\sim}{a}}$ and $\varphi(\underset{\sim}{c}^{(4)}) \in \mathcal{C}^{\underset{\sim}{b}}$. Thus $\underset{\sim}{e}$ the base of our cone of fixed points is recursive in the join of $(\underset{\sim}{c}^{(4)} \vee \varphi^{-1}(\underset{\sim}{d}^{(4)})) \in \mathcal{C}^{\underset{\sim}{a}}$ and $(\underset{\sim}{d}^{(4)}) \vee \varphi(\underset{\sim}{c}^{(4)})) \in \mathcal{C}^{\underset{\sim}{b}}$.

Our basic result on automorphisms now follows by setting $\underset{\sim}{a} = \underset{\sim}{0} = \underset{\sim}{b}$.

Theorem 4.5. Any automorphism of $\mathfrak{D}$ is the identity on a cone with base in $\mathbf{C} \subseteq \aleph$ and so every degree above all those in $\mathbf{C} \subseteq \aleph$ is fixed by every automorphism.

Note that in general Theorems 4.1 and 4.4 combine to show that if $\varphi: \mathfrak{D}(\geq \underset{\sim}{a}) \cong \mathfrak{D}(\geq \underset{\sim}{b})$ then $\varphi(\underset{\sim}{x}) = \underset{\sim}{x}$ for every $\underset{\sim}{x}$ above some degree hyperarithmetic in $\underset{\sim}{a}$ (and so also in $\underset{\sim}{b}$). The natural question now, of course, asks if there are any nontrivial automorphisms of $\mathfrak{D}$ or indeed any non-trivial isomorphisms of $\mathfrak{D}(\geq \underset{\sim}{a})$ onto $\mathfrak{D}(\geq \underset{\sim}{b})$ for any $\underset{\sim}{a}$ and $\underset{\sim}{b}$.

REFERENCES

Addison, J. W. [1959] Some consequences of the axiom of constructibility, Fund. Math. 46, 337-357.

Balbes, R. and Dwinger, P. [1974] Distributive lattices, University of Missouri Press, Columbia, Missouri.

Church, A. and Quine, W. V. [1952] Some theorems on definability and decidability, J. Symb. Logic 17, 179-187.

Epstein, R. [1979] Degrees of unsolvability: Structure and theory, Lect. Notes in Math. 759, Springer-Verlag, Berlin.

Ershov, Yu. L., Lavrov, I. A., Taimanov, A. D. and Taitslin, A. M. [1965] Elementary theories, Russian Mathematical Surveys 20, 35-105.

Feiner, L. [1970] The strong homogeneity conjecture, J. Symb. Logic 35, 375-377.

Friedberg, R. M. [1957] The fine structure of degrees of unsolvability of recursively enumerable sets, seminars of Cornell Institute for Symbolic Logic, pp.404-406.

Friedberg, R. M. [1957a] A criterion for completeness of degrees of unsolvability, J. Symb. Logic 22, 159-160.

Harding, C. J. [1974] Forcing in recursion theory, Thesis, University College of Swansea.

Harrington, L. and Kechris, A. [1975] A basis result for Σ_3^0 sets of reals with an application to minimal covers, Proc. Am. Math. Soc. 53, 445-448.

Harrington, L. and Shore, R. A. [1981] Definable degrees and automorphisms of $\mathfrak{D}$, to appear.

Jockusch, C. G. Jr. [1973] An application of Σ_4^0 determinacy to the degrees of unsolvability, J. Symb. Logic 38, 293-294.

Jockusch, C. G. Jr. [1974] Review of Selman [1972] in Math Reviews 47 #3155.

Jockusch, C. G. Jr. and Simpson, S. G. [1976] A degree theoretic definition of the ramified analytical hierarchy, Ann. Math.Logic, 10,1-32.

Jockusch, C. G. Jr. and Soare, R. I. [1970] Minimal covers and arithmetical sets, Proc. Am. Math. Soc. 25, 856-859.

Jockusch, C. G. Jr. and Solovay, R. M. [1977] Fixed points of jump preserving automorphisms of degrees, Israel J. Math. 26, 91-94.

Lachlan, A. H. [1968] Distributive initial segments of the degrees of unsolvability, Z. f. Math. Logik und Grund. d. Math, 14, 457-472.

Lachlan, A. H. [1969] Initial segments of one-one degrees, Pacific J. Math. 29, 351-366.

Lachlan, A. H. [1970] Initial segments of many-one degrees, Can. J. Math. 22, 75-85.

Lavrov, I. A. [1963] Effective inseparability of the sets of identically true formulae and finitely refutable formulae for certain elementary theories, Alg. i. Logika 2, 5-18.

Lerman, M. [1978] Initial segments of the degrees below $\underset{\sim}{0}'$, Notices Am. Math. Soc. 25, A-506.

Lerman, M., Shore, R. A. and Soare, R. I. [1981] The elementary theory of the recursively enumerable degrees is not $\aleph_0$-categorical, to appear.

Moschovakis, Y. N. [1973] Uniformization in a playful universe, Bull. Am. Math. Soc. 77, 731-736.

Nerode, A. and Shore, R. A. [1979] Second order logic and first order theories of reducibility orderings, in <u>The Kleene Symposium</u>, J. Barwise, J. Keisler and K. Kunen eds. North-Holland, Amsterdam.

Nerode, A. and Shore, R. A. [1981] Reducibility orderings: Theories, definability and automorphisms, Ann. Math. Logic 18, 61-89.

Rabin, M. [1965] A simple method for undecidability proofs and some applications, <u>Logic Methodology and Philosophy of Science, Proc. 1964 Int. Congress</u>, Bar-Hillel ed., North-Holland, Amsterdam, 58-68.

Rabin, M. and Scott, D. [n.d.] The undecidability of some simple theories, mimeographed notes.

Richter, L. S. C. [1979] On automorphisms of the degrees that preserve jumps, Israel J. Math. 32, 27-31.

Rogers, H. Jr. [1967] <u>Theory of Recursive Functions and Effective Computability</u>, McGraw Hill Co., New York.

Rogers, H. Jr. [1967a] Some problems of definability in recursive function theory, in <u>Sets Models and Recursion Theory</u>, Proc. of the Summer School in Mathematical Logic and Tenth Logic Colloquium Lancaster August-September 1965, Crossley, J. N. ed., North-Holland, Amsterdam.

Sacks, G. E. [1966] <u>Degrees of Unsolvability</u>, Am. Math. Studies no.55, 2nd ed. Princeton University Press, Princeton, N.J.

Selman, A. L. [1972] Applications of forcing to the degree theory of the arithmetic hierarchy, Proc. London Math. Soc. (3), 25, 586-602.

Shore, R. A. [1979] The homogeneity conjecture, Proc. Nat. Ac. Sci. 76, no.9, 4218-4219.

Shore, R. A. [1980] The degrees recursively enumerable in a degree $\mathbf{d}$. Abstracts Am. Math. Soc. 1, no.6.

Shore, R. A. [1981] On homogeneity and definability in the first order theory of the Turing degrees, J. Symb. Logic, to appear.

Shore, R. A. [1981a] The theory of the degrees below $\mathbf{0}'$, J. London Math. Soc., to appear.

Simpson, S. G. [1977] First order theory of the degrees of recursive unsolvability, Ann. Math. (2) 105, 121-139.

Spector, C. [1956] On degrees of recursive unsolvability, Ann. Math. 64, 581-592.

Yates, C. E. M. [1970] Initial segments of the degrees of unsolvability Part II: Minimal degrees, J. Symb. Logic 35, 243-266.

Yates, C. E. M. [1972] Initial segments and implications for the structure of degrees, in Conference in Mathematical Logic, London 1970. Hodges ed. Lecture Notes in Mathematics no.255, Springer-Verlag, Berlin, 305-335.

Two Theorems on Autostability in p-Groups[1]

Rick L. Smith[2]
Department of Mathematics
The Ohio State University
Columbus, Ohio 43210

1. Summary. Mal'cev [6] introduced the notion of an autostable structure as a recursive structure where an isomorphism with another recursive structure can be replaced by a recursive isomorphism. Aside from the simplicity and naturalness of this notion, it has fundamental importance since effectiveness questions about algebraic properties are determined in an autostable structure. We will call the problem of characterizing the autostable structures the autostability problem. At present there is no general model theoretic solution to the autostability problem. There are solutions in specific categories. The most general was provided by Nurtazin [8] for the category of decidable models. As a sample application of Nurtazin's Theorem one obtains the following: an algebraically closed field is autostable iff it has finite transcendence degree over its prime field. LaRoche [4] has shown that a Boolean algebra is autostable iff it has only finitely many atoms. Theorem 1 of this paper is classification of the autostable p-groups.

The autostability problem is altered when new relations are added to a structure and solutions to the problem acquire a different flavor. Suppose F is a recursive algebraically closed field and $K \subset F$ is an r.e. subfield with F algegraic over K. Now form the recursive structure $(F,a)_{a \in K}$ i.e. we include constants from K in the new structure. The autostability problem for this type of structure is the same as the problem of characterizing those fields K which have a recursively unique algebraic closure. Metakides and Nerode [7] and Smith [11] have shown that this is equivalent to the existence of a splitting algorithm for the separable polynomials over K. In Theorem 2 we classify those p-groups which have a recursively unique divisible closure.

Theorem 2 presupposes that every recursive p-group has a recursive divisible closure. This is proven in Proposition 1. In fact, we show that every recursive abelian group has a recursive divisible closure. We know of no proof in the literature which is sufficiently effective to give this result. The proof given here is drawn from ideas in Rabin [9].

---

[1] Section 3 of this paper is part of the author's doctoral thesis written under Stephen G. Simpson. Section 4 was motivated by conversations with Anil Nerode.

[2] Supported by an AMS Postdoctoral Research Fellowship and NSF Grant MCS-79-23743.

2. Notation. The set of natural numbers is denoted ω. All groups in this paper are abelian. The order of an element $g \in G$ is written $o(g)$. G is a p-group if $o(g)$ is always some power of p. The easiest examples of p-groups are the cyclic groups of order p^n, written $\mathbb{Z}(p^n)$. The direct limit of these is the quasi-cyclic group, $\mathbb{Z}(p^\infty) = \varinjlim \mathbb{Z}(p^n)$. $G \oplus H$ indicates the direct sum of G and H, $\oplus_m G$ is m-copies of G, and $\oplus_\omega G$ is the weak direct sum of countably many copies of G. We use the standard notations, $pG = \{px: x \in G\}$ and $G[p^k] = \{x \in G: p^k x = o\}$. A p-group has bounded order if $G[p^k] = G$ for some k. For $g \in G$, we say g has finite height if there is an n such that $p^{n+1}x \neq g$ for all $x \in G$, and if g has finite height, then the height of g is the least such n. A p-group is divisible if no element has finite height and reduced if (o) is the only divisible subgroup. $\langle X \rangle$ is the subgroup generated by $X \subseteq G$. Fuchs [2] and Kaplansky [3] are references for this material.

A group is recursive if its domain is a recursive set of natural numbers and the group operations are recursive functions. The eth partial recursive function is written $\{e\}$ and its restriction to computations of $\leq s$ steps is written $\{e\}_s$. Rogers [10] is a general reference for recursion theory.

3. Autostable p-Groups

Definition 1: Let G be a recursive p-Group. G is autostable if for all recursive $H \cong G$ there is a recursive $\varphi: H \cong G$.

The next theorem classifies the autostable p-groups.

Theorem 1: Let G be a recursive p-group. G is autostable iff either $G = \oplus_\omega \mathbb{Z}(p^\infty) \oplus F$ or $G = \oplus_m \mathbb{Z}(p^\infty) \oplus \oplus_\omega \mathbb{Z}(p^n) \oplus F$ where F is a finite p-group and $m,n \in \omega$.

Proof: If G has one of these forms, then it is easy to build a recursive isomorphism with an effective back and forth procedure. Thus we will assume for the remainder of the proof that G is autostable.

Lemma 1: If G has bounded order, then $G = \oplus_\omega \mathbb{Z}(p^n) \oplus F$ for some $n \in \omega$ and some finite F.

Proof: Suppose G is infinite, then there is an n such that $\oplus_\omega \mathbb{Z}(p^n)$ is a summand of G. Let n be least so that

$$G = \oplus_\omega \mathbb{Z}(p^n) \oplus M \oplus F_o$$

where F_o is finite and every summand of F_o has size less than p^n and every summand of M has size greater than p^n. We may assume that M is infinite. Let $A \subseteq \omega$ be any r.e. nonrecursive set. We will build an $H \cong G$ such that

$$\{x \in H: o(x) = p \ \& \ \exists y\ (p^n y = x)\}$$

has the same Turing degree as A.

H will be a direct limit of a recursive sequence of finite groups, $\{H_s\}_{s<\omega}$. Let $\{c_n\}_{n<\omega}$ be a recursive basis for $G[p]$ and $\{G_s\}_{s<\omega}$ a recursive sequence of finite subgroups of G such that $G = \bigcup_s G_s$ and $G_s = \bigoplus_{i\leq s} C_{i,s}$ where $C_{i,s}$ is cyclic and $c_i \in C_{i,s}$. Thus $\lim_s C_{i,s} = C_i$. In the construction of H_s, we will have $H_s = \bigoplus_{i\leq n(s)} B_{i,s}$ where $B_{i,s} \subseteq B_{i,s+1}$ and $B_{i,s}$ is cyclic. At the same time we define a bijection η: $\{i: C_i$ has size $> p^n\} \to A$ in stages, $\eta = \bigcup_s \eta_s$, $\eta_s \subseteq \eta_{s+1}$, η_s: $\{i: C_{i,s}$ has size $> p^n\} \to A$. Let $a_0, a_1, \ldots$ be a recursive enumeration of A.

The Construction

Stage 0: Let $H_0 = (o)$ and η_0 be undefined.

Stage s+1: Let $H_{s+1} = H_s$ unless there are j,k such that $a_k \notin$ ran (η_s) and $j \notin$ dom (η_s) but $C_{j,s+1}$ has size $> p^n$. Pick the least j and the least k and define $\eta_{s+1} = \eta_s \cup \{(j,a_k)\}$, $r(s+1) = \max\{r(s), a_k\}$. For $i \in$ dom (η_{s+1}) and $\eta_{s+1}(i) = \ell$ define $B_{\ell,s+1} \supseteq B_{\ell,s}$ and $B_{\ell,s+1} \cong C_{i,s+1}$. For $i \leq r(s+1)$ and $i \notin$ ran (η_{s+1}) define $B_{i,s+1} = \mathbb{Z}(p^n)$.

End of Construction

Let $H = \varinjlim H_s$. Clearly dom $(\eta) = \{i: C_i$ has size $> p^n\}$ and ran $(\eta) = A$. For $i \in$ dom(η) $C_i \cong B_{\eta(i)} = \bigcup_s B_{\eta(i),s}$ and for $i \notin$ dom(η) $C_i \cong \mathbb{Z}(p^n)$. If $i \notin$ ran (η), $B_i \cong \mathbb{Z}(p^n)$. Thus $H \cong G$. Let $X = \{x \in H: o(x) = p \;\&\; \exists y(p^n y = x)\}$.

Claim: X has the same Turing degree as A.

By the construction $i \in A$ iff B_i has size $> p^n$. A is reducible to X, for given any $i < \omega$, pick $x_i \in B_i$ such that $o(x_i) = p$. Then $i \in A$ iff $x_i \in X$. X is reducible to A, for given any $x = (x_0, \ldots, x_n) \in H$, $x \in X$ iff $\forall i \leq n$ $(i \notin A \to x_i = o)$. □

<u>Remark</u>: Suppose $G = \bigoplus_\omega \mathbb{Z}(p^\infty) \oplus B$ where B has bounded order. The proof of Lemma 1 can be used to show that B is finite.

<u>Lemma 2</u>: G does not have arbitrarily large cyclic summands.

If R is a reduced p-group which does not have arbitrarily large cyclic summands, then R has bounded order. Before proving Lemma 2 we show how the theorem follows.

Let $G = D \oplus R$ where D is divisible and R is reduced. By Lemma 2, R has bounded order. Now if $D = \oplus_\omega \mathbb{Z}(p^\infty)$, then by the Remark, R is finite. Suppose $D = \oplus_m \mathbb{Z}(p^\infty)$. Let k such that $p^k R = (o)$, so that $G[p^k] = F_o \oplus R$ where $F_o = D \cap G[p^k]$. By Lemma 1, it suffices to show that $G[p^k]$ is autostable. Suppose $F_1 \oplus R_1 \cong F_o \oplus R$ where $F_1 \cong F_o$ and $R_1 \cong R$. Build a divisible tower over F_1 to get a recursive $H \cong G$ where $H[p^k] = F \oplus R_1$. Since G is autostable, there is a recursive $\varphi: G \cong H$. The restriction of φ to $G[p^k]$ maps $F_o \oplus R \cong F_1 \oplus R_1$.

<u>Proof of Lemma 2</u>: Suppose G has arbitrarily large cyclic summands. We will build an $H \cong G$ which is not recursively isomorphic to G. As in Lemma 1, H is a direct limit of a recursive sequence $\{H_s\}_{s<\omega}$ of finite groups. Let $g_o g_1, \ldots$ be a recursive enumeration of H. At each stage s we will define a subgroup G_s of G and an isomorphism $\varphi_s: G_s \cong H_s$. For each s, $g_s \in G_{2s}$ so that $G = \bigcup_s G_s$. To insure that $G \cong H$ we will satisfy the negative requirements

$$N_e: \lim_s \varphi_s(g_e) \text{ and } \lim_s \varphi_s^{-1}(h_e) \text{ exist.}$$

Let $G - ht(x)$ be the height of x in G and let $G - ht_s(x)$ be the height of x in G_s. Similarly for $H - ht$ and $H - ht_s$. To insure that G is not recursively isomorphic to H we will satisfy the positive requirements

$$P_e: G - ht(b) \text{ is finite and } G - ht(b) < H - ht(\{e\}(b))$$
$$\text{for some } b \in G[p]$$

To witness the element b which satisfied P_e we will use a system of markers, $b(e,s)$, and require that $\lim_s b(e,s)$ exists. To satisfy N_e and protect higher priority requirements we will preserve the subgroup $K(e,s)$ which is generated by g_i, $\varphi_s^{-1}(h_i)$ where $i \leq e$ and all $x \in G[p]$ such that $x \leq b(i,s)$ for some $i \leq e$. Suppose $b = b(e,s)$, it is possible that $G - ht_t(b) < H - ht_t(b)$ for all t but $G - ht(b) = H - ht(b) = \omega$, so that b will not witness P_e. To avoid this problem we introduce the function

$$t(e,s) = \text{least } t \text{ such that for all } k$$
$$\text{if } t \leq k \leq s, \text{ then } b(e,k) = b(e,s),$$

and commit ourselves to attempting to move $b(e,s)$ when $G - ht_s\,(b,e,s)) \neq G - ht_t\,(b(e,s))$ for $t = t(e,s)$. As an aid in our search for a final resting place for

$b(e,s)$ we use the function

$$n(e,s) = \text{the number of attempts to satisfy } P_e$$

together with a recursive function $r: \omega \to \omega$ with the property that for each y there are infinitely many x such that $r(x) = y$.

P_e <u>is satisfied at stage</u> s if for $b = b(e,s)$ and $t = t(e,s)$, $G\text{-}ht_s(b) < H\text{-}ht_s(\{e\}(b))$ and $G\text{-}ht_s(b) = G\text{-}ht_t(b)$.

P_e <u>requires attention</u> at stage s+1 if

(i) P_e is not satisfied at stage s

(ii) G_s and H_s can be decomposed as $G_s = M \oplus \langle y \rangle$ where $M \cong N$, $K(e,s) \subsetneq M$ and $\varphi_s[K(e,s)] \subseteq N$. There is a $b \in \langle x \rangle \cap G[p]$, $b \neq 0$, $b \geq r(n(e,s))$ such that $\{e\}_s$ is defined on all elements $g \in G[p]$, where $g \leq b$, and $\{e\}(b) \in \langle y \rangle \cap G[g]$, $\{e\}(b) \neq 0$.

(iii) e is least with respect to (i) and (ii).

The Construction

At even stages of the construction let $G_{2s} = \langle G_{2s-1}, g_s \rangle$ and define $\varphi_{2s} \supseteq \varphi_{2s-1}$ and H_{2s} so that $\varphi_{2s}: G_{2s} \cong H_{2s}$. At odd stages we work on the positive requirements. Suppose s+1 is an odd stage. If no $e \leq s$ requires attention at stage s+1, let $G_{s+1} = G_s$, $H_{s+1} = H_s$ and $\varphi_{s+1} = \varphi_s$. Suppose e requires attention and $b \in G[p]$ is the least element which satisfies clause (ii). Since G has arbitrarily large cyclic summands, we can find a $z \in G$ such that $G_s \cap \langle z \rangle = (o)$ and $o(z) > o(x)$. Let $G_{s+1} = G_s \oplus \langle z \rangle$. Build $H_{s+1} = N \oplus \langle v \rangle \oplus \langle w \rangle$ where $y \in \langle v \rangle$, $o(v) = o(z)$ and $o(w) = o(x)$. Define $\varphi_{s+1} = \varphi_s$ on N, $\varphi_{s+1}(x) = w$, $\varphi_{s+1}(z) = v$, $n(e,s+1) = n(e,s)+1$, and $b(e,s+1) = b$. Notice that $G\text{-}ht_{s+1}(b) < H\text{-}ht_{s+1}(\{e\}(b))=$

End of Construction

<u>Sublemma</u> 1: For all e, $\lim_s b(e,s)$ exists and N_e is satisfied.

<u>Proof</u>: Suppose e is least such that $\lim_s b(e,s)$ does not exist and let t be a stage where $b(i,s) = b(i,t)$ for all $s \geq t$, $i < e$. We may also assume that P_i does not require attention after stage t for $i < e$. Thus $K(e,s) = K(e,t)$ for all $s \geq t$ and N_e is satisfied. Let $b \in G[p]$, $b \neq o$, be any element of finite height

such that $G = A \oplus \langle x \rangle$ where $b \in \langle x \rangle$ and $K(e,t) \subseteq A$. There are infinitely many s for which $b \geq r(n(e,s))$ is the least witness to clause (ii) and e requires attention at stage $s+1$. At some stage $s \geq t$, $G\text{-}ht(b) = G\text{-}ht_s(b)$ and $b(e,s) = b$. P_e will be satisfied at all stages after s, thus $b(e,s)$ never moves again which is a contradiction. □

<u>Sublemma</u> 2: For all e, $\{e\}: G \not\cong H$.

<u>Proof</u>: Suppose e is least such that $\{e\}: G \cong H$. Let t be a stage where $b(i,s) = b(i,t)$ for all $s \geq t$ and $i \leq e$. Hence for $s \geq t$, P_e does not require attention at stage s. Now if $G\text{-}ht(b) < \omega$ where $b = b(e,t)$ then at some stage $s \geq t$ $G\text{-}ht(b) = G\text{-}ht_s(b) = H\text{-}ht_s(\{e\}(b))$. If the $G\text{-}ht(b)$ is infinite, then at some stage $s > t$, $G\text{-}ht_s(b) > G\text{-}ht_t(b)$. In either case there is a s age $s \geq t$ where P_e is not satisfied at stage s. Now the argument used in Sublemma 1 can be applied to show that P_e requires attention at some stage $s > t$, so that $b(e,s) \neq b$ which is a contradiction. □

A result of Lin [5] states that for all recursive ordinals $\alpha \geq \omega$ there is a recursive reduced p-group of length α which is not autostable. An immediate corollary of Theorem 1 is that no reduced p-group of infinite length is autostable. The reader can consult Feferman [1] for a general method of constructing p-groups of specified length.

4. <u>Divisible Closures of p-Groups</u>

Let D be an abelian group. D is <u>divisible</u> if for all $m \geq o$ and $d \in D$ there is a $d' \in D$ such that $md' = d$. If G is an abelian group, a pair (D,φ) is a <u>divisible closure</u> of G if D is divisible, $\varphi: G \hookrightarrow D$, and for all $d \in D^{d \neq o}$, there is an $m \geq o$ and a $g \in G$, $g \neq o$ such that $md = g$. (D,φ) is a <u>recursive divisible closure</u> of G if (D,φ) is a divisible closure and both D and φ are recursive.

<u>Proposition 1</u>: Every recursive abelian group has a recursive divisible closure.

<u>Proof</u>: Let $G = F/K$ where F is the free abelian group on the generators $x_o, x_1, \ldots$ and K is a recursive subgroup of F. Let D be the divisible group of all finite formal sums $\Sigma\, r_i x_i$ where $r_i \in \mathbb{Q}$. F is a recursive subgroup of D, thus there is a

recursive $\alpha: D \to \omega$ defined by $\alpha(d)$ = least k such that $kd \in F$.

Lemma 3: Suppose N is a subgroup of D maximal with respect to $N \cap F = K$, then

(1) If $d \in N$, then $\alpha(d) \cdot d \in K$

(2) If $\varphi: F/K \to D/N$ is the canonical map, then $(D/N, \varphi)$ is a recursive divisible closure of F/K.

Proof: (1) If $d \in N$, $\alpha(d) \cdot d \in N \cap F = K$.

(2) φ is a recursive embedding. Suppose $d \in D - N$, then by the maximality of N there is an $a \in F \cap \langle N, d \rangle$ such that $a \notin K$, so $a \notin D$. It follows that $md \equiv a \pmod N$ for some $m \geq 1$. □

In view of Lemma 3 we need only construct a subgroup N of D maximal with respect to $N \cap F = K$.

Definition 2: Let $a_o, \ldots, a_s \in D$ and suppose $\alpha(a_i) \cdot a_i \in K$ for $o \leq i \leq s$. (a $(a_o, \ldots, a_s)$ is consistent over K if $\langle K, a_o, \ldots, a_s \rangle \cap F = K$.

Lemma 4: $\{(a_o, \ldots, a_s): \alpha(a_i) \cdot a_i \in K$ for $o \leq i \leq s$ and $(a_o, \ldots, a_s)$ is consistent over $K\}$ is recursive.

Proof: It sufficies to show $b \in F \cap \langle K, a_o, \ldots, a_s \rangle$ iff $b \equiv \Sigma\, m_i a_i \pmod K$ where $-\alpha(a_i) \leq m_i \leq \alpha(a_i)$ and $\Sigma\, m_i a_i \in F$. Suppose $b = \Sigma\, n_i a_i + c$ where $n_i \in \mathbb{Z}$ and $c \in K$. Now $n_i a_i \equiv m_i a_i \pmod K$ where $-\alpha(a_i) \leq m_i \leq \alpha(a_i)$. Thus $b \equiv \Sigma\, m_i a_i \pmod K$ and if $b \in F$, then $\Sigma\, m_i a_i \in F$. The converse is trivial. □

We are now ready to construct N. Let $a_o, a_1, \ldots$ be a recursive enumeration of those elements of D such that $\alpha(a_i)a_i \in K$.

Stage o: Let $N_o = \begin{cases} \{a_o\} & \text{if } a_o \text{ is consistent over } K \\ \emptyset & \text{otherwise} \end{cases}$

Stage s+1: Let $N_{s+1} = \begin{cases} N_s \cup \{a_{s+1}\} & \text{if } (N_s a_{s+1}) \text{ is consistent over } K \\ N_s & \text{otherwise} \end{cases}$

Let $N = \bigcup N_s$. By Lemma 4, N is recursive and maximal with respect to every finite subset being consistent over K. Easily $N \cap F = K$ and by the maximality of N, N is a subgroup of D. □

The divisible closure of an abelian group G is unique up to a G-isomorphism. That is, if (D_1,φ_1) and (D_2,φ_2) are two divisible closure of G, then there is a $\Psi: D_1 \cong D_2$ such that $\Psi \circ \varphi_1(g) = \varphi_2(g)$ for all $g \in G$. We now address the effective analog of this question for p-groups.

Definition 3: G has a recursively unique divisible closure if for all recursive divisible closures (D_1,φ_1), (D_2,φ_2) there is a recursive $\Psi: D_1 \cong D_2$ such that $\Psi \circ \varphi_1(g) = \varphi_2(g)$ for all $g \in G$.

Theorem 2: G has a recursively unique divisible closure iff pG is recursive.

Proof: Suppose that pG is recursive and let (D,φ) be a recursive divisible closure.

Lemma 5: $\varphi[G]$ is a recursive subgroup of D.

Proof: $\varphi[G]$ is r.e., so it sufficies to show that $D\backslash\varphi[G]$ is r.e. Given a recursive enumeration $g_o, g_1, \ldots$ of G define the following sets.

$$A_s = \begin{cases} \{d \in D: p^n d = \varphi(g_s) \text{ for some } n \geq 1\} \text{ if } g_s \notin pG \\ \varphi \qquad\qquad\qquad\qquad\qquad\qquad \text{otherwise} \end{cases}$$

Let $A = \bigcup_s A_s$. A is r.e. and we claim $A = D\backslash\varphi[G]$. If $d \in A_s$, then $p^n d = \varphi(g_s)$ for some $n \geq 1$. Now if $d \in \varphi[G]$, then $g_s \in pG$ which is false since $A_s \neq \emptyset$. Suppose $d \in D\backslash\varphi[G]$ and let $n \geq 1$ be the least integer such that $p^n d \in \varphi[G]$. Thus $p^n d = \varphi(g_s)$ for some s. Suppose $g_s \in pG$, then $p^n d = \varphi(g)$ for some $g \in G$. Thus $p^{n-1} d - \varphi(g)$ has order p. Since (D,φ) is a divisible closure of G, the elements of D of order p are in $\varphi[G]$. It follows that $p^{n-1} d \in \varphi[G]$ which contradicts the choice of n, thus $g_s \in pG$ and $d \in A_s \subset A$. □

We now indicate how to set up an effective back and forth procedure. Suppose $\Psi: \varphi[G] \hookrightarrow D'$ where D' is divisible and $p\, d \in \varphi[G]$ but $d \notin \varphi[G]$. Select $d' \in D'$ such that $pd' = \Psi(pd)$. The map $\Psi'(\varphi(g) + kd) = \Psi \circ \varphi(g) + kd'$ where $o \leq k < d$ is an extension of Ψ to $<\varphi[G], d>$. □

Conversely, suppose that G has a recursively unique divisible closure. Let (D, id) be a recursive divisible closure of G where id is the inclusion map of

G into D. The strategy is to build a recursive $\varphi : G \hookrightarrow D$ so that (D,φ) is a recursive divisible closure and at the same time satisfy the requirements.

$$R_e: \{e\}: D \cong D \Rightarrow \{e\} \neq \varphi \text{ on } G$$

The failure to meet some requirement R_e will permit us to decide $g \in pG$.

Let $g_o, g_1, \ldots$ be a recursive enumeration of G with the property that for $G_o = G[p]$ and $G_{s+1} = \langle G[p], g_o, \ldots, g_s \rangle$, $pG_{s+1} \subseteq G_s$. Let $W_e^s = \{d \in D: d \leq s$ and $\{e\}_s(d)$ is defined$\}$. We construct φ in stages, $\varphi = \bigcup_s \varphi_s$ where $\varphi_s : G_s \hookrightarrow D$. R_e <u>requires attention at stage</u> $s+1$ if there is a $g \in G_{s+1} \setminus G_s$ such that $pg, g \in W_e^s$ and $\{e\}_s(g') = \varphi_s(g')$ for all $g' \in W_e^s$ and $\{e\}_s(g') = \varphi_s(g')$ for all $g' \in W_e^s \cap G_s$.

The Construction

Stage o: $\varphi_o : G_o \hookrightarrow D$ via $\varphi_o(g) = g$ for all $g \in G_o$.

Stage s+1: If there is no $e \leq s$ such that R_e requires attention at stage $s+1$, extend φ_s to φ_{s+1} in any fashion so that $\varphi_{s+1} : G_{s+1} \hookrightarrow D$. Otherwise let e be least such that R_e requires attention. Pick $g \in G_{s+1} \setminus G_s$ such that $pg, g \in W_e^s$. Let $d \in D$ such that $pd = \{e\}(pg) = \varphi_s(pg)$ but $d \neq \{e\}(g)$. Extend φ_s to φ_{s+1} by mapping $g \to d$.

End of Construction.

<u>Lemma 6</u>: pG is recursive.

<u>Proof</u>: Let e be least such that $\{e\}: D \cong D$ and $\{e\}(g) = \varphi(g)$ for all $g \in G$. Let t be a stage where i never requires attention after stage t for $i < e$. Given $r \geq t$ to decide $g_r \in pG$. Let $d \in D$ such that $pd = g_r$ and find a stage $s \geq t$ where $d, pd \in W_e^s$. Claim

$$g_r \in gG \text{ iff } g_r \in p\, G_s$$

If $g_r \notin p\, G_s$ and $g_r \in pG$, then $d \neq G_s$ but $d \in G$. Thus there will be a stage $s' > s$ where $d \in G_{s'+1} \setminus G_{s'}$ and R_e requires attention at stage $s'+1$, which is a contradiction. □

References

[1] Feferman, S., Impredicativity of the existence of the largest divisible subgroup, in "Model Theory and Algebra: A Memorial Tribute to Abraham Robinson", Springer Lecture Notes 498, Springer-Verlag, Berlin-New York, 1975.

[2] Fuchs, L. "Infinite Abelian Groups" v.1, Academic Press, New York, 1970.

[3] Kaplansky, I., "Infinite Abelian Groups", University of Michigan Press, Ann Arbor, Michigan, 1969.

[4] LaRoche, P., Contributions to recursive algebra, Ph.D. Dissertation, Cornell University, August, 1978.

[5] Lin, C., The effective content of Ulm's Theorem, Ph.D. Dissertation, Cornell University, August, 1977.

[6] Mal'cev, A. I., On recursive abelian groups, Soviet Math. Dokl. 3 (1962), 1431-1434.

[7] Metakides, G. and Nerode, A., Effective content of field theory, Ann Math. Logic 17 (1979), 289-320.

[8] Nurtazin, A. T., Strong and weak constructivizations and computable families, Algebra and Logic 13 (1974), 177-184.

[9] Rabin, M. O., Computable algebra, general theory and theory of computable fields, Trans. Amer. Math. Soc. 95 (1960), 341-360.

[10] Rogers, H., Jr., "Theory of Recursive Functions and Effective Computability", McGraw-Hill, New York, 1968.

[11] Smith, R. L., Effective valuation theory, to appear in "Aspects of Effective Algebra", Upside Down A Book Company Yarra Glen, Victoria, Australia, 1980.

CONSTRUCTIVE AND RECURSIVE SCATTERED ORDER TYPES

Richard Watnick
The University of Connecticut, Stamford, Connecticut

1. Introduction and Summary

In this paper we generalize Kleene's work on constructive ordinals by offering several possible definitions of a constructive scattered linear ordering, which lead to the classes $\mathcal{O}$, $\mathcal{A}$, $\mathcal{J}$, $\mathcal{J}'$, and $\mathcal{N}$ of constructive scattered linear orderings. In [7], we began by closely following the treatment of constructive ordinals found in Rogers [5]. The development is technical and leads to universal systems of notation. For the purposes of this paper we will begin with these universal systems and use them as definitions. We will develop some terminology which has proven useful in studying these order types and show how we arrive at the result $R \supseteq \mathcal{J} \supsetneq \mathcal{O} \supsetneq \mathcal{A} \supsetneq \mathcal{N}$, and $\mathcal{J} = \mathcal{J}'$, where R is the class of recursive scattered linear orderings.

The original motivation for this project was the formalization of a priority argument which extends Tennenbaum's result (see [7]) on the existence of infinite recursive linear orderings which contain no infinite recursive ascending or descending sequence. (The definition of what I call $\mathcal{A}$ above also appears in Pinus [4].)

2. Terminology and Notation

Our universe is the rational numbers, Q, together with the usual ordering $<$. All our sets will be subsets of Q, which inherit the usual ordering $<$. It is well known (due to Cantor) that the order types represented by subsets of Q are precisely all the countable order types. The effective version of this theorem is that an order type is recursive iff there is a recursive subordering of Q of that order type.

N will denote the strictly positive integers. The symbol "n," besides denoting the natural number n or 0, will also denote the order type represented by a set of n linearly ordered elements.

Recall that if α and β are order types of sets A and B respectively, then $\alpha+\beta$ denotes the order type of a set which is a copy of A followed by a copy of B, and $\alpha\cdot\beta$ denotes the order type of a set which is β copies of α. For example, $\alpha+\alpha$ is $\alpha\cdot 2$, but is not $2\cdot\alpha$.

Again, suppose we are given an order type α, and a pairwise disjoint collection of sets $\{B(\beta) | \beta \in \alpha\}$ where $B(\beta)$ has order type

$\gamma(\beta)$. Let C be the ordered set whose elements are $\bigcup_{\beta\in\alpha}(B(\beta))$, where for $x,y \in C$, $x < y$ if $x \in B(\beta)$, $y \in B(\beta')$, and either $\beta < \beta'$ or $\beta = \beta'$ and $x < y$ in $B(\beta)$. We will denote the order type of C by $\sum_{\beta\in\alpha}\gamma(\beta)$.

If α is an order type represented by $\langle A, < \rangle$, then the opposite order type, represented by $\langle A, > \rangle$, will be denoted by α^*. Z will denote the order type of the integers as well as the integers themselves. That is, $Z = \{\cdots -2,-1,0,1,2,\cdots\}$ and also $Z = \omega^*+\omega$. The order type of Q will be denoted by η.

Our universe is countable, and all objects we deal with will be countable. For example, if we say let α be an order type, we are assuming that α is a countable order type.

We say that a linear ordering A is scattered if it has no subset of order type η. An order type α is said to be scattered, if every (or equivalently, if some) set of order type α is scattered. Hausdorff [2] first showed that any order type α can be expressed as a dense sum of scattered order types. That is, $\alpha = \sum_D H_d$, where D is dense and H_d is scattered for all d in D.

There is an alternate inductive characterization of scattered order types equivalent to the definition above. (Also due to Hausdorff.) Let 0 and 1 be in S_0. For α a successor ordinal, if $\{\mu_i\}$ is a sequence of order types in $S_{\alpha-1}$, then $\sum_\omega \mu_i$, $\sum_{\omega^*}\mu_i$, and $\sum_Z \mu_i$ are in S_α. For α a limit ordinal, if $\{\mu_i\}$ is a sequence of order types in $\bigcup_{\beta<\alpha} S_\beta$, then $\sum_\omega \mu_i$, $\sum_{\omega^*}\mu_i$ and $\sum_Z \mu_i$ are in S_α. Then, a countable order type α is scattered iff there is an ordinal β such that $\alpha \in S_\beta$. We say that $r(\alpha) = \beta$, or that α has rank β, or α is scattered of rank β, if $\beta = \min\{\gamma \mid \alpha \in S_\gamma\}$.

γ is a recursive $(\Sigma_n,\pi_n,\Delta_n)$ order type if there is a recursive $(\Sigma_n,\pi_n,\Delta_n)$ set of order type γ. ϕ_e will denote the eth partial recursive function, as is customary.

3. Definitions of $\mathcal{O}$ and $\mathcal{A}$, $\mathcal{A} \subseteq \mathcal{O} \subseteq \mathcal{R}$

Let S be the collection of countable scattered order types. We will define $\mathcal{O}$ and $\mathcal{A}$, where $\mathcal{O}$ is the collection of Original constructive scattered order types, and $\mathcal{A}$ is the collection of Alternate constructive scattered order types. For any family F of scattered order types, let F_α be the collection of order types in F with rank less than or equal to α.

If $\gamma \in S_\alpha$, then $\gamma = \sum_Z \gamma_i$, where $\{\gamma_i\}_Z \subseteq \bigcup_{\beta<\alpha} S_\beta$. Intuitively, γ will be in $\mathcal{O}_\alpha$ iff there is an algorithm Φ, which effectively lists a two-way infinite sequence of elements in $\bigcup_{\beta<\alpha} \mathcal{O}_\beta$, say $\{\gamma_i\}_Z = \{\Phi(i)\}_Z$, such that $\gamma = \sum_Z \gamma_i$. We will define $\mathcal{O}$ so that if $\gamma \in \mathcal{O}$, then we can say not only that γ is constructed by building sums of smaller order types, using 0 and 1 as the foundation, but that this construction is effective.

Scattered order types can similarly be said to be built up from the single building block, 1. To build S using only 1, and not using 0, we have to allow ω-sums, ω^*-sums, and finite sums, in addition to allowing Z-sums. This is precisely what we will do in A, again requiring the construction to be effective.

Now let us show how this can be formalized.

<u>DEFINITION 1 $\mathcal{O}$</u> <u>The O-system; $\nu\mathcal{O}$: N $\to$ scattered order types</u>

$\nu\mathcal{O}(2) = 0$, $\nu\mathcal{O}(4) = 1$. If $\gamma = \sum_{n\in\omega^*} \nu\mathcal{O}\phi_y(n) + \sum_{n\in\omega} \nu\mathcal{O}\phi_z(n)$, and $r(\nu\mathcal{O}\phi_y(n)) < r\gamma$ and $r(\nu\mathcal{O}\phi_z(n)) < r\gamma$ for n in N, then $\nu\mathcal{O}(3^y \cdot 5^z) = \gamma$.

<u>DEFINITION 2 $\mathcal{O}$</u> $\underline{\mathcal{O}}$ = range of $\nu\mathcal{O}$

<u>DEFINITION 1 A</u> <u>The A-system νA: N $\to$ scattered order types</u>

$\nu A(1) = 1$.

If $\gamma = \sum_{i=1}^{n+1} \gamma(i)$, and $r(\gamma(i)) < r\gamma$, for $1 \le i \le n+1$, and $\nu A(y(i)) = \gamma(i)$, then $\nu A(2^n \cdot 3^{y(1)} \cdot 5^{y(2)} \ldots (P_{n+1})^{y(n+1)}) = \gamma$, where P_i is the ith odd prime. If $\gamma = \sum_{n\in\omega} \nu A\phi_y(n)$, and for all n in N, $r(\nu A\phi_y(n)) < r(\gamma)$, then $\nu A(3^y) = \gamma$. If $\gamma = \sum_{n\in\omega^*} \nu A\phi_y(n)$, and for all n in N, $r(\nu A\phi_y(n)) < r(\gamma)$, then $\nu A(5^y) = \gamma$. If $\gamma = \sum_{n\in\omega^*} \nu A\Phi_y(n) + \sum_{n\in\omega} \nu A\phi_z(n)$, and for all n in N, $r(\nu A\phi_y(n)) < r(\gamma)$ and $r(\nu A\phi_z(n)) < r(\gamma)$, then $\nu A(7^y \cdot 11^z) = \gamma$.

DEFINITION 2 A $\underline{A}$ = range of $\nu A \cup \{0\}$.

LEMMA 3 $A \subseteq O$.

Sketch of proof: Intuitively, A should be contained in O, since if $\gamma \in A$ is not coded by a Z-sum, we need merely tack on infinitely many 0-summands to transform it into a Z-sum, and any function eventually constant, eventually always equal to the code for 0, must be recursive. This idea can be carried out using the fixed point theorem. We first define mappings ψ_z, our first approximations of mappings from the domain of νA to the domain of νO. For example, $\psi_z(3^u) = 3^v \cdot 5^w$, where $\phi_v \equiv$ code for 0, and $\phi_w = \phi_z \circ \phi_u$. Once ψ_z is defined, we let $\phi_{f(z)} = \psi_z$, and find a number m such that $\psi_m = \phi_{f(m)} = \phi_m$, using the fixed point theorem. Then we can establish by induction on $r\nu A(x)$, that $\nu O \phi_m(x) = \nu A(x)$, for all x in the domain of νA. For example, $\nu O \phi_m(3^u) = \nu O(3^v \cdot 5^w) = 0 + \sum_{\omega} \nu O \phi_w(n) = \sum_{\omega} \nu O \phi_m \phi_u(n)$, which by the induction hypothesis would $= \sum_{\omega} \nu A \phi_u(n) = \nu A(3^u)$.

LEMMA 4 $A \subseteq O \subseteq R$.

Before proving lemma 4, we will develop some further terminology. Let $\nu O(x) = \gamma = \sum_{\omega^*} \gamma(i) + \sum_{\omega} \gamma(i)$, where $x = 3^y \cdot 5^z$, $\phi_y(i)$ codes $\gamma(-i)$ and $\phi_z(i)$ codes $\gamma(i)$. Then, we will say that $\underline{x \text{ presents } \gamma \text{ as}}$ $\sum_{Z} \gamma(i)$. Let $\nu A(x) = \gamma = \sum_{\omega} \gamma(i)$, where $x = 3^y$, and $\phi_y(i)$ codes $\gamma(i)$. Then, we will say that $\underline{x \text{ presents } \gamma \text{ as}}$ $\sum_{\omega} \gamma(i)$. $\underline{x \text{ presents } \gamma}$ is defined similarly for the other cases in A. In general, x presents γ as $\sum_{\beta} \gamma(i)$, if $\nu s(x) = \gamma$ and x describes γ as a β-sum of order types $\{\gamma(i)\}_{i \in \beta}$.

Let us return to O. Suppose $x = 3^y \cdot 5^z$ presents γ as $\sum_{Z} \gamma(i)$, then $\phi_y(i)$ is a $\underline{\text{code for } \gamma(-i) \text{ supplied by } x}$, which we will refer to as $\underline{x(-i)}$. Also, $\phi_z(i)$ is a $\underline{\text{code for } \gamma(i) \text{ supplied by } x}$, which we will refer to as $\underline{x(i)}$. Suppose further that for some i, x(i) presents $\gamma(i)$ as $\sum_{j \in Z} (\gamma(i))(j)$. We let $\gamma(i,j) = (\gamma(i))(j)$, and let $x(i,j) = (x(i))(j)$ be a $\underline{\text{code for } (\gamma(i))(j) \text{ supplied by } x(i)}$, and also a $\underline{\text{code for } \gamma(i,j) \text{ supplied by } x}$. We can continue in this way forming the following two trees:

$T_1(0,x)$

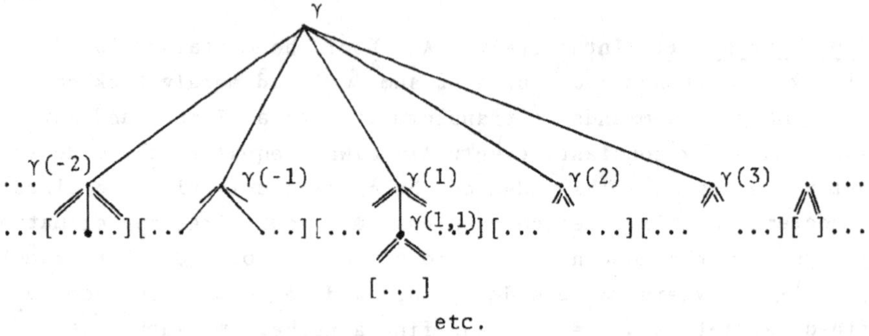

$T_2(0,x)$

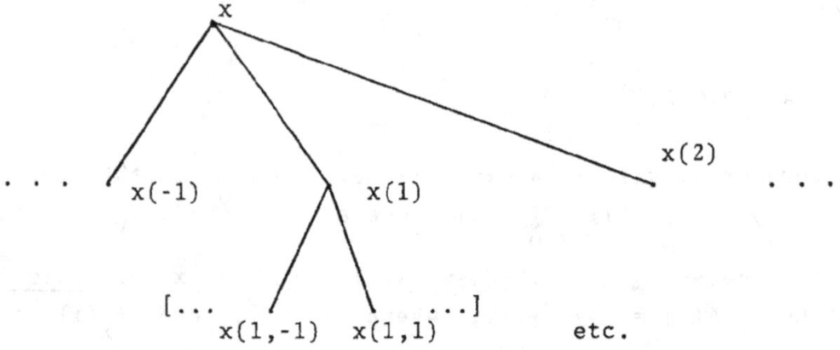

Let $\vec{\alpha}$ be a finite sequence of numbers in Z. $x(\vec{\alpha})$ is a node of $T_2(0,x)$ iff $\gamma(\vec{\alpha})$ is a node of $T_1(0,x)$ iff $x(\vec{\alpha})$ is a code for $\gamma(\vec{\alpha})$ supplied by x. Either $\gamma(\vec{\alpha}) = 0$ and $x(\vec{\alpha}) = 2$, or $\gamma(\vec{\alpha}) = 1$ and $x(\vec{\alpha}) = 4$, or $\gamma(\vec{\alpha})$ and $x(\vec{\alpha})$ are undefined and do not appear as nodes, or $x(\vec{\alpha})$ presents $\gamma(\vec{\alpha})$ as $\sum_{i \in Z} (\gamma(\vec{\alpha}))(i)$. In the last of these cases $x(\vec{\alpha},i)$ and $\gamma(\vec{\alpha},i)$ are defined for all i in $Z-\{0\}$.

Note that all paths in $T_1(0,x)$ and $T_2(0,x)$ are finite because $r(\gamma(\vec{\alpha},i)) \lneq r(\gamma(\vec{\alpha}))$. Furthermore, $x(\vec{\alpha})$ can be effectively found. Therefore, we know when $\gamma(\vec{\alpha})$ or $x(\vec{\alpha})$ is a terminal node of the tree. $\gamma(\vec{\alpha})$ is a terminal node of T_1 iff $x(\vec{\alpha})$ is a terminal node of T_2 iff $x(\vec{\alpha}) = 2$ or 4.

In A we can establish the same terminology. However, $\gamma(\vec{\alpha})$ is a tip iff $x(\vec{\alpha})$ is a tip iff $\gamma(\vec{\alpha}) = 1$ iff $x(\vec{\alpha}) = 1$. $T_1(A,x)$ and $T_2(A,x)$ will have the same properties as above, except some of the nodes of the trees may be finitely branching. That is, $x(\vec{\alpha})$ may present $\gamma(\vec{\alpha})$ as a finite sum. In general, the two trees together display how γ is built using the code x. Note that we have adopted the convention that $0 \notin N$, and for instance, in O, $x = 3^y \cdot 5^z$ supplies codes $\{x(i)\}_{i\in\omega^*+\omega} = \{\phi_y(i)\}_{i=1}^{\infty} \cup \{\phi_z(i)\}_{i=1}^{\infty}$. There is no $x(0)$. In general, if $x(\vec{\alpha})$ or $\gamma(\vec{\alpha})$ exists, then 0 does not appear in $\vec{\alpha}$.

We are now ready to prove lemma 4.

Sketch of proof. Let $\nu O(x) = \gamma$, we must produce a recursive set A of order type γ. This is done by using trees $T_1(O,x)$ and $T_2(O,x)$. We associate with $\vec{\alpha}$, not only $x(\vec{\alpha})$ and $\gamma(\vec{\alpha})$, but also an interval in $\mathbb{Q}$, $I(\vec{\alpha})$. We do this in such a way that $\vec{\alpha}$ is an initial segment of $\vec{\beta}$ iff $I(\vec{\beta}) \subseteq I(\vec{\alpha})$. Also, if neither $\vec{\alpha}$ nor $\vec{\beta}$ is an initial segment of the other, then $I(\vec{\alpha}) \cap I(\vec{\beta}) = \emptyset$ and for all p in $I(\vec{\alpha})$, q in $I(\vec{\beta})$, $p \lneq q$ iff $\vec{\alpha} \lneq \vec{\beta}$ in the dictionary ordering iff $x(\vec{\alpha})$ is to the left of $x(\vec{\beta})$ on $T_1(O,x)$. Then, we define what proves to be the desired set A by letting $q \in A$ iff q is the first element of $I(\vec{\alpha})$ from a fixed enumeration of $\mathbb{Q}$, where $x(\vec{\alpha}) = 4$ and $\gamma(\vec{\alpha}) = 1$.

This concludes section 3. We have defined two notions of constructive scattered linear orderings O and A. We have that $A \subseteq O \subseteq R$.

4. Nice Presentations and $O \neq A$

If $\gamma = Z+c(1)+Z+c(2)+\ldots = \nu A(x)$, $c(i) \in N$, then x may code γ as $Z+3+2+\omega^*+5+1+1+\omega+\ldots$. However, in [7], we show that $C = \{c(i)\}$ is a recursively enumerable set and that there must be a "nicer" code for γ. That is, there is an x' which presents γ as $Z+c(1)+Z+c(2)+\ldots$. This suggests the following.

DEFINITION 5 A Let $\nu A(x) = \gamma$, $r(\gamma)$ finite. Suppose that x presents γ as $\sum_{\beta}\gamma(i)$. We will say that x is an almost nice code for γ if whenever i and j are consecutive elements of β, $r(\gamma(i)+\gamma(j)) = r(\gamma)$.

In $\mathcal{O}$, we are allowed to code the 0 ordering. In fact, since we only have Z-sums, to code essentially finite sums we must use the 0 ordering and could never meet the requirements of the above definition. Therefore, we propose the following alternative.

DEFINITION 5 $\mathcal{O}$ Suppose $r(\gamma)$ is finite, $\nu\mathcal{O}(x) = \gamma$, and $x = 3^y \cdot 5^z$. We will say that x is an almost nice code for γ if $[(\phi_y \equiv 2$ and $\phi_z(i) \neq 2$ for $i \in N)$ or $(\phi_y \equiv 2$ and $\{i \mid \phi_z(i) \neq 2\}$ is an initial segment of $N)$ or $(\phi_y(i) \neq 2$ for $i \in N$ and $\phi_z \equiv 2)$ or (for $i \in N$ $\phi_y(i) \neq 2$ and $\phi_z(i) \neq 2)]$ and [for all consecutive pairs of elements i,j in Z, $\gamma(i) = 0$, or $\gamma(j) = 0$, or $r(\gamma(i)+\gamma(j)) = r(\gamma)$.]

DEFINITION 6 In a system s, we will say that an almost nice code x is a nice code if all nodes of $T_2(s,x)$ are almost nice.

We observe below that $A \cap S_i = R \cap S_i$ for $i = 0,1$ and that all codes for $\gamma \in S_1$ in A are nice. Recall that for a family of order types F, $F_\alpha = F \cap S_\alpha$.

$A_0 = \{0,1\} = R_0$. There is no code for 0 in A, and 1 is the only code for 1 in A. $A_1 = N \cup \{0,\omega,\omega^*,Z\} = R_1$. $2^{n-1} \cdot 3 \cdot 5 \ldots P_n$ is the only code for n in A, $n > 1$. $\{3^u \mid \phi_u \equiv 1\}$ are the only codes for ω in A. $\{5^u \mid \phi_u \equiv 1\}$ are the only codes for ω^* in A. $\{7^u \cdot 11^u \mid \phi_u \equiv 1\}$ are the only codes for Z in A. All codes in A of order types in $A_1 = R_1$ are nice. This also implies that every almost nice code for $\gamma \in A_2$ is also nice. On the other hand, there are codes for $\gamma \in \mathcal{O}_2$ which are almost nice but not nice. There are, of course, codes for ω, ω^*, and Z in $\mathcal{O}$ which are not nice.

We will now show that if $\gamma \in A_2$, then, as for Z-representations, at least one of the codes for γ in A is nice. On the other hand, we will show that not all elements of $\mathcal{O}_2$ have nice codes in $\mathcal{O}$. These two facts will imply that $\mathcal{O} \neq A$.

LEMMA 7 If $\gamma \in A_2$, then there is a nice code for γ in A.

Sketch of proof. All codes in A for order types in A_1 are nice. Therefore, any almost nice code for γ in A_2 is a nice code for γ. We must show that γ has an almost nice code. For example, suppose $\gamma = \nu A(3^x) = \sum_{i \in \omega} \nu A \phi_x(i)$. The only way 3^x could fail to be almost nice is for there to be one or more pairs (n,m) such that

$r(\sum_{i=n}^{m} \nu A\phi_x(i)) = 1$ or $0 < 2$, where $1 \leq n < m \leq \infty$. For example, if $\nu A\phi_x(1) = \omega^*$, $\nu A\phi_x(2) = 5$, and $\nu A\phi_x(3) = Z$. Then $\nu A\phi_x(1)+\nu A\phi_x(2) = \omega^* \in A_1$. We would let $f(1)$ code ω^* which equals ω^*+5, and let $f(2)$ code Z. We attempt to continue in this way supplying an algorithm for a recursive function f. 3^y will be a nice code for $\gamma = \nu A(3^x)$, where $\phi_y = f$. There is only one obstruction; if $m = \infty$ for one of the bad pairs (n,m). In this case, we define $f(1)$ through $f(j)$, $j \leq n$, but never define $f(j+1)$ because $r(\sum_{i=n}^{\infty} \nu A\phi_x(i)) = 1$ or 0. However, in this case $\gamma = \nu Af(1)+\ldots+\nu Af(j)+\nu A(w)$, where w is a code for $\sum_{i=n}^{\infty} \nu A\phi_x(i)$. Therefore γ has a nice code which presents γ as a finite sum. In any case, γ does have a nice code.

Notice that in the above proof, we never can tell if f will end up a total function or if we never define $f(K+1)$ for some K. In the latter case all we really need is a finite sum. For this reason, the above proof does not supply a uniform procedure which takes an arbitrary code to a nice code. It describes a procedure which will succeed if it has to, but never tells us which case we are in. With a uniform procedure we could proceed by induction to A_3, A_4, etc. However, we cannot improve upon the above proof, and in fact, there is not necessarily a nice code for γ in A_3, [7].

LEMMA 8 There is a $\gamma \in \mathcal{O}_2$ with no nice code in $\mathcal{O}$.

Proof. We define a recursive operator ψ. Suppose $\phi_e(1),\ldots,\phi_e(K)$, and $\psi\phi_e(1),\ldots,\psi\phi_e(n)$ have been evaluated and actual values are found for each of them. Perform one more stage of the computation of $\phi_e(K+1)$. If at this point a value for $\phi_e(K+1)$ is found, then define $\psi\phi_e(n+1)$ to be 4, (the code for 1 in $\mathcal{O}$). Otherwise, define $\psi\phi_e(n+1)$ to be 2, (the code for 0 in $\mathcal{O}$). Let f be a recursive function such that $\psi\phi_e = \phi_{f(e)}$. Let $\phi_u \equiv 2$, $\phi_v \equiv 4$, and define $\phi_z(K) =$

$$\begin{cases} 3^v \cdot 5^v & \text{if } K = 2n,\ n \geq 1 \\ 3^v \cdot 5^{f(n)} & \text{if } K = 2n-1,\ n \geq 1 \end{cases}$$

Let $x = 3^u \cdot 5^z$, then x is in the domain of $\mathcal{O}$.

$$\nu O(x) = \sum_{K\in\omega^*} \nu O(\phi_u(K)) + \sum_{K\in\omega} \nu O\phi_z(K) = \sum_{K\in\omega} \nu O\phi_z(K)$$

$$= \nu O(3^v \cdot 5^{f(1)}) + Z + \nu O(3^v \cdot 5^{f(2)}) + Z + \ldots$$

$$= (\omega^* + \sum_{i\in\omega} \nu O\phi_{f(1)}(i)) + Z + (\omega^* + \sum_{i\in\omega} \nu O\phi_{f(2)}(i)) + Z + \ldots$$

$$= (\omega^* + \sum_{i\in\omega} \nu O\psi\phi_1(i)) + Z + (\omega^* + \sum_{i\in\omega} \nu O\psi\phi_2(i)) + Z + \ldots .$$

If ϕ_e is total, then $\psi\phi_e(i) = 4$, coding 1, for infinitely many i. Otherwise, $\psi\phi_e(i) = 4$ for only finitely many i. Therefore, $\nu O(x) = (\omega^* + a(1)) + Z + (\omega^* + a(2)) + Z + \ldots$, where $a(i)$ is finite iff ϕ_i is not recursive, and $a(i)$ is ω otherwise.

$\nu O(x) \in O_2$. Suppose it has a nice code $x' = 3^u \cdot 5^\eta$ for some η. Then, $\nu O\phi_\eta(2e-1) = Z$ iff ϕ_e is total, and ω^* otherwise. Therefore, ϕ_e is recursive iff $\phi_\eta(2e-1) = 3^a \cdot 5^b$ and $\phi_b(1) \neq 2$, since ϕ_e is recursive if $\phi_\eta(2e-1)$ is a (nice) code for Z, and not ω^*. This contradiction completes the proof.

COROLLARY 9 $A \subsetneq O$.

Proof. Let γ be the element of O_2 with no nice code given above in the proof of Lemma 8. If γ were in A_2, then γ would have a nice code in A, by lemma 7. If x were such a nice code in A for γ, then $x = 3^n$ and $7 | \phi_n(2e-1)$ iff ϕ_e is total. This is impossible. Therefore $\gamma \notin A$. ∎

5. $O \neq R$

LEMMA 10 If $\gamma \in O$ such that $\gamma = \sum_\omega (Z \cdot a(i) + c(i))$, where $a(i), c(i) \in N$, then $C = \{c(i)\}$ is a Σ_3 set.

Sketch of proof. First we must verify that even for the most general presentation of $\gamma = \nu O(3^y \cdot 5^z)$, $x \in C$ iff there are a group of finite summands in the presentation adding up to x, and surrounded by summands they cannot be absorbed into, (i.e., $\omega^* + 3 + 2 + \omega = \omega^* + \omega = Z$). Then we must verify that this holds iff

$\exists i \exists j \;\; i < j \;\&$

(1) The ith summand in the presentation ends up with ω (actually is ω or Z). &

(2) The jth summand in the presentation begins with ω^* (actually is ω^* or Z). &

(3) All summands strictly between the ith and the jth in the presentation are finite (including possibly 1 or 0). &

(4) The total sum of all summands, strictly between the ith and j^{th}, is x.

(1) and (2) can be expressed as a Π_2 relation $R(i,j)$, and (3) and (4) as a Σ_2 relation $T(i,j,x)$. $x \in C$ iff $\exists i \exists j$ $(i < j \cap R(i,j) \cap T(i,j,x))$. Therefore, C is a Σ_3 set.

COROLLARY 11 If $c(i)$ from lemma 10 is a strictly increasing function, then C is a Δ_3 set.

Proof. Let $S(i,j,x) = R(i,j) \,\&\, T(i,j,x)$. Then, S is $\Sigma_2 \wedge \pi_2$, and $x \in C$ iff $\exists i \exists j\ (i < j \,\&\, S(i,j,x))$. Here we know also that $[S(i,j,x)$ and $S(u,v,w)] \rightarrow (i \leq u \leftrightarrow j \leq v \leftrightarrow x \leq w)$. Therefore, to determine whether $x \in C$, we could search through all the triples (u,v,w) with $u < v$, until we find $S(u,v,w)$ holds where $w \geq x$. (This assumes that $|C|$ is infinite, otherwise, C is trivially Σ_3.) Then $x \in C$ iff $S(i,j,x)$ holds for some $i < j$ where $i \leq u$ and $j \leq v$. Therefore, there is a procedure, recursive in S which is $\Sigma_2 \wedge \pi_2$, determining membership in C. Therefore, C is a Δ_3 set. ∎

COROLLARY 12 $\mathcal{O} \neq \mathcal{R}$.

Proof. It is known [6 and 1] that if C is a Σ_3 set, then there is a $\gamma \in \mathcal{R}$ such that $\gamma = \sum_{\omega} (Z \cdot a(i)+c(i))$, and $c(i) < c(j)$ iff $i < j$, $C = \{c(i)\}$. This fact combined with the previous corollary yields the desired result.

6. $\mathcal{R} \supseteq \mathcal{J} \supsetneq \mathcal{O} \supsetneq \mathcal{A} \supsetneq \mathcal{N}$, $\mathcal{J} = \mathcal{J}'$

We now know that $\mathcal{A} \subsetneq \mathcal{O} \subsetneq \mathcal{R}$. Three of the concepts and features used to reach the above conclusion were as follows: nice codes, the fact that summands in a presentation must have lower rank than their sum, and the restriction of using only total recursive functions for forming codes.

In this section, we will use the concepts mentioned above to indicate new possible directions. We will start by defining $\mathcal{J}$-constructive scattered order types, where we will not require that presentations show a jump in rank from the summands to the sum.

The definition of $\mathcal{J}$ and νJ is then the same as $\mathcal{O}$ and νO, except from definition 1 $\mathcal{O}$ we drop the requirement that $r\nu J\phi_e(n) < r\gamma$, for $e = y,z$.

The notion of rank cannot play the role in $\mathcal{J}$ that it did in $\mathcal{O}$. Instead we define $\mathcal{J}$-rank on codes. Let $JR_0 = \{2,4\}$. Suppose JR_β is defined for all $\beta < \alpha$. Let JR_α be the collection of codes in $\mathcal{J}$, $x = 3^a \cdot 5^b$, such that $a,b \in \bigcup_{\beta<\alpha} JR_\beta$. Let $jr(x)$ = the $\mathcal{J}$-rank of x = the least ordinal α such that $\gamma \in JR_\alpha$. In $\mathcal{J}$, it is easier to perform a proof by induction on $\mathcal{J}$-rank than by induction on rank.

Clearly, for all x in the domain of νO, $\nu O(x) = \nu J(x)$. Also we can prove $\rho\nu J(x) \subseteq R$, just as we proved $\rho\nu O(x) \subseteq R$, using trees T_1 and T_2. Therefore, $\mathcal{O} \subseteq \mathcal{J} \subseteq R$. We will now show that $\mathcal{O} \neq \mathcal{J}$, by showing that the order type from corollary 12 is in $\mathcal{J}$.

<u>LEMMA 13</u> If C is Σ_3, then there is a $\gamma \in \mathcal{J}$ such that $\gamma = Za_1+c_1+Za_2+c_2+\ldots$, where $c_i < c_j$ iff $i < j$, and where $C = \{c_i\}$.

<u>Proof</u>. We will adjust the construction of a recursive set A of order type γ, referred to in corollary 12, to a construction of a code for γ in $\mathcal{J}$. To do this we will apply the method used to prove lemma 8. Let $n \in C$ iff $\exists x \forall y \exists z\ S(x,y,z,n)$ iff $\exists! x \forall y \exists z\ S(x,y,z,n)$. For each $n \in N$, let gn be defined by $gn(3K-2) = 3^m \cdot 5^{u(n,K)}$, $\nu J(gn(3K-1)) = n$, and $gn(3K) = 3^{u(n,K)} \cdot 5^m$, where $\phi_m \equiv 2$, so $\phi_m(i)$ codes 0 for all i in N, and where $u(n,K)$ codes f_n^K, defined as follows. Look for a $Z(1)$ such that $S(K,1,Z(1),n)$ holds. Define $f_n^K(t) = 2$ until such a $Z(1)$ is found at time $t(1)$. Let $f_n^K(t(1)) = 4$, which codes 1. Starting with $t(1)+1$, do the same for $S(K,2,Z(2),n)$, etc. Let v be a code for h where h is defined by $\nu J(h(2i-1)) = \omega^*$, and $h(2i) = 3^m \cdot 5^{a(i)}$ where $\phi_{a(i)} = gi$. Let $\phi_v = h$, then $\nu J(3^m \cdot 5^v) = 0 + \sum_{n\in\omega} \nu J(\phi_v(n)) = \sum_{n\in\omega} \nu J(h(n)) =$ $\sum_{n\in\omega} [\omega^* + \sum_{K\in\omega} \nu J(gn(K))] = \sum_{n\in\omega} [\omega^* + \sum_{K\in\omega} (\alpha(K,n)+n+\alpha(K,n)^*)]$, where

$$\alpha(K,n) = (0+0+\ldots+0+\underset{t(1)}{1}+0+0+\ldots+0+\underset{t(2)}{1}+0+0+\ldots+0+\underset{t(3)}{1}+0+0+\ldots).$$

If $n \in C$, then $\alpha(K,n) = \omega$ for a single K, otherwise, $\alpha(K,n)$ is finite. Therefore, for each n,

$$\sum_{K\in\omega} (\alpha(K,n)+n+\alpha(K,n)^*)$$

$$= \begin{cases} (\text{finite})+\ldots+(\text{finite})+\omega+n+\omega^*+(\text{finite})+(\text{finite})+\ldots \\ = \omega+n+Z & \text{if } n \in C \\ (\text{finite})+(\text{finite})+\ldots = \omega & \text{if } n \notin C \end{cases}$$

Therefore, for each n, $\omega^* + \sum_{K\in\omega} (\alpha(K,n)+n+\alpha(K,n)^*)$

$$= \begin{cases} Z+n+Z & \text{if } n \in C \\ Z & \text{iff } n \notin C. \end{cases}$$

Therefore, $\nu J(3^m \cdot 5^v)$ is $\sum_{i\in\omega} [Z(c(i)-c(i-1)+1)+c(i)]$. ∎

<u>COROLLARY 14</u> $\mathcal{O} \subsetneq J$.

<u>Proof</u>. This follows from corollary 11 and lemma 13.

Next, we will generalize J to J' in an attempt to include more of R. We allow the use of partial functions to form codes. Note that if we define $\nu J'$(undefined) to be a nonzero order type, then we cannot hope to have $J' \subseteq R$.

<u>DEFINITION 15</u> $\nu J'(2) = 0$, $\nu J'(4) = 1$, $\nu J'(3^a \cdot 5^b) = \sum_{\omega^*} \nu J'\phi_a(i) + \sum_{\omega} \nu J'\phi_b(i)$, where $\sum_{D} \alpha(i)$ is interpreted as $\sum_{D'} \alpha(i)$ where $D' = \{i \in D \mid \alpha(i) \text{ is defined}\}$.

Note, that by omitting undefined summands we are essentially adding $\nu J'$ (undefined) $= 0$ to the definition of νJ. We define J'-rank analogous to J-rank.

<u>DEFINITION 16</u> J' = range of $\nu J'$.

LEMMA 17 $J' = J$.

Proof. $J \subseteq J'$, because $\nu J'(x) = \nu J(x)$ whenever $\nu J(x)$ is defined. To show that $J' \subseteq J$ we will establish the existence of ϕ_m such that $\nu J\phi_m(x) = \nu J'(x)$ whenever $\nu J'(x)$ is defined. As in the proof of lemma 3 we start by defining a collection of mappings ψ_z which try to be ϕ_m, and let the fixed point theorem tell us which ψ_z works. Essentially, ψ_z stalls by using zeros waiting to see if anything is being coded. If nothing ever happens we end up with zero. If a value is produced, we use it.

Let $\phi_u \equiv 2$, let $\psi_z(2) = 2$, $\psi_z(4) = 4$, $\psi_z(3^a \cdot 5^b) = 3^x \cdot 5^y$, where $\phi_x(i) = 3^{p(i)} \cdot 5^u$, $\phi_y(i) = 3^u \cdot 5^{q(i)}$, where $\phi_{p(i)}(j) = 2$ for all j if searching for $\phi_a(i)$ never yields any value, and $\phi_{p(i)}(j) = 2$ for $j \neq k(i)+1$, and $\phi_{p(i)}(k(i)+1) = \phi_z\phi_a(i)$, if the search for $\phi_a(i)$ terminates successfully at stage $k(i)$ of the search. $\phi_{q(i)}$ and $L(i)$ are defined using ϕ_b in the same way as $\phi_{p(i)}$ and $k(i)$ were defined using ϕ_a. That is, $\{\phi_{q(i)}(j)\}_{j\in\omega} = \{\underbrace{2,\ldots,2}_{L(i)\text{ times}},\phi_z\phi_b(i),2,2,2,\ldots\}$, if $\phi_b(i)$ is defined, and $\phi_q(i) \equiv 2$, if $\phi_b(i)$ is not defined.

Let $\psi_z = \phi_{f(z)}, \phi_m = \phi_{f(m)} = \psi_m$, using the fixed point theorem. We claim that $\nu J\phi_m(x) = \nu J'(x)$, whenever $\nu J'(x)$ is defined. Assume $3^a \cdot 5^b = x$ is a counterexample of minimal J'- rank. This leads to a contradiction as follows. $\nu J\phi_m(3^a \cdot 5^b) = \nu J(3^x \cdot 5^y) = \sum_{\omega^*} \nu J\phi_x(i) + \sum_{\omega} \nu J\phi_y(i) = \sum_{\omega^*} \nu J(3^{p(i)} \cdot 5^u) + \sum_{\omega} \nu J(3^u \cdot 5^{q(i)}) = \sum_{i\in\omega^*} [\sum_{j\in\omega^*} \nu J\phi_{p(i)}(j) + \sum_{j\in\omega} \nu J\phi_u(j)] + \sum_{i\in\omega} [\sum_{j\in\omega^*} \nu J\phi_u(j) + \sum_{j\in\omega} \nu J\phi_{q(i)}(j)] =$

$$\sum_{i\in\omega^*} [\cdots \underbrace{\nu J(2)+\ldots+\nu J(2)}_{K(i)\text{ times}}+\nu J(2) \text{ or } \nu J\phi_m\phi_a(i)+\nu J(2)+\ldots+\nu J(2)+ \sum_{j\in\omega} 0]$$

$$+ \sum_{i\in\omega} [(\sum_{j\in\omega^*} 0) + \underbrace{\nu J(2)+\ldots+\nu J(2)}_{L(i)\text{ times}}+\nu J(2) \text{ or } \nu J\phi_m\phi_b(i)+\nu J(2)+\ldots+\nu J(2)+\ldots].$$

By the induction hypothesis, height $T_1(J', \phi_{\substack{b\\a}}(i)) <$ height $T_1(J', 3^a \cdot 5^b)$ implies $\nu J\phi_m\phi_{\substack{b\\a}}(i) = \nu J'\phi_{\substack{b\\a}}(i)$, so the above $= \sum_{\omega^*} \nu J'\phi_a(i) + \sum_{\omega} \nu J'\phi_b(i) = \nu J'(3^a \cdot 5^b)$. This contradicts that $x = 3^a \cdot 5^b$ is a counterexample.

In another direction, we could define N to be the collection of order types in A with nice codes. This is equivalent to using O instead of A since an order type has a nice code in A iff it has a nice code in O. The nice constructive order types have already played an important role in our discussion and provide our strongest notion of effectively constructed order types. However, we have defined nice codes only for order types of finite rank. In [7], we discuss why a satisfactory extension of this concept to infinite ranks seems unlikely. (The explanation refers to the actual systems of notation.) In any case, we already know that $N_2 = A_2$, $N_3 \subsetneq A_3$, $N \subsetneq A$.

We have been interested only in scattered order types. It would take only slight modifications to extend our definitions to non-scattered order types. Pinus does not restrict his definition in [4] to scattered order types.

We will close by stating, as a theorem, a summary of the relationships between the alternate definitions of constructive order types.

<u>THEOREM</u> 1. $R \supseteq J \supsetneq O \supsetneq A \supsetneq N$.

2. $A_2 = N_2$, but $A_3 \supsetneq N_3$.

3. $R_1 = J_1 = O_1 = A_1 = N_1$, but $J_2 \supsetneq O_2 \supsetneq A_2$.

4. $J' = J$.

We wish to use these new concepts to settle questions about recursive order types. As explained in the introduction, one such use has already been made, [7]. One goal is to find a constructive definition for R. If not all of R, then perhaps we could find a constructive definition for a family of order types in R sharing a certain property. For example, if $\gamma \in A$, then there is a set A of order type γ such that A has a recursive successor function, [7]. Can A be characterized by this property? On the other hand, nonconstructive elements of R may be a source of counterexamples.

Bibliography

1. S. Fellner, Recursiveness and Finite Axiomatizability of Linear Orderings, Ph.D. thesis, Rutgers University, New Brunswick, New Jersey (1976).

2. F. Hausdorff, "Grundzüge einer Theorie der Geordnete Mengen," Math. Ann., 65 (1908), 435-505.

3. L. Hay, A. B. Manaster, and J. G. Rosenstein, "Small recursive ordinals, many-one degrees and the arithmetical difference hierarchy," Annals of Math. Logic 8 (1975), 297-343.

4. A. G. Pinus, "Efficient Linear Orderings," Siberian Math. J. V16 No. 6, Plenum Press, New York (1975), 956-962.

5. H. Rogers, Jr., Theory of Recursive Functions and Effective Computability, McGraw-Hill, New York, 1967.

6. J. G. Rosenstein, Linear Orderings: An Introduction, Academic Press, 1980.

7. R. M. Watnick, Recursive and Constructive Linear Orderings, Ph.D. thesis, Rutgers University, New Brunswick, New Jersey (1979).

Vol. 700: Module Theory, Proceedings, 1977. Edited by C. Faith and S. Wiegand. X, 239 pages. 1979.

Vol. 701: Functional Analysis Methods in Numerical Analysis, Proceedings, 1977. Edited by M. Zuhair Nashed. VII, 333 pages. 1979.

Vol. 702: Yuri N. Bibikov, Local Theory of Nonlinear Analytic Ordinary Differential Equations. IX, 147 pages. 1979.

Vol. 703: Equadiff IV, Proceedings, 1977. Edited by J. Fábera. XIX, 441 pages. 1979.

Vol. 704: Computing Methods in Applied Sciences and Engineering, 1977, I. Proceedings, 1977. Edited by R. Glowinski and J. L. Lions. VI, 391 pages. 1979.

Vol. 705: O. Forster und K. Knorr, Konstruktion verseller Familien kompakter komplexer Räume. VII, 141 Seiten. 1979.

Vol. 706: Probability Measures on Groups, Proceedings, 1978. Edited by H. Heyer. XIII, 348 pages. 1979.

Vol. 707: R. Zielke, Discontinuous Čebyšev Systems. VI, 111 pages. 1979.

Vol. 708: J. P. Jouanolou, Equations de Pfaff algébriques. V, 255 pages. 1979.

Vol. 709: Probability in Banach Spaces II. Proceedings, 1978. Edited by A. Beck. V, 205 pages. 1979.

Vol. 710: Séminaire Bourbaki vol. 1977/78, Exposés 507–524. IV, 328 pages. 1979.

Vol. 711: Asymptotic Analysis. Edited by F. Verhulst. V, 240 pages. 1979.

Vol. 712: Equations Différentielles et Systèmes de Pfaff dans le Champ Complexe. Edité par R. Gérard et J.-P. Ramis. V, 364 pages. 1979.

Vol. 713: Séminaire de Théorie du Potentiel, Paris No. 4. Edité par F. Hirsch et G. Mokobodzki. VII, 281 pages. 1979.

Vol. 714: J. Jacod, Calcul Stochastique et Problèmes de Martingales. X, 539 pages. 1979.

Vol. 715: Inder Bir S. Passi, Group Rings and Their Augmentation Ideals. VI, 137 pages. 1979.

Vol. 716: M. A. Scheunert, The Theory of Lie Superalgebras. X, 271 pages. 1979.

Vol. 717: Grosser, Bidualräume und Vervollständigungen von Banachmoduln. III, 209 pages. 1979.

Vol. 718: J. Ferrante and C. W. Rackoff, The Computational Complexity of Logical Theories. X, 243 pages. 1979.

Vol. 719: Categorial Topology, Proceedings, 1978. Edited by H. Herrlich and G. Preuß. XII, 420 pages. 1979.

Vol. 720: E. Dubinsky, The Structure of Nuclear Fréchet Spaces. V, 187 pages. 1979.

Vol. 721: Séminaire de Probabilités XIII. Proceedings, Strasbourg, 1977/78. Edité par C. Dellacherie, P. A. Meyer et M. Weil. VII, 647 pages. 1979.

Vol. 722: Topology of Low-Dimensional Manifolds. Proceedings, 1977. Edited by R. Fenn. VI, 154 pages. 1979.

Vol. 723: W. Brandal, Commutative Rings whose Finitely Generated Modules Decompose. II, 116 pages. 1979.

Vol. 724: D. Griffeath, Additive and Cancellative Interacting Particle Systems. V, 108 pages. 1979.

Vol. 725: Algèbres d'Opérateurs. Proceedings, 1978. Edité par P. de la Harpe. VII, 309 pages. 1979.

Vol. 726: Y.-C. Wong, Schwartz Spaces, Nuclear Spaces and Tensor Products. VI, 418 pages. 1979.

Vol. 727: Y. Saito, Spectral Representations for Schrödinger Operators With Long-Range Potentials. V, 149 pages. 1979.

Vol. 728: Non-Commutative Harmonic Analysis. Proceedings, 1978. Edited by J. Carmona and M. Vergne. V, 244 pages. 1979.

Vol. 729: Ergodic Theory. Proceedings, 1978. Edited by M. Denker and K. Jacobs. XII, 209 pages. 1979.

Vol. 730: Functional Differential Equations and Approximation of Fixed Points. Proceedings, 1978. Edited by H.-O. Peitgen and H.-O. Walther. XV, 503 pages. 1979.

Vol. 731: Y. Nakagami and M. Takesaki, Duality for Crossed Products of von Neumann Algebras. IX, 139 pages. 1979.

Vol. 732: Algebraic Geometry. Proceedings, 1978. Edited by K. Lønsted. IV, 658 pages. 1979.

Vol. 733: F. Bloom, Modern Differential Geometric Techniques in the Theory of Continuous Distributions of Dislocations. XII, 206 pages. 1979.

Vol. 734: Ring Theory, Waterloo, 1978. Proceedings, 1978. Edited by D. Handelman and J. Lawrence. XI, 352 pages. 1979.

Vol. 735: B. Aupetit, Propriétés Spectrales des Algèbres de Banach. XII, 192 pages. 1979.

Vol. 736: E. Behrends, M-Structure and the Banach-Stone Theorem. X, 217 pages. 1979.

Vol. 737: Volterra Equations. Proceedings 1978. Edited by S.-O. Londen and O. J. Staffans. VIII, 314 pages. 1979.

Vol. 738: P. E. Conner, Differentiable Periodic Maps. 2nd edition, IV, 181 pages. 1979.

Vol. 739: Analyse Harmonique sur les Groupes de Lie II. Proceedings, 1976–78. Edited by P. Eymard et al. VI, 646 pages. 1979.

Vol. 740: Séminaire d'Algèbre Paul Dubreil. Proceedings, 1977–78. Edited by M.-P. Malliavin. V, 456 pages. 1979.

Vol. 741: Algebraic Topology, Waterloo 1978. Proceedings. Edited by P. Hoffman and V. Snaith. XI, 655 pages. 1979.

Vol. 742: K. Clancey, Seminormal Operators. VII, 125 pages. 1979.

Vol. 743: Romanian-Finnish Seminar on Complex Analysis. Proceedings, 1976. Edited by C. Andreian Cazacu et al. XVI, 713 pages. 1979.

Vol. 744: I. Reiner and K. W. Roggenkamp, Integral Representations. VIII, 275 pages. 1979.

Vol. 745: D. K. Haley, Equational Compactness in Rings. III, 167 pages. 1979.

Vol. 746: P. Hoffman, τ-Rings and Wreath Product Representations. V, 148 pages. 1979.

Vol. 747: Complex Analysis, Joensuu 1978. Proceedings, 1978. Edited by I. Laine, O. Lehto and T. Sorvali. XV, 450 pages. 1979.

Vol. 748: Combinatorial Mathematics VI. Proceedings, 1978. Edited by A. F. Horadam and W. D. Wallis. IX, 206 pages. 1979.

Vol. 749: V. Girault and P.-A. Raviart, Finite Element Approximation of the Navier-Stokes Equations. VII, 200 pages. 1979.

Vol. 750: J. C. Jantzen, Moduln mit einem höchsten Gewicht. III, 195 Seiten. 1979.

Vol. 751: Number Theory, Carbondale 1979. Proceedings. Edited by M. B. Nathanson. V, 342 pages. 1979.

Vol. 752: M. Barr, *-Autonomous Categories. VI, 140 pages. 1979.

Vol. 753: Applications of Sheaves. Proceedings, 1977. Edited by M. Fourman, C. Mulvey and D. Scott. XIV, 779 pages. 1979.

Vol. 754: O. A. Laudal, Formal Moduli of Algebraic Structures. III, 161 pages. 1979.

Vol. 755: Global Analysis. Proceedings, 1978. Edited by M. Grmela and J. E. Marsden. VII, 377 pages. 1979.

Vol. 756: H. O. Cordes, Elliptic Pseudo-Differential Operators – An Abstract Theory. IX, 331 pages. 1979.

Vol. 757: Smoothing Techniques for Curve Estimation. Proceedings, 1979. Edited by Th. Gasser and M. Rosenblatt. V, 245 pages. 1979.

Vol. 758: C. Năstăsescu and F. Van Oystaeyen; Graded and Filtered Rings and Modules. X, 148 pages. 1979.

Vol. 759: R. L. Epstein, Degrees of Unsolvability: Structure and Theory. XIV, 216 pages. 1979.

Vol. 760: H.-O. Georgii, Canonical Gibbs Measures. VIII, 190 pages. 1979.

Vol. 761: K. Johannson, Homotopy Equivalences of 3-Manifolds with Boundaries. 2, 303 pages. 1979.

Vol. 762: D. H. Sattinger, Group Theoretic Methods in Bifurcation Theory. V, 241 pages. 1979.

Vol. 763: Algebraic Topology, Aarhus 1978. Proceedings, 1978. Edited by J. L. Dupont and H. Madsen. VI, 695 pages. 1979.

Vol. 764: B. Srinivasan, Representations of Finite Chevalley Groups. XI, 177 pages. 1979.

Vol. 765: Padé Approximation and its Applications. Proceedings, 1979. Edited by L. Wuytack. VI, 392 pages. 1979.

Vol. 766: T. tom Dieck, Transformation Groups and Representation Theory. VIII, 309 pages. 1979.

Vol. 767: M. Namba, Families of Meromorphic Functions on Compact Riemann Surfaces. XII, 284 pages. 1979.

Vol. 768: R. S. Doran and J. Wichmann, Approximate Identities and Factorization in Banach Modules. X, 305 pages. 1979.

Vol. 769: J. Flum, M. Ziegler, Topological Model Theory. X, 151 pages. 1980.

Vol. 770: Séminaire Bourbaki vol. 1978/79 Exposés 525–542. IV, 341 pages. 1980.

Vol. 771: Approximation Methods for Navier-Stokes Problems. Proceedings, 1979. Edited by R. Rautmann. XVI, 581 pages. 1980.

Vol. 772: J. P. Levine, Algebraic Structure of Knot Modules. XI, 104 pages. 1980.

Vol. 773: Numerical Analysis. Proceedings, 1979. Edited by G. A. Watson. X, 184 pages. 1980.

Vol. 774: R. Azencott, Y. Guivarc'h, R. F. Gundy, Ecole d'Eté de Probabilités de Saint-Flour VIII-1978. Edited by P. L. Hennequin. XIII, 334 pages. 1980.

Vol. 775: Geometric Methods in Mathematical Physics. Proceedings, 1979. Edited by G. Kaiser and J. E. Marsden. VII, 257 pages. 1980.

Vol. 776: B. Gross, Arithmetic on Elliptic Curves with Complex Multiplication. V, 95 pages. 1980.

Vol. 777: Séminaire sur les Singularités des Surfaces. Proceedings, 1976-1977. Edited by M. Demazure, H. Pinkham and B. Teissier. IX, 339 pages. 1980.

Vol. 778: SK_1 von Schiefkörpern. Proceedings, 1976. Edited by P. Draxl and M. Kneser. II, 124 pages. 1980.

Vol. 779: Euclidean Harmonic Analysis. Proceedings, 1979. Edited by J. J. Benedetto. III, 177 pages. 1980.

Vol. 780: L. Schwartz, Semi-Martingales sur des Variétés, et Martingales Conformes sur des Variétés Analytiques Complexes. XV, 132 pages. 1980.

Vol. 781: Harmonic Analysis Iraklion 1978. Proceedings 1978. Edited by N. Petridis, S. K. Pichorides and N. Varopoulos. V, 213 pages. 1980.

Vol. 782: Bifurcation and Nonlinear Eigenvalue Problems. Proceedings, 1978. Edited by C. Bardos, J. M. Lasry and M. Schatzman. VIII, 296 pages. 1980.

Vol. 783: A. Dinghas, Wertverteilung meromorpher Funktionen in ein- und mehrfach zusammenhängenden Gebieten. Edited by R. Nevanlinna and C. Andreian Cazacu. XIII, 145 pages. 1980.

Vol. 784: Séminaire de Probabilités XIV. Proceedings, 1978/79. Edited by J. Azéma and M. Yor. VIII, 546 pages. 1980.

Vol. 785: W. M. Schmidt, Diophantine Approximation. X, 299 pages. 1980.

Vol. 786: I. J. Maddox, Infinite Matrices of Operators. V, 122 pages. 1980.

Vol. 787: Potential Theory, Copenhagen 1979. Proceedings, 1979. Edited by C. Berg, G. Forst and B. Fuglede. VIII, 319 pages. 1980.

Vol. 788: Topology Symposium, Siegen 1979. Proceedings, 1979. Edited by U. Koschorke and W. D. Neumann. VIII, 495 pages. 1980.

Vol. 789: J. E. Humphreys, Arithmetic Groups. VII, 158 pages. 1980.

Vol. 790: W. Dicks, Groups, Trees and Projective Modules. IX, 127 pages. 1980.

Vol. 791: K. W. Bauer and S. Ruscheweyh, Differential Operators for Partial Differential Equations and Function Theoretic Applications. V, 258 pages. 1980.

Vol. 792: Geometry and Differential Geometry. Proceedings, 1979. Edited by R. Artzy and I. Vaisman. VI, 443 pages. 1980.

Vol. 793: J. Renault, A Groupoid Approach to C^*-Algebras. III, 160 pages. 1980.

Vol. 794: Measure Theory, Oberwolfach 1979. Proceedings 1979. Edited by D. Kölzow. XV, 573 pages. 1980.

Vol. 795: Séminaire d'Algèbre Paul Dubreil et Marie-Paule Malliavin. Proceedings 1979. Edited by M. P. Malliavin. V, 433 pages. 1980.

Vol. 796: C. Constantinescu, Duality in Measure Theory. IV, 197 pages. 1980.

Vol. 797: S. Mäki, The Determination of Units in Real Cyclic Sextic Fields. III, 198 pages. 1980.

Vol. 798: Analytic Functions, Kozubnik 1979. Proceedings. Edited by J. Ławrynowicz. X, 476 pages. 1980.

Vol. 799: Functional Differential Equations and Bifurcation. Proceedings 1979. Edited by A. F. Izé. XXII, 409 pages. 1980.

Vol. 800: M.-F. Vignéras, Arithmétique des Algèbres de Quaternions. VII, 169 pages. 1980.

Vol. 801: K. Floret, Weakly Compact Sets. VII, 123 pages. 1980.

Vol. 802: J. Bair, R. Fourneau, Etude Géometrique des Espaces Vectoriels II. VII, 283 pages. 1980.

Vol. 803: F.-Y. Maeda, Dirichlet Integrals on Harmonic Spaces. X, 180 pages. 1980.

Vol. 804: M. Matsuda, First Order Algebraic Differential Equations. VII, 111 pages. 1980.

Vol. 805: O. Kowalski, Generalized Symmetric Spaces. XII, 187 pages. 1980.

Vol. 806: Burnside Groups. Proceedings, 1977. Edited by J. L. Mennicke. V, 274 pages. 1980.

Vol. 807: Fonctions de Plusieurs Variables Complexes IV. Proceedings, 1979. Edited by F. Norguet. IX, 198 pages. 1980.

Vol. 808: G. Maury et J. Raynaud, Ordres Maximaux au Sens de K. Asano. VIII, 192 pages. 1980.

Vol. 809: I. Gumowski and Ch. Mira, Recurences and Discrete Dynamic Systems. VI, 272 pages. 1980.

Vol. 810: Geometrical Approaches to Differential Equations. Proceedings 1979. Edited by R. Martini. VII, 339 pages. 1980.

Vol. 811: D. Normann, Recursion on the Countable Functionals. VIII, 191 pages. 1980.

Vol. 812: Y. Namikawa, Toroidal Compactification of Siegel Spaces. VIII, 162 pages. 1980.

Vol. 813: A. Campillo, Algebroid Curves in Positive Characteristic. V, 168 pages. 1980.

Vol. 814: Séminaire de Théorie du Potentiel, Paris, No. 5. Proceedings. Edited by F. Hirsch et G. Mokobodzki. IV, 239 pages. 1980.

Vol. 815: P. J. Slodowy, Simple Singularities and Simple Algebraic Groups. XI, 175 pages. 1980.

Vol. 816: L. Stoica, Local Operators and Markov Processes. VIII, 104 pages. 1980.